Statistics for the Environment

Statistics for the Environment

Proceedings of the SPRUCE Conference held in the Calouste Gulbenkian Centre, Lisbon, Portugal, 7–10 April 1992.

SPRUCE is an international initiative concerned with *Statistics in Public Resources, Utilities and in Care of the Environment.*

Sponsored by:

COMMISSION
OF THE
EUROPEAN COMMUNITIES

DIRECTORATE-GENERAL
FOR SCIENCE, RESEARCH
AND DEVELOPMENT
STD_3
LIFE SCIENCE AND TECHNOLOGIES
FOR DEVELOPING COUNTRIES

Statistics for the Environment

Edited by

Vic Barnett

Rothamsted Experimental Station, UK

and

K. Feridun Turkman

Centre of Statistics, University of Lisbon, Portugal

JOHN WILEY & SONS

Chichester · New York · Brisbane · Toronto · Singapore

Other Wiley Editorial Offices

John Wiley & Sons, Inc., 605 Third Avenue,
New York, NY 10158-0012, USA

Jacaranda Wiley Ltd, G.P.O. Box 859, Brisbane,
Queensland 4001, Australia

John Wiley & Sons (Canada) Ltd, 22 Worcester Road,
Rexdale, Ontario M9W 1L1, Canada

John Wiley & Sons (SEA) Pte Ltd, 37 Jalan Pemimpin #05-04,
Block B, Union Industrial Building, Singapore 2057

Library of Congress Cataloging-in-Publication Data

Statistics for the environment / edited by Vic Barnett and K. Feridun Turkman.
p. cm.
'Proceedings of the SPRUCE Conference held in the Calouste Gulbenkian Centre, Lisbon, Portugal from 7–10 April 1992, sponsored by Commission of the European Communities'—P.
Includes bibliographical references (p.) and index.
ISBN 0 471 93467 4
1. Environmental sciences—Statistical methods—Congresses.
I. Barnett, Vic. II. Turkman, K. Feridun.
GE45.S73S83 1993 92-21239
CIP

British Library Cataloguing in Publication Data.

A catalogue record for this book is available from the British Library

ISBN 0 471 93467 4

Typeset by Thomson Press (India) Ltd, New Delhi
Printed and bound in Great Britain by Biddles Ltd, Guildford, Surrey

To Antónia

Contents

Foreword *J. Sant'Ana Calazans* xv

Preface *V. Barnett and K. Feridun Turkman* xvii

PART I ENVIRONMENTAL MONITORING AND SAMPLING 1

1 **The Active Mixing Volume: A New Concept in Modelling Environmental Systems** 3
P. C. Young and M. Lees

1.1 Introduction 3
1.2 A motivational example: The ADZ model for transport and dispersion of solutes in stream channels 6
1.2.1 Derivation of the aggregated dead zone (ADZ) model 6
1.2.2 Transfer function representation of the ADZ model 11
1.3 Identification and estimation of TF models 14
1.3.1 Recursive parameter estimation 14
1.3.2 TF model order identification 18
1.3.3 Identifiability 20
1.4 The ADZ model revisited: Practical results 21
1.4.1 Wider implications of the ADZ model 23
1.5 Other practical examples of the ADZ-type model 23
1.5.1 Solute transport in soils: I 24
1.5.2 Solute transport in soils: II 26
1.6 A generalisation of the ADZ modelling concept: The active mixing volume (AMV) 29
1.6.1 An AMV model for heat flow in soils 30
1.7 Conclusions 38
Acknowledgments 39
Appendix 1.1 Transfer function models 39

2 **Encounter Sampling and Modelling in Ecological and Environmental Studies Using Weighted Distribution Methods** 45
G. P. Patil, C. Taillie and S. Talwalker

2.1 Background and introduction 45
2.2 Extra-distance sources of encounter bias in transect sampling 46
2.2.1 Deep-sea red crab 47
2.2.2 Bivariate sighting-functions 48
2.2.3 Random sighting-functions do not cause overdispersion 51
2.3 Informativeness of weighted distributions 52
2.3.1 Species abundance 53
2.4 Overdispersion and heterogeneity of teratological binary data 58
2.4.1 Comparison of beta-binomial and double binomial as models for overdispersed binary data 61
2.4.2 Accuracy of Efron's approximation formulae 61
2.5 Modelling bivariate overdispersion 62
2.5.1 A bivariate form of the Diaconis–Efron DEF 64
2.5.2 Double bivariate normal distribution 66
2.5.3 A weighted bivariate normal correlation model 67
2.5.4 Relative informativeness of the weighted observations 69

3 **Comparing Sampling Designs for Monitoring Ecological Status and Trends: Impact of Temporal Patterns** 71
N. S. Urquhart, W. S. Overton and D.S. Birkes

3.1 Introduction 71
3.2 Modelling approach 74
3.3 Results 79
3.4 Discussion 84
Acknowledgments 85

PART II MEASURING LEVELS AND CONSEQUENCES OF POLLUTION AND CONTAMINATION 87

4 **Point Process Modelling in Environmental Epidemiology** 89
P. J. Diggle

4.1 Introduction 89
4.2 Point processes and their statistical analysis 91
4.2.1 First- and second-moment properties 91
4.2.2 Estimation of the first- and second-moment properties of a bivariate stationary, isotropic point process 95
4.2.3 Models 96
4.2.4 Data-perturbation 97
4.2.5 The practical relevance of stationarity 99
4.3 Spatial clustering 100
4.4 Space–time interaction 102
4.5 Raised incidence near one or more point sources 104
4.6 Splancs: Software for environmental epidemiology 107
4.7 Discussion 109

5 Estimation of Quantiles in Airborne Pollution; Measuring Pollution around Gas Furnaces **111**
G. Lindgren, L. Zetterqvist and G. Holmstedt

5.1 Introduction 111
5.2 Normal and extreme NO_χ-emission 112
5.2.1 Background 112
5.2.2 The measuring system 112
5.2.3 Correction for drift 113
5.2.4 Data 114
5.2.5 Influence of wind direction 115
5.3 A statistical model 115
5.3.1 The model 115
5.3.2 Analysis 116
5.4 Quantile relations in disturbed distributions 118
5.4.1 Log-normal case 119
5.4.2 Application to NO_χ data 120
5.5 Conclusions 121

6 A Stochastic Model for Times of Exposures to Air Pollution from a Point Source **123**
A. C. Davison and N. I. Ramesh

6.1 Introduction 123
6.2 The MESOS model and data 124
6.3 Markov-modulated Poisson processes 127
6.3.1 Model 127
6.3.2 Second-order properties 129
6.4 Data analysis 131
6.4.1 Separate analyses 131
6.4.2 Joint analysis 131
6.5 Discussion 138
Acknowledgments 138

PART III CLIMATOLOGICAL AND METEOROLOGICAL ISSUES 139

7 Long-range Dependence and Global Warming **141**
R. L. Smith

7.1 Introduction 141
7.2 Global warming and the detection of trends in temperature series 142
7.3 Long-range dependence and the estimation of a linear trend 143
7.4 Central England temperature data 147
7.5 Long-range dependence in English data 149
7.6 US data 155
7.7 Global data 157
7.8 Discussion 160
Acknowledgments 161

8 Environmental Factors Affecting Reservoir Safety **163**
C. W. Anderson and S. Nadarajah

8.1 Introduction 163
8.2 Reservoir safety and the current method of assessment 164
8.3 Strategy of the joint probability studies 166
8.4 Statistical methodology 167
8.4.1 Primary and secondary variables 167
8.4.2 Dependence between extremes of primary variables 168
8.4.3 Inclusion of secondary variables 171
8.4.4 Further aspects of modelling 171
8.4.5 Declustering 172
8.5 Some preliminary results 173
8.5.1 Dependence between rain and wind 173
8.5.2 Wind speed and direction 180
8.6 Conclusions 180
Acknowledgments 180
Appendix 8.1 A distributional overview 181
Appendix 8.2 Dependence models 181

9 Rainfall Depth-Duration-Frequency Curves; A Problem of Dependent Extremes **183**
T. A. Buishand

9.1 Introduction 183
9.2 Annual maxima at De Bilt 184
9.3 Estimation of the correlation between parameter estimates 186
9.3.1 Smith's method 187
9.3.2 The jack-knife 188
9.3.3 The linear jack-knife 189
9.3.4 The bootstrap 190
9.3.5 The estimated correlation between $\hat{\tau}$ and $\hat{c}$ 191
9.4 Mathematical description of the change of the Gumbel parameters with duration 192
9.5 Quantile estimation 195
9.6 Conclusions 196
Acknowledgment 197

PART IV WATER RESOURCES 199

10 Water Resources and Statistics: Past, Present and Future **201**
V. Yevjevich

10.1 Introduction 201
10.2 Characteristics of past applications of statistics in solutions of water resources problems 204
10.3 The present state of application of advanced statistics in hydrology and water resources 210
10.4 Expected future applications of statistics in water resources 219

11 **On the Estimation of Parameters of Rare Events in Environmental Time Series** **225**
M. I. Gomes

11.1 Introduction and preliminaries 225
11.2 The asymptotic behaviour of exceedances of high levels and of upper order statistics 226
11.3 The influence of the extremal index on the estimation of upper quantiles 229
11.4 Comparison of estimators of upper quantiles 231
11.5 The estimation of the extremal index 236
11.6 An example of application—levels of the River Douro at Fraga hydrometric station 239
Acknowledgments 241

12 **Extreme Sea Levels** **243**
J. A. Tawn

12.1 Introduction 243
12.2 Background to extreme-value modelling 245
12.3 Separate site methods 247
12.3.1 Direct methods 247
12.3.2 Indirect methods 250
12.4 Spatial modelling 252
12.5 Application of the spatial model 254
12.5.1 UK data analysis 254
12.5.2 Australian data analysis 258
12.6 Incorporating waves in the analysis 260
12.7 Further work 263
Acknowledgments 263

PART V DYNAMICS OF FISH POPULATIONS 265

13 **Modelling and Managing a Salmon Population** **267**
T. Speed

13.1 Introduction 267
13.2 Background 268
13.3 Models and methods 271
13.4 Discussion 289
Acknowledgments 292

14 **Ambiguous Inferences from Fisheries Data** **293**
J. T. Schnute

14.1 Introduction 293
14.2 A simple example 294
14.3 Chinook hatchery data 297
14.4 Serial correlation 301

14.5 Stock and recruitment 307
14.6 Conclusions 309

15 Recapture Data on Tagged Fish: Some Questions of Analysis and Interpretation **311**
R. M. Cormack

15.1 Introduction 311
15.2 Approaches to recapture data 312
15.3 Multinomial models 313
15.4 Generalized linear models 315
15.5 Example 1: Effect of experimental dam manipulation on local survival 319
15.6 Example 2: Commercial recaptures of chinook in south-eastern Alaska waters 323
15.7 Extensions 329
Appendix 15.1 Complete GLIM input for data and analysis of Example 1 330

PART VI FORESTRY: SUPPLY AND CONSERVATION 333

16 Statistical Analysis of Climatological and Ecological Factors in Forestry: Spatial and Temporal Aspects **335**
W. Urfer

16.1 Introduction 335
16.2 Data analysis in the process of scientific investigation: longitudinal studies 336
16.3 Analysis of categorical data in longitudinal studies by a marginal model 337
16.4 Analysis of categorical data in longitudinal studies by a logistic regression model 339
16.5 Application of weighted least squares and logistic regression analysis on data from permanent observation plots 341
16.6 Further aspects of forest ecosystem measurements 344

17 Uncertainty and the Conservation or Destruction of Natural Forests and Wilderness **347**
W. J. Reed

17.1 Introduction 347
17.2 A single-stand model for the harvest–conservation decision problem 354
17.2.1 The cost–benefit prescription 356
17.2.2 The optimal harvest rule 357
17.2.3 Suboptimality of the cost–benefit procedure 359
17.2.4 The certainty-equivalence procedure 364
17.3 A model for conservation–development decisions for a large area of wilderness land, allowing for partial development 367

17.3.1 The model 367
17.3.2 The cost–benefit prescription 369
17.3.3 The optimal development rule 371
17.3.4 The suboptimality of the cost–benefit procedure 373
17.3.5 The certainty-equivalence procedure 373
17.3.6 Size-independent land-service flows 375
17.4 Final remarks 376

18 Quantitative Assessment of Growth in the Tropical Rainforests 379
S. Nokoe

18.1 Introduction 379
18.1.1 The tropical forest 379
18.1.2 The need for growth studies 381
18.2 Quantitative assessments 382
18.2.1 Spatial distribution 382
18.2.2 Species diversity 383
18.2.3 Frequency distribution 383
18.2.4 Growth and yield 385
18.2.5 Stand dynamics 386
18.3 Illustration 388
18.3.1 The problem 388
18.3.2 The data and initial computations 388
18.3.3 Final species structure 388
18.3.4 Determining time-period equivalent in years 391
18.3.5 Assessment of unregulated interference and interventions 391
18.4 Conclusions 394
Acknowledgments 395
Appendix 18.1 Basic program for stable structure simulation 395

References 401

Index 419

Foreword

Major policy reforms and institutional changes are required to bring about long-term economic development and better environmental management and the current time is unprecedented in its potential for change. The growing recognition of the importance of environmental concerns at national, regional and global level facilitates, and benefits from, increased scientific and technological cooperation between industrialised and developing countries (DCs) in solving problems associated with our sustainable presence on planet Earth.

The United Nations Conference on Environment and Development (UNCED), held in June 1992, provided a formal starting point for an in-depth reappraisal of key issues affecting sustainable development. This will undoubtedly lead to the identification of areas in which our knowledge of the environment and the way it is impacted by development activities is not sufficient to supply an objective base for correct policy formulation.

Scientific research has an irreplaceable role in analysing ways of reconciling long-term economic growth with the conservation of renewable natural resources such as water, forests, soils and living organisms, the preservation of human health and of natural aesthetic values. Scientific and Technological Cooperation with Developing Countries, currently embodied in the Science and Technology for Development (STD) Programme of the Community (RTD) Framework Programme, constitutes a privileged and unique instrument to promote the awareness of problems posed by development to scientific communities in DCs and the EC. Scientific and Technological Cooperation reflects the Community's solidarity with DCs in their efforts to improve their quality of life but also its perception of the nature of problems that must be addressed and solved through joint efforts.

STD's support of the first SPRUCE conference aims at strengthening the links between the scientific communities of DCs and of the European Community while providing a sharper and scientifically objective focus on environmental conservation and development issues. We trust that these two objectives have been amply achieved.

J. Sant'Ana Calazans
Scientific and Technological Cooperation with DCs
DG XII/G-4
Science, Research and Development
Commission of the European Communities

Preface

Environmental issues will be at the forefront of concern of governments as we move towards the 21st century. A proper understanding of mechanisms and relationships will be vital if we are to reconcile the economic and environmental conflicts. Statistics has a major role to play in such areas as quantifying effects, assessing consequences, measuring risks, interpreting the evidence. Already *Environmental Statistics* is becoming an important, high-profile, specified theme throughout the world.

Many organisations are signalling their commitment to this theme; by including within their present research, education and training portfolios a specific place for 'environmental statistics'; for example, we have universities and colleges establishing departments or centres for 'environmental statistics'. New international coordinated themes are being set up to harness this growing interest and activity through conferences, journals, books and joint research programmes.

One such initiative, which has had a major impact over the few years of its existence, is SPRUCE: *Statistics in Public Resources, Utilities and in Care of the Environment*. This was set up initially as a part of the collaboration between the statistics departments of the Universities of Lisbon and Sheffield. Its aims were to bring together statisticians, probabilists and other scientists from both the developed and the developing world in the fields of protection and conservation of the environment, the safeguarding and garnering of resources, and the supply of energy in its various forms in a safe and reliable mode. The first tangible product of the SPRUCE initiative was a major International Conference in the Calouste Gulbenkian Centre in Lisbon from 7 to 10 April 1992, which attracted more than 200 delegates from 30 or so countries.

The Executive Committee for the Conference consisted of V. Barnett (*Chairman*), M. A. Amaral Turkman, C. W. Anderson, I. R. Dunsmore, K. Feridun Turkman (*Local Organiser*), and F. Rosado.

The keynote and invited speakers represented the highest levels of endeavour throughout the world in the quantitative and stochastic study of environmental matters and included C. W. Anderson, T. A. Buishand, R. M. Cormack, A. C. Davison, P. J. Diggle, M. I. Gomes, G. Lindgren, S. Nokoe, G. P. Patil, W. J.

Reed, J. T. Schnute, R. L. Smith, T. Speed, J. A. Tawn, W. Urfer, N. S. Urquhart, V. Yevjevich and P. C. Young, whose papers are presented as chapters in this volume.

The material has been presented in an integrated and cross-edited manner to highlight the important links and themes. In this regard it has been categorised into the topics:

Part I Environmental Monitoring and Sampling
Part II Measuring Levels and Consequences of Pollution and Contamination
Part III Climatological and Meteorological Issues
Part IV Water Resources
Part V Dynamics of Fish Populations
Part VI Forestry: Supply and Conservation

which constitute the distinct parts of these published proceedings.

The importance of the theme of the conference was highlighted by the remarks of a number of eminent national (Portuguese) and EC speakers in their opening addresses. For example, Eng. Flores (Ministry of the Environment and National Resources, Portugal) remarked as follows:

> Degradation of the environment has been the reverse face of unrestricted economic growth in the second half of the 20th century ... This model of society, where unruled growth was tolerated, is no longer acceptable. Similarly, we can no longer accept the first generation of environmental policies, based on systematic opposition to development ... To implement an integrated environmental policy in any particular domain is not, however, an easy task ... Uncertainty cannot be avoided or bypassed in the analysis of any natural system ... In this setting, the statistical tools that are the object of SPRUCE play a most vital and irreplaceable role in the analysis, systematisation and modelling of environmental data; Statistics affords a simple and intelligible interpretation of symptoms and evaluation of the health condition of our planet, better understanding of causes of its illnesses, deeper knowledge of its resources and potentialities ... International conferences sharing the results of research in this field are therefore of the utmost importance, bringing together nations, contributing to overcome world problems.

Dr Vaz (Portuguese Information Office of the EC) re-echoed this theme, but with stress on the problems of the developing world. He reminded the conference of the need for

> further strengthening of research capacity in both the developing countries and in the Community Member States in areas defined as having priority for Third World development (agriculture, health, nutrition and the environment in tropical and subtropical areas), by means of joint research projects.

and concluded by saying that

> the results of this conference, with the participation of so many experts coming from all over the world, will contribute to a better comprehension of control mechanisms and analysis of the environmental degradation in the world.

It is fitting that Dr Vaz should concentrate on the needs of developing countries in the light of the generous support given by DGXII (G-4) of the CEC to the SPRUCE conference which enabled us to invite 15 or so scientists (involved in environment issues) from various parts of the developing world to participate in the conference. They came from Africa, the Far East, the Middle East and Latin America, and added a richness to the interchange of ideas. We are grateful to DGXII for this support, and in particular for the financial assistance in the publication of this volume and for contributing the Foreword statement

Environmental statistics is as yet far from being a well-defined methodological theme. The objective of this wide-ranging first SPRUCE conference was to review different fields of study to see in each case if any common methodological threads were evident. What was clear was the highly demanding nature of the problems in terms of the statistical methods needed (e.g. sophistated spatial/temporal processes, extreme-value methods for dependent systems) and the commonly encountered experience that standard methodology might not readily transfer to the environmental problem: idiosyncratic, tailored, methods were frequently needed. Future SPRUCE conferences will concentrate on more specific topics (SPRUCE II, Rothamsted Experimental Station, will be on **Statistics of Water**: resources, quality, power) but the first SPRUCE conference in Lisbon in 1992 has provided a wealth of ideas and methods reflected in the pages of this book.

We hope that you find it a valuable and stimulating resource.

V. Barnett and **K. Feridun Turkman**
July 1992

PART I

Environmental Monitoring and Sampling

1

The Active Mixing Volume: A New Concept in Modelling Environmental Systems

Peter C. Young and Matthew Lees,
University of Lancester, UK

1.1 INTRODUCTION

The following is a quotation from a United Kingdom Treasury Committee Report on Policy Optimisation set up to investigate economic modelling techniques:

> One central problem is that the economist, unlike the physical scientist, must by and large, draw conclusions about economic relationships from a continuous stream of behaviour, rather than being able to conduct controlled experiments that are possible under laboratory conditions. He has at his disposal statistcal tools which are used to attempt to analyse historical information about the economy and decompose it into sets of meaningful economic relationships. These inter-relationships of economic data make it hard to establish the direction of causation and to distinguish between causal and coincidental associations.

In this paragraph, the authors neatly encapsulate some of the central problems of modelling from economic data. At the same time, however, they seem unware that not all 'physical scientists' have the luxury of being able to perform tightly planned experiments under laboratory conditions. Certainly many environmental scientists would describe their modelling problems in a similar manner

Statistics for the Environment. Edited by Vic Barnett and K. Feridun Turkman

to that of the Treasury mandarins: namely, they wish to relate theoretical dynamic models of the environment, which are normally obtained by reference to physical and chemical laws, to a 'continuous stream' of passively monitored, time-series data. In other words, the environmental modeller, like his econometric counterpart, is confronted by a very difficult problem of statistical inference; one which is quite unlike that of the laboratory scientist and, as we shall see in this chapter, presents a formidable philosophical and methodological challenge.

Not surprisingly, given the nature of the subject, environmental scientsts tend to have a wide variety of different academic origins. It is quite understandable, therefore, that they should consider the problem of model building from rather different standpoints. In simple terms, however, we can classify environmental modellers in two main groups. By far the largest category are those who favour *mechanistic* models which reflect the perceived complexity of environmental systems and have a structure which resembles closely the physical, biological and chemical structure of the real world *as it is understood by the scientist*, given the contemporary state of scientific knowledge about the system.

In such mechanistic models, the equations are obtained in various ways, e.g. from dynamic conservation equations (mass, energy, momentum, etc.), which have a direct physical interpretation. They are then in the form of *simulation* models, which are normally solved either by writing a special-purpose computer program to solve the equations; or, more likely these days, by using a general-purpose computer simulation language. The parameters in these models are obtained either directly, under the assumption that they represent a physical 'reality' which can be measured or inferred from measurements; or they are estimated from monitored data using some form of model 'fitting' or 'optimisation', often with constraints on assumed known ('measured') parameters.

Scientists and engineers who favour mechanistic models of this type can be contrasted with those who have a more statistical turn of mind and tend to warn of the dangers inherent in any approach which relies too strongly on the perception of the model builder and so constrains the model to have a structure which conforms with this prior 'prejudice'. This second, smaller group of scientists tends to favour 'data-based' procedures, where the model structure is inferred, and the model parameters are estimated, directly by reference to experimental data, using more objective, and often statistically based methods.

Data-based models of this type are often simple in structure, inherently stochastic in form, and are characterised by only as many parameters as can be justified by the information content of the available data (i.e. they are 'parametrically efficient' or 'parsimonious'; see Box and Jenkins, 1970). In contrast, mechanistic models (sometimes referred to as *physically based* or *structural* models) are normally fairly complex in structure, deterministic in form, and are often characterised by many parameters, some of which are constrained by prior assumptions to fixed 'known' values, based on their physical interpretation (e.g. volumes, reaction rates, etc.).

Although it clarifies the nature of the controversy, this dichotomy is too simplistic: as in most generalisations, the lines of argument are drawn too rigidly and the way in which environmental models are constructed in practice is rarely as clearly defined as this interpretation would suggest. For example, if the available data are severely limited, then the more speculative, mechanistic models become almost *sine qua non* for any exercise in model building and an almost essential extension of the 'mental models' of the scientist. Moreover, when accompanied by intelligent exercises in sensitivity analysis (e.g. Hornberger and Spear, 1983; Young, 1983), they can prove extremely useful in examining ideas about alternative mechanisms and their relative importance in defining the behaviour of the model system. At the other extreme, where plentiful supplies of data are available, even the truly 'black box' time-series models can prove very useful in understanding more about the nature of the data and allowing for exercises such as seasonal adjustment, interpolation over gaps in the data, and forecasting.

But there is an intermediate type of model which we discuss in this chapter: one which exploits the availability of time-series data in statistical terms but which overtly attempts to produce models which, although they are much more parametrically efficient than the normal mechanistic model, have a sensible physical interpretation. In these *data-based, mechanistic models*, the structure is obtained by some form of objective statistical inference applied to the time-series data and based on a given, general class of time-series model. But the resulting model equations are only considered fully acceptable if, in addition to explaining the data well, they also provide a description which has relevance to the physical reality of the system under study. This general approach to environmental modelling has been evolved by the first author over many years and has been adumbrated in previous publications on the subject (Young, 1978, 1983).

The present chapter starts with a motivational example based on the modelling of pollutant transport and dispersion in river systems. This example introduces the more general reader to the transfer function (*TF*) description of a dynamic process, which has long been utilised successfully by engineers and applied systems analysts; and to the related methodological tools which provide the basis for the identification and estimation of *TF* models from time-series data. Other practical examples, which exploit a similar methodological approach and are concerned with solute transport in soils, reinforce the efficacy of this data-based mechanistic approach to environmental modelling. But these examples are merely a prelude to the main object of the chapter, which is to introduce a new concept in modelling environmental systems; one which is inherent in all the examples but has wider connotations: namely, the *active mixing volume (AMV)*. This concept is illustrated further by a final practical example concerned with the modelling of heat flow and temperature variations in a soil column; an example which also demonstrates the importance of *identifiability* concepts in environmental systems modelling.

1.2 A MOTIVATIONAL EXAMPLE: THE ADZ MODEL FOR TRANSPORT AND DISPERSION OF SOLUTES IN STREAM CHANNELS

Classical hydrodynamic analysis associated with dispersion in flowing media, such as the seminal work of G. I. Taylor (1954) on flow in pipes, often provides the starting point for modelling the transport and dispersion of solutes in stream channels. In the purely longitudinal case, this involves a mathematical representation in the form of the following single-dimensional, partial differential equation (see e.g. Fischer *et al*, 1979; Orlob, 1983; Henderson-Sellars *et al*, 1990; Young and Wallis, 1992), usually known as the Fickian diffusion equation or the advection dispersion equation (*ADE*),

$$\frac{\partial x(s,t)}{\partial t} + U\frac{\partial x(s,t)}{\partial s} = D\frac{\partial^2 x(s,t)}{\partial s^2}, \tag{1.1}$$

where $x(s, t)$ is the concentration of the solute (typically a pollutant) at spatial location s and time t; U is the cross-sectional average longitudinal velocity; and D is the longitudinal dispersion coefficient. The solution to equation (1.1) for an impulsive ('gulp') input of solute with finite mass M is, with suitable choice for the origins of s and t, given by

$$x(s,t) = \frac{M}{2A(\pi Dt)^{0.5}}\exp\left\{-\frac{(s-ut)^2}{4Dt}\right\}. \tag{1.2}$$

Although this represents a Gaussian distribution in space, when viewed in the time domain at a fixed longitudinal location, it yields a mildly skewed distribution, because the solute cloud continues to evolve as it passes the fixed observation point.

Observations of dispersing solute clouds in rivers have revealed persistent deviations from the behaviour predicted by this classical Fickian theory and the measured concentration profiles rarely, if ever, attain the theoretical Gaussian distribution in space, except at very large distances from the injection point, and the skewness of the temporal profiles measured in practice is much larger than that predicted by the Fickian theory. The reasons for this are essentially twofold: firstly, observations are rarely made at long enough times after injection for the Gaussian distribution to evolve; and, secondly, real stream channels are nonuniform, and this nonuniformity leads to rather different patterns and mechanisms of mixing than those predicted by the pure Fickian theory.

1.2.1 Derivation of the the aggregated dead zone (ADZ) model

The limitations of the *ADE* have stimulated scientists to either modify the Fickian-inspired theory or to develop alternatives. Of these, one approach which has been particularly successful depends on the assumption that 'dead zones' play an important role in the dispersion process. These zones are traditionally associated with side pockets, bad irregularities and other 'roughness' elements in the stream channel, elements which tend to create relatively slower moving regions or even areas of 'backflow'. The casual observer who throws a stick into a river may often observe that, near to the bank, the stick can get caught in such a dead zone and may often be seen to move in an opposite direction to the main stream flow. Solute which is entrained within these 'backflow' areas in then released back into the main flow relatively slowly and at diluted concentrations, due to the 'eddy-like' flow structure which tends to characterise such regions. This induces a distributed time delay effect and tends to raise significantly the tail of the concentration profile which, for an impulsive input of solute, has a noticeably skewed distribution, as shown in Figure 1.1(a), which gives the results obtained from a typical tracer experiment using the fluorescent dye Rhodamine WT carried out on the River Conder in north-west England (see Young and Wallis, 1986; Wallis *et al*, 1987, 1989). This experiment is discussed in more detail later in the chapter.

Various dead-zone analyses (e.g. Valentine and Wood, 1979; Bencala and Walters, 1983) have shown that dead zones can account for some of the observed deviations from the traditional ADE model behaviour. Until recently, however, most of these dead-zone models have been based on an extension of the ADE model, in which the dead zones are characterised explicity as a first-order differential equation for the dead-zone effect, adjoined to the classical partial differential equation (1.1) of the ADE. In the past few years, however, some of the concepts of systems theory have been brought to bear on the problem and this has led to a rather different model formulation: the *aggregated dead zone* (*ADZ*), where the model is in the form of a lumped parameter, ordinary differential-delay equation for the changes in concentration occurring over a reach of specified length.

This ADZ approach to dispersion modelling began in the early 1970s (Beck and Young, 1975, 1976) and came to fruition in the 1980s (Beer and Young, 1983; Young and Wallis, 1986; Wallis *et al*, 1989). Here, the direct analysis of experimental tracer data such as that shown in Figure 1.1(a), using the recursive estimation procedures described later, is utilised to statistically estimate the parameters in a simple TF model, of low or minimal dimension, which is almost always able to explain the measured concentration profile much better than previous conventional models, such as the ADE. The outputs of two such models are shown in (ii) of Figure 1.1(a).

This *data-based ADZ* approach can be contrasted to more conventional *mechanistic* modelling procedures, such as the ADE, where a model structrue is derived by physico-chemical reasoning and then utilised as the basis for exercises in model 'fitting' or 'optimisation'. Whereas the data-based approach advocated here allows the data themselves to expose the possible model structure within the larger general class of TF models, the conventional approach tends to rely on the adequacy of the analyst's preconceptions about the behavioural mechanisms of the system under study.

Although the ADZ model is obtained from an objective exercise in statistical identification and parameter estimation, as described below, it can be interpreted best, in physical terms, from its formulation as a continuous-time, first-order, differential delay equation obtained from mass-conservation arguments. Let us consider a reach of a river system, in which the flow through a reach with total water volume V_a is denoted by Q. Suppose the water is contaminated by a dissolved pollutant which is known to decay over time with a decay rate k which is proportional to the mass of dissolved pollutant in the reach at any time. There are many approaches to formulating a dynamic model for such a system but we will choose one of the simplest and assume that the solute is '*imperfectly mixed*' in the reach, so that the relationship between the measured

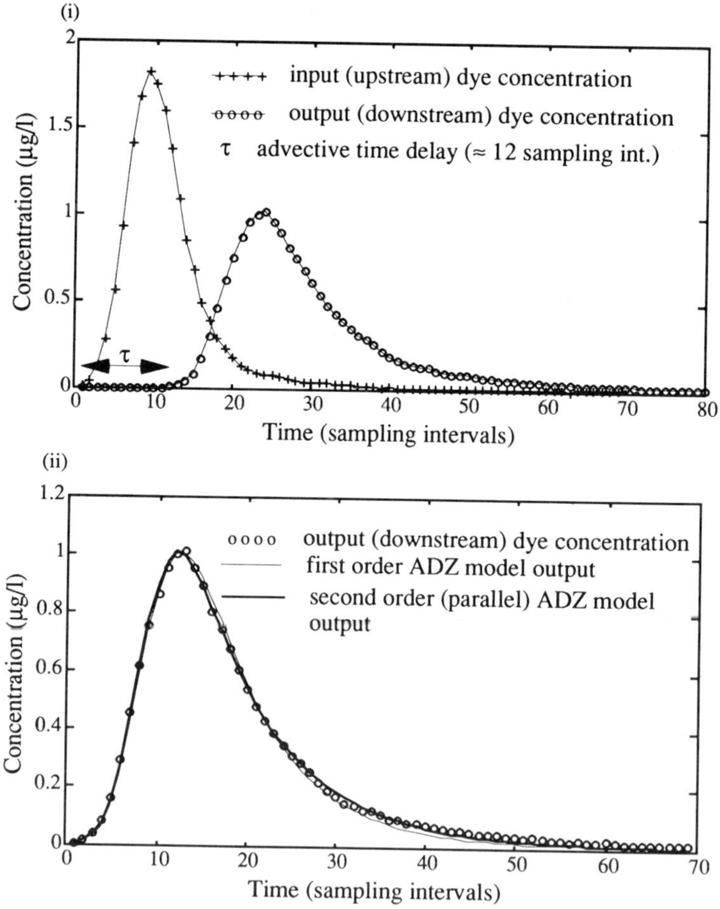

(a)

Figure 1.1(a) Typical dye-tracer experimental results from the River Conder in north-west England: (i) dye tracer experiment data; (ii) comparison of estimated ADZ model outputs and measured downstream tracer concentration

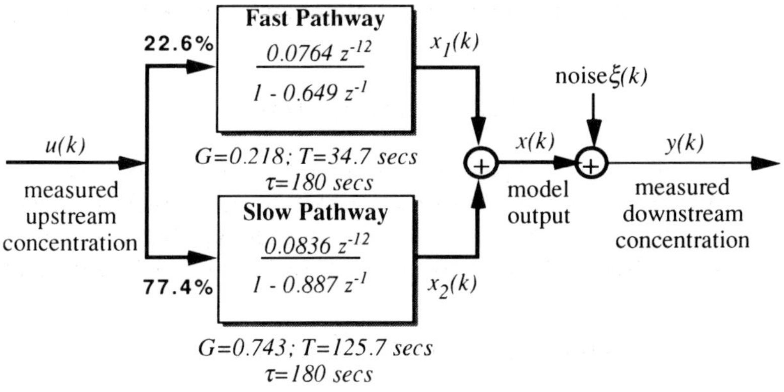

(b)

Figure 1.1(b) TF block diagram of the second-order model in Figure 1(a) showing that it has a parallel structure with one pathway faster than the other but carrying less tracer material

concentration $x(t)$ of the pollutant at the output (or downstream) location in the reach to the measured concentration $u(t)$ at the input (or upstream) location, can be obtained from the simplest dynamic mass-conservation considerations as

change of mass per unit time		mass flow in per unit time		mass flow out per unit time		mass lost due to decay per unit time
$\dfrac{d[V_a y(t)]}{dt}$	$=$	$Q\exp(-k\tau)u(t-\tau)$	$-$	$Qx(t)$	$-$	$k[V_a y(t)]$

(1.3)

where $y(t)$ is the average concentration of pollutant in the reach and τ is the *advective time-delay* introduced to allow for the translational effects of the unidirectional river flow.

Referring to the results of an impulsive tracer experiment, such as that shown in Figure 1.1(a), the term $\exp(-k\tau)u(t-\tau)$ allows for the decay of the solute over the advective time-delay period τ, while the term $k[V_a y(t)]$ accounts for the decay within the 'actively mixed' parts of the reach, as represented by the dispersed response which follows the advective delay. What we are postulating here is that while, at the detailed level, the mixing processes in a natural channel are complex and non-stationary, the results of tracer experiments, such as those shown in Figure 1.1(a), suggest that the overall effect of such processes yields a much simpler and more predictable pattern of aggregative behaviour which we believe can be modelled well by a lumped parameter differential equation such as (1.3).

In order to convert equation (1.3) to a differential equation relating $u(t)$ to $x(t)$ directly, it is necessary to make some assumption about the relationship between $x(t)$ and $y(t)$. Here, we will make the simplest such assumption, namely that they are linearly related by a factor D_f, which we will term the *dispersive fraction* for reasons that will become obvious below. In other words,

$$y(t) = D_f x(t)$$

so that,

$$\frac{d[Vx(t)]}{dt} = Q\exp(-k\tau)u(t-\tau) - Qx(t) - k[Vx(t)] \tag{1.4}$$

where $V = D_f V_a$. Now, if V_a and Q are assumed constant (i.e. steady flow conditions), then equation (1.4) can be written as

$$\frac{dx(t)}{dt} = -Ax(t) + Bu(t-\tau) \tag{1.5}$$

where,

$$A = \frac{Q}{V} + k \quad \text{and} \quad B = \frac{Q\exp(-k\tau)}{V}.$$

Alternatively, the equation can be written in the form

$$T\frac{dx(t)}{dt} = -[1 + kT]x(t) + gu(t-\tau) \tag{1.6}$$

where $g = \exp(-k\tau)$; while $T = V/Q$ can be considered as a *residence time* parameter for the ADZ, in the sense that it provides a measure of the time which the solute, with concentration $x(t)$, resides in the volume $V = D_f V_a$. Another useful parameter in the model (1.6) is the *steady state gain* (*SSG*), usually denoted by G and defined as $G = B/A$: this determines the relationship between the equilibrium value of the output $x(t)$, for $t \to \infty$, when the input is maintained at a steady value, say $u(\infty)$.

The unit impulse response of equation (1.5) (i.e. temporal response to a unit or 'gulp' input of pollutant applied at $t = 0$) is given by

$$x(t) = 0 \quad \text{for } t < \tau; \quad x(t) = B\exp\{-A(t-\tau)\} \quad \text{for } t \geq \tau. \tag{1.7}$$

In other words, the impulsive input is transformed by the model to yield a concentration profile at the output of the reach which, following the advective time delay of τ time units, jumps to a level B and then decreases exponentially with time constant $1/A = T/(1 + kT)$ time units. This exponential decrease is clearly a measure both of the dispersive nature of the model, as defined by the residence time parameter T, and of the natural decay of solute, as defined by the decay rate k. But if the solute is conservative, as in the case of the fluorescent tracer Rhodamine WT mentioned above in connection with Figure 1.1(a), then $k = 0$, and the parameter $T = 1/A$ completely defines the dispersive characteristics of the reach (in this conservative situation, note also that $A = B$). This suggests an obvious experimental approach to characterising the river reach in dispersive terms: first, input–output data from a conservative tracer experiment, such as that shown in Figure 1.1(a), is used to estimate the parameters A and B; then the inverse of the parameter A yields a direct estimate of the residence time T.

In such analysis, however, it is important to note that T is associated with the volume $V = D_f V_a$ and *not* the total reach volume V_a; so that if $D_f < 1.0$ then the *effective residence time* associated with dispersion processes is less than the residence time of the reach as a whole. Bearing in mind our hypothesis that it is dead zones that provide the main mechanism for dispersion, therefore, it is reasonable to associate the volume V with the aggregated effect of all the dead-zone behaviour in the reach and refer to it as the *ADZ volume*. In this manner, we see that the *ADZ* volume V is *not* the total reach volume V_a and will normally be much less than V_a; indeed, as we shall see, the dispersive fraction $D_f = V/V_a$, is normally in the range 0.3 to 0.4 for medium-sized natural streams (see later and Young and Wallis, 1986; Wallis *et al*, 1989)

One very interesting and significant aspect of the ADZ modelling studies is the manner in which the model characteristics change with river discharge. In particular, while both the ADZ volume V and the total volume of water V_a in the reach vary over a wide range, as might be expected, the fractional volume of the reach which appears attributable to ADZ effects, i.e. the dispersive fraction D_f, is approximately constant over the whole range of flows. For example, in the River Conder, $D_f = 0.37$; in other words, only 37% of the river volume appears primarily responsible for introducing the pollutant dispersion effects, with the remainder functioning in a largely advective role.

Finally, it is worth emphasising that the name 'aggregated dead zone' is rather misleading in its physical connotations. It is a name that was coined originally to draw attention to the relationship between the previous dead-zone research and the ADZ model. But the ADZ is clearly not a 'dead zone' in any real sense of the term; rather it is a lumped approximation to the many 'active mixing zones' that characterise the reach and lead to the dispersion of the pollutant. For this reason, it might be better to refer to the aggregated dispersive effect modelled by the residence time T as the *active mixing zone* (*AMZ*) and the associated volume V as the *active mixing volume* (*AMV*). Indeed, these rather more physically appropriate terms will be used later when we suggest that models of this type have a potentially much wider significance in many different areas of environmental systems modelling.

1.2.2 Transfer function representation of the ADZ model

A general introduction to transfer function (TF) models is given in *Appendix 1.1*. In the present first-order differential-equation context, the time-derivative

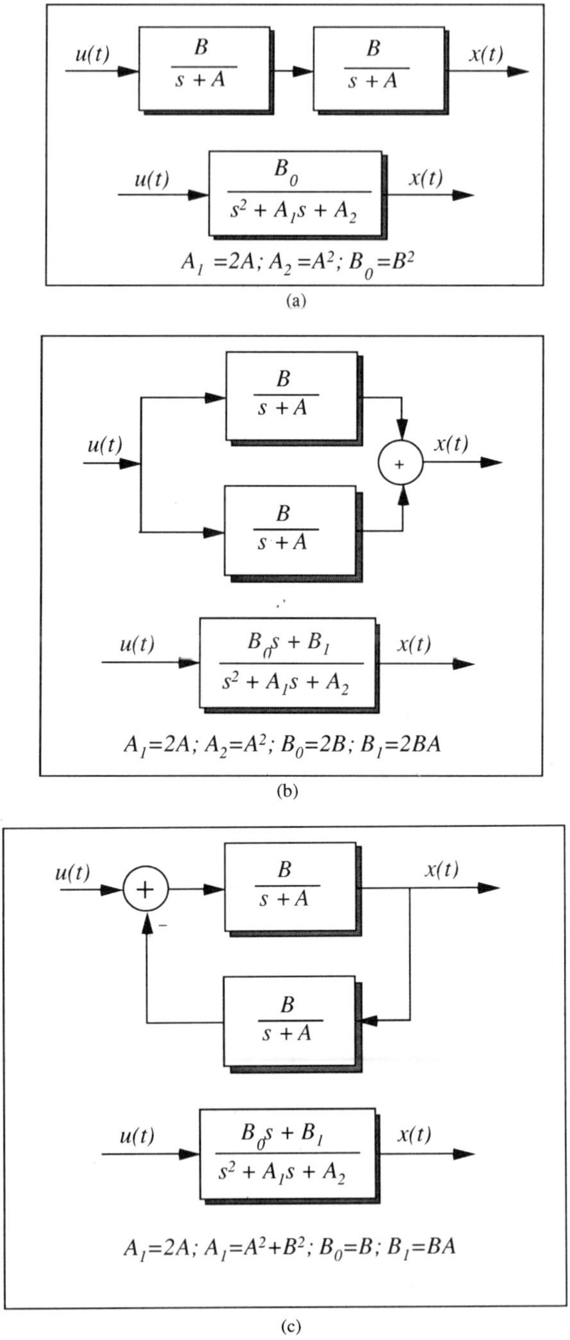

Figure 1.2 Serial, parallel and feedback connection of first-order transfer function models: (a) serial connection; (b) parallel connection; (c) feedback connection (single equivalent, second-order block diagram shown below in each case)

(*s*) operator TF representation of equation (1.5) is obtained simply by setting $d/dt = s$ and rearranging the equation to yield,

$$x(t) = \frac{B}{s + A} u(t - \tau) \quad \text{or} \quad x(t) = \frac{G}{1 + Ts} u(t - \tau) \tag{1.8}$$

where $G = B/A$ and $T = 1/A$. The associated delta (δ) and backward shift (z^{-1}) operator TF forms can then be obtained straightforwardly for any defined sampling interval Δt, as described in *Appendix 1.1*.

In general, for an arbitrarily selected stretch of river, we cannot assume that a *single* ADZ element such as this will be able to model fully the dispersive properties; it could well be that several elements such as this would be required to model a reasonably long length of river. Depending upon the nature of the river network, this higher-order model could be made up of serial, parallel, or feedback connections of first-order elements, as shown diagrammatically in Figures 1.2(a), (b) and (c). These figures demonstrate how such connections of two identical, first-order ADZ elements described by models of similar form to equation (1.8) lead to overall second-order transfer functions in the *s* operator. It is quite obvious, therefore, that networks involving *n* ADZ elements can be represented by a general *n*th-order transfer function of the form

$$x(t) = \frac{B_0 s^m + B_1 s^{m-1} + \cdots + B_m}{s^n + A_1 s^{n-1} + \cdots + A_n} u(t - \tau) \tag{1.9}$$

as described in *Appendix 1.1*. Moreover, as the *appendix* points out, this same higher-order model can be considered in its equivalent δ and z^{-1} operator TF

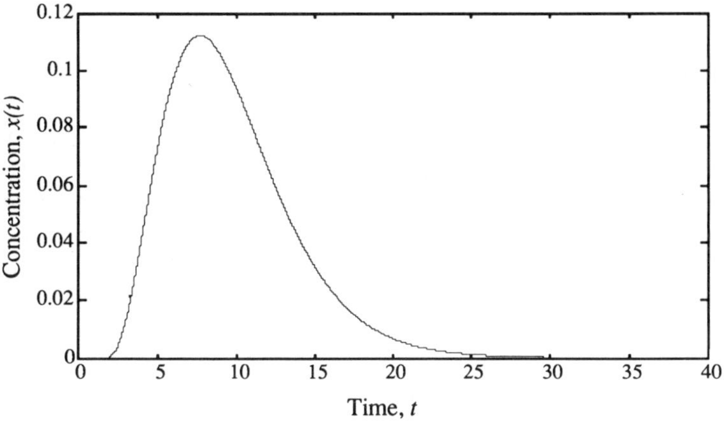

Figure 1.3 Impulse response of four identical conservative ADZ element in series.

forms, which may be more appropriate if data on the system are available in sampled form.

The need to specify a general model of this type which allows for the interconnections of first-order ADZ elements, arises from physical considerations. In a single uniform stretch of river, for instance, we might assume that a serial connection of ADZ elements, each defined by a first-order TF such as (1.8), could describe the dispersive characteristics. An example of this is given in *Appendix 1.1*, with four conservative ADZ elements in series, each with $T = 2$ time units ($A = B = 0.5$), and pure time delay $\tau = 0.5$ time units. The impulse response of this serial combination is shown in Figure 1.3. The general similarity between this response and the measured tracer experiment responses in Figure 1.1(a) is clear on inspection. If we allow for more complicated river networks, or more complex multi-layer structure in each reach of the river channel, however, then serial and parallel connections are both possible (see later and Young and Wallis, 1992).

1.3 IDENTIFICATION AND ESTIMATION OF TF MODELS

In the previous section, we have suggested that the transport and dispersion of a solute in a river system can be described by combinations of elemental first-order ADZ models, as represented by multi-order transfer functions in continuous or discrete time. Later, we will see that such TF models can be applied generally in environmental systems modelling and provide an essential tool for data-based applied systems analysis. The problem of identifying the appopriate order of TF model, and then estimating the parameters that characterise this model, is one of *time-series analysis*. In this section, therefore, we will outline a particular approach to time-series analysis based on recursive parameter estimation which has proven particularly effective in the environmental context.

1.3.1 Recursive parameter estimation

Of the various methods available for time-series analysis, recursive estimation (see Young, 1984) is probably the most flexible method of modelling TF models from time-series data. This approach, which involves sequential updating of the TF model parameter estimates whilst processing the data in a serial manner, has several theoretical and practical advantages over the alternative *en bloc* approaches, such as the better-known maximum likelihood methods proposed by Box and Jenkins (1970). In particular, the recursive formulation means that the estimation algorithms can be extended quite easily to handle *nonstationary* models, i.e. models which are characterised by *time-variable parameters* (*TVP*).

The inherent nonstationarity of many environmental time-series is widely acknowledged and the relevance of such TVP methods in environmental systems analysis is obvious. In particular, the algorithms can be used in an *on-line* form as an aid in the *adaptive* management of environmental systems. Typical examples of this currently being developed at Lancaster University are an adaptive flood-warning system for the Solway River Purification Board; an adaptive flow-management model of the River Derwent for the National Rivers Authority; and self-adaptive controllers for greenhouse environmental systems, in association with the Agriculture and Food Research Council (AFRC) Silsoe Research Institute.

Of the many recursive methods that are now available (see e.g. Ljung and Soderstrom, 1983; Young, 1984; Norton, 1986; Soderstrom and Stoica, 1989) for the estimation of z^{-1} operator transfer-function model parameters, only one can also be applied directly to δ and s operator models. This is the *simplified refined instrumental variable* (*SRIV*) procedure (Young, 1984, 1985) which exploits special adaptive prefiltering, both to achieve good estimation performance and, in the δ and s operator cases, to avoid numerical differentiation. As such, the SRIV algorithm provides the most obvious practical vehicle for identification and estimation of TF models and it is a central tool in the *microCAPTAIN* time-series analysis computer program package developed at Lancaster and used to generate all the results presented later in this chapter.

Time-series analysis is normally posed in discrete-time, stochastic terms. In order to introduce SRIV estimation, therefore, let us consider the stochastic form of the z^{-1} operator TF (cf equation (A1.3) in *Appendix 1.1*),

system:
$$x(k) = \frac{B(z^{-1})}{A(z^{-1})} u(k-\Delta), \tag{1.10}$$

output equation:
$$y(k) = x(k) + e(k) \tag{1.11}$$

or, combining these two equations,

$$y(k) = \frac{B(z^{-1})}{A(z^{-1})} u(k-\Delta) + e(k) \tag{1.12}$$

where $e(k)$ is a zero-mean, serially uncorrelated sequence of random variables with variance σ^2, which is introduced to allow for any unmeasurable stochastic inputs, disturbances, or measurement noise that may affect the system between the input $u(t)$ and the 'noisy' output measurement $y(t)$. This equation can be written in the following alternative vector form, which is *linear-in-the-parameters* $\{a_i, b_j\}$ of the TF model,

$$y(k) = \mathbf{z}(k)^{\mathrm{T}}\mathbf{a} + \eta(k), \tag{1.13}$$

where

$$\mathbf{z}(k)^{\mathrm{T}} = [-y(k-1), \ldots, -y(k-n)u(k-\Delta), u(k-\Delta-1), \ldots, u(k-\Delta-m)], \tag{1.14}$$

$$\mathbf{a} = [a_1, a_2, \ldots, a_n, b_0, b_1, \ldots, b_m]^{\mathrm{T}}$$

and $\eta(k)$ is a coloured noise variable defined as follows in relation to the original white noise $e(k)$,

$$\eta(k) = e(k) + a_1 e(k-1) + \cdots + a_n e(k-n).$$

Note that here, for generality, we have assumed that the numbers of a and b parameters are different (n and m respectively). The orders n and m, as well as the time delay Δ, will depend upon the nature of the connection between the elements. In later analysis, it is convenient to use the shorthand notation $[n, m+1, \Delta]$ to refer to such a model.

Most estimation problems are posed in a manner such that the variable to be minimised has white-noise properties. Thus, a sensible error function is the *response* or *prediction error*, $\hat{e}(k)$,

$$\hat{e}(k) = y(k) - \frac{\hat{B}(z^{-1})}{\hat{A}(z^{-1})} u(k-\Delta)$$

where $\hat{B}(z^{-1})$ and $\hat{A}(z^{-1})$ are estimates of the TF polynomials $A(z^{-1})$ and $B(z^{-1})$. Unfortunately, this is nonlinear in the unknown parameters and so the estimation problem cannot be posed directly in simple, linear estimation terms. However, it becomes *linear-in-the-parameters* if we assume prior knowledge of $A(z^{-1})$ in the form of an estimate $\hat{A}(z^{-1})$: then the error equation can be written in the form

$$\hat{e}(k) = \frac{1}{\hat{A}(z^{-1})} [\hat{A}(z^{-1}) y(k) - \hat{B}(z^{-1}) u(k-\Delta)],$$

which can be rewritten as

$$\hat{e}(k) = \hat{A}(z^{-1}) y^*(k) - \hat{B}(z^{-1}) u^*(k-\Delta),$$

where

$$y^*(k) = \frac{1}{\hat{A}(z^{-1})} y(k); \quad u^*(k) = \frac{1}{\hat{A}(z^{-1})} u(k)$$

are 'prefiltered' variables, obtained by passing $y(k)$ and $u(k)$ through the prefilter $F(z^{-1})$, where

$$F(z^{-1}) = \frac{1}{\hat{A}(z^{-1})}.$$

It will be noted that, if a method for defining $\hat{A}(z^{-1})$ can be defined and the resultant estimation algorithm converges, then this prefilter should have a similar frequency passband to that of the system itself: as a result, it performs the intuitively reasonable function of passing those parts of the input and output signals that contain information on the system, whilst filtering off higher-frequency 'noise' components that will have a deterious effect on the parameter estimation.

With the above reasoning in mind, the ordinary IV algorithm (e.g. Young, 1984) can be applied *iteratively* to estimate the model parameter vector **a**, with the variables $y(k)$, $u(k)$ and the instrumental variable $\hat{x}(k)$ replaced, at each iteration, by their adaptively prefiltered equivalents $y^*(k)$, $u^*(k)$ and $\hat{x}^*(k)$, respectively, and with the prefilter parameters based on the parameter estimates obtained at the previous iteration (see Young, 1984, 1985). The main recursive part of the SRIV algorithm takes the form

$$\hat{\mathbf{a}}(k) = \hat{\mathbf{a}}(k-1) + \mathbf{g}(k)\{y^*(k) - \mathbf{z}^*(k)^{\mathrm{T}}\hat{\mathbf{a}}(k-1)\} \qquad \text{(i)}$$

where,

$$\mathbf{g}(k) = \mathbf{P}(k-1)\hat{\mathbf{x}}^*(k)[1 + \mathbf{z}^*(k)^{\mathrm{T}}\mathbf{P}(k-1)\hat{\mathbf{x}}^*(k)]^{-1}. \qquad \text{(ii)} \qquad (1.15)$$

and

$$\mathbf{P}(k) = \mathbf{P}(k-1) + \mathbf{g}(k)\mathbf{z}^*(k)^{\mathrm{T}}\mathbf{P}(k-1) \qquad \text{(iii)}$$

where $\mathbf{P}(k)$, which is the inverse of the instrumental cross-product matrix (*ICPM*; see e.g. Young *et al*, 1980), i.e.,

$$\mathbf{P}(k) = \left[\sum_{i=1}^{i=k} \hat{\mathbf{x}}^*(i)^{\mathrm{T}}\mathbf{z}^*(i)\right]^{-1} \qquad (1.16)$$

is also related to the covariance matrix $\mathbf{P}^*(k)$ of the estimated parameter vector $\hat{\mathbf{a}}(k)$ by the equation

$$\mathbf{P}^*(k) = \sigma^2\mathbf{P}(k), \quad \text{where } \mathbf{P}^*(k) = E\{\tilde{\mathbf{a}}(k)\tilde{\mathbf{a}}^T(k)\} \qquad (1.17)$$

and $\tilde{\mathbf{a}}$ is the parameter estimation error, defined by

$$\tilde{\mathbf{a}}(k) = \hat{\mathbf{a}}(k) - \mathbf{a}.$$

An estimate $\hat{\sigma}^2$ of the variance σ^2 can be obtained from an additional recursive equation based on the squared values of a suitably normalised recursive innovation sequence (see Young, 1984, p. 100; Young *et al*, 1991). The standard errors on the parameter estimates are computed from the diagonal elements of $\mathbf{P}^*(k)$; so we see that, at every sample in time, the recursive algorithm (1.15) provides both the currently best SRIV estimate of the model parameter vector **a** and an estimate of the uncertainty associated with this estimate.

The δ and s operator versions of the SRIV algorithm (1.15) are quite similar in algorithmic terms but applied to the δ and s operator equivalents of the equation (1.13). The δ operator algorithm and its use in various automatic control and environmetric applications are described fully in Young *et al.* (1991). The continuous-time versions of the SRIV algorithm, which are closely related to the δ operator versions, are described in Young and Jakeman (1980) and Young (1981); however, they constitute theoretically justified versions of heuristic estimation algorithms developed by Young in the 1960s (Young, 1965, 1970). Also, an alternative approach to s operator TF estimation, which exploits numerical differentiation based on recursive fixed-interval smoothing, has been suggested recently by Young (1992). This latter approach is utilised later in the chapter to estimate the parameters in a model of temperature variations in a soil column.

1.3.2 TF model order identification

The ability to estimate the parameters in a transfer function model represents only one side of the time-series analysis problem. Equally important is the related problem of *model order identification*: namely the identification of the most appropriate values for the orders n and m (or p) of the TF polynomials and the size of any accompanying pure time delay which affects the input–output relationship. This is not a simple problem and there is, as yet, no foolproof and entirely objective method for solving it in purely statistical terms. The process of identification can, however, be assisted by the use of well-chosen statistical measures which indicate the presence of overparametrisation; i.e. they indicate when the model appears to be characterised by more parameters than are really necessary to explain the observed input–output behaviour.

A reasonably successful identification procedure, which exploits certain useful properties of IV estimation (see Young, 1989), is to choose the model which minimises the following identification statistic,

$$\mathrm{YIC} = \ln\left\{\frac{\sigma^2}{\sigma_y^2}\right\} + \ln\{\mathrm{NEVN}\} \tag{1.18}$$

where,

σ^2 is the sample variance of the model residuals $e(k)$,
σ_y^2 is the sample variance of the measured system output $y(k)$ about its mean value;

while NEVN is the 'normalised error variance norm' (Young *et al*, 1980) defined as

$$\mathrm{NEVN} = \frac{1}{np}\sum_{i=1}^{np}\frac{\hat{\sigma}^2 p_{ii}}{\hat{a}_i^2}. \tag{1.19}$$

Here, in relation to the TF models (A1.3) and (A1.4), np is the total number of parameters estimated, i.e. $n+m+1$ for model (1) and $2p+1$ for model (2); $\hat{a}_i^2$ is the estimate of the ith parameter in the parameter vector $\mathbf{a}$; p_{ii} is the ith diagonal element of the $\mathbf{P}(N)$ matrix, where N is the sample size (so that $\hat{\sigma}^2 p_{ii}$ is an estimate of the error variance associated with the ith parameter estimate after N samples).

This YIC statistic is heuristic but can be justified on two grounds: statistical and numerical. From the statistical standpoint, the first term in (1.18) provides a normalised measure of how well the model explains the data: the smaller the variance of model residuals in relation to the variance of the measured output, the more negative the first term becomes. Similarly, the second term is a normalised measure of how well the parameter estimates are defined for the npth order model: clearly the smaller the relative error variance, the better defined are the parameter estimates in statistical terms, and this is once more reflected in a more negative value for the term. Thus the model which minimises the YIC should provide a good compromise between model fit and parametric efficiency: as the model order is increased, so the first term tends to decrease; while the second term tends to decrease at first and then to increase quite markedly when the model becomes overparametrised and the standard errors on its parameter estimates become large in relation to their estimated values (in this connection, note that the square root of $\hat{\sigma}^2 p_{ii}/\hat{a}_i^2$ is simply the relative standard error on the ith parameter estimate).

From a numerical analysis standpoint, the NEVN is a normalised measure of the magnitude of the $\mathbf{P}(N)$ matrix which, it will be recalled from equation (1.16), is the inverse of the instrumental cross-product matrix (ICPM). It can be shown (see Young *et al*, 1980) that, in theory, if the order of the system is greater than that justified by the data, then the ICPM will become singular. In practice, complete singularity will rarely occur, and so this problem is revealed by the poor conditioning of the ICPM, with the result that the elements of its inverse $\mathbf{P}(k)$ can become very large. In this situation, the NEVN, whose value is strongly dependent upon $\mathbf{P}(k)$, will also increase markedly in value, indicating the poor conditioning of the IPCM and associated poor definition (i.e. high variance) of the TF model parameters.

In practical applications, of course, the minimisation of YIC will not, in itself, guarantee that the 'best' model has been identified: it is, after all, only one possible, heuristic, statistic which is naturally dependent upon the quality and size of the time-series data set. Consequently, inadequate or very noisy data can lead to a YIC identified model structure which may not be acceptable for some good physical reasons. In other words, the YIC should be used, together with other *physically motivated* considerations, such as the nature of the model response characteristics and any *a priori* information on the system dynamics, as a *guide* to the selection of the most appropriate model order.

For example, the YIC can tend to under identify the model order when the input signal is limited for some reason; e.g. in tracer experiments where it is restricted to an impulsive 'gulp' or single-step excitation. In such circumstances, where the input is not *persistently exciting* (see e.g. Young, 1984), it may well be that higher-order (and usually better-fitting) models, with somewhat larger YIC values, will have certain practical advantages. It would than be advisable to select one of these higher-order models, particularly if *a priori* considerations suggest that its superior fitting ability outweighs the possibility of overparametrisation. Practical examples of such discrimination are discussed in the next sections of the chapter.

1.3.3 Identifiability

Later, when we consider the more general concept of an *active mixing volume*, it is necessary to examine the relationship between the data-based models, as identified and estimated from experimental or monitored data, and mechanistic models, in the form of theoretical differential equations, obtained via conservation arguments. As we shall see, this can introduce certain **identifiability** problems, in which the unique estimation of the physically significant mechanistic model parameters from the data-based model parameters proves impossible because of ambiguity in the equations relating the parameters. This problem of identifiability is well known in various areas of science and social science. For instance, Godfrey and DiStefano (1985) present a useful survey of the subject in relation to problems encountered by scientists, engineers and medical researchers; where the latter are concerned with 'compartmental models' that are rather similar to the ADZ and AMV models considered here. Econometricians are also confronted by similar problems when considering the relationship between their 'reduced form', data-based models and the 'structural' models which are assumed to reflect economic theory (see e.g. Johnston, 1963, p. 240 *et seq*:).

In order to introduce the topic, let us consider the ADZ model (1.5). The problem of identifiability in this case concerns the relationship between the data-based model parameters $\{A, B\}$ and the mechanistic model parameters $\{Q, V, k\}$, as defined by the equations

$$A = \frac{Q}{V} + k \quad \text{and} \quad B = \frac{Q \exp(-k\tau)}{V}.$$

(Econometricians would call this a *structural* model.)

The two models are obviously *observationally equivalent*, in the sense that, for a given input $u(t)$, the output of equation (1.4) will be identical to the output of (1.5). But this does not mean that all the parameters in (1.4) can be unambiguously defined from such input–putput data; or that they can be derived from knowledge of the parameters in (1.5).

For instance, suppose we obtain estimates of $\{A, B, \tau\}$ using TF model identification and estimation, and then attempt to derive estimates of $\{Q, V, k\}$ from these relationships. It is clearly not possible to obtain unique values for $\{Q, V, k\}$

in this manner, since there are only two equations in the three unknown parameters. Thus the parameters $\{Q, V, k\}$ are said to be *not fully identifiable* from the parameters $\{A, B, \tau\}$ (or *underidentified* in the semantics of econometrics). However, if we only wish to estimate the ADZ residence time $T = V/Q$, then the parameters T and k are identifiable from $\{A, B\}$, since we now have two equations in two unknowns. And if we assume knowledge of the flow rate Q, then the parameters $\{V, k\}$ are also identifiable from $\{A, B\}$.

In this ADZ case, the identifiability restrictions do not cause any real practical problems, since knowledge of the mechanistic model parameters T or V (given the flow rate Q) satisfy our requirements for estimating physically meaningful parameters. But the situation may not always be so fortuitous: in the later example concerned with the modelling of temperature variations in a soil column, for example, the mechanistic model parameters are not uniquely determined from the TF model parameter estimates and the system is not fully identifiable in this sense.

1.4 THE ADZ MODEL REVISITED: PRACTICAL RESULTS

The elucidation of a method-theory, and development of associated methodological tools, are only justified if the approach can be shown to work in practical applications. Before proceeding to consider the more general implications of the methodological approach described in previous sections of this chapter, therefore, let us consider first how it has been applied successfully in the case of our motivational example; namely, to the ADZ modelling of pollutant transport and dispersion in a river system. In particular, we will present some of the results obtained in an NERC-sponsored research programme carried out a few years ago by the first author and his colleagues at Lancaster (Young and Wallis, 1986; Wallis *et al*; 1989).

The study involved four different stream channels in the Lancaster area, but we will consider here only those obtained from a dye-tracer experiment carried out on a short reach of the River Conder, a medium-sized stream in north-west England, the results of which are shown in Figure 1.1(a). These typify the results obtained on the other stream channels, as well as those from many subsequent studies carried out on other rivers and channels. The reach was 5 metres wide and 116 metres in length, characterised by one major pool-riffle structure with a fine cobble bed and samples were taken every 15 seconds. A number of different order TF models were estimated from these data using the SRIV method of recursive estimation in the microCAPTAIN microcomputer program (see Young and Benner, 1991). Table 1.1 compares the results obtained from this exercise and shows that the best identified model in terms of the YIC identification statistic is the first-order (i.e. $[1, 1, 12]$ or $n = 1$; $m + 1 = 1$; $\Delta = 12$) model, which can be interpreted in physical terms as a single ADZ element with

Table 1.1 SRIV identification results

Demominator order (n)	Numerator order ($m+1$)	Advective delay δ (sampling interval)	R_T^2	YIC orderimg
1	1	12	0.9933	**1**
2	1	11	0.9923	4
2	2	12	0.9978	**2**
3	1	11	0.9971	5
3	2	11	0.9981	3

a residence time of 65 seconds and an advective time-delay of 12 sampling intervals (i.e. 180 seconds).

An alternative model is the second-order model [2, 2, 12], which can be interpreted as two ADZ elements with residence times of 34.7 and 125.7 seconds respectively. This second-order model has a somewhat larger YIC value but it fits the data rather better, with a coefficient of determination $R_T^2 = 0.9978$ compared with 0.9933 for the first-order model (i.e. 99.78% of data explained), particularly in the important tail of the concentration profile.

Clearly, from a purely objective YIC standpoint, we should accept the first-order model as the most appropriate representation of the data, with the better fit provided by the higher-order model not fully justified in statistical terms by its increased parametrisation. However, we see that the second-order model is quite well identified, being the second choice in YIC terms, and the improvement in the explanation of the elevated tail seems to be visually and, as we discuss below, physically significant.

This improved modelling of the tail portion of the concentration profile is particularly important in the present context since it is a direct product of the model's structure. If we assume that the second-order model is the best candidate, then it can be decomposed, quite unambiguously, into two ADZ elements which have rather different residence times and are arranged in a *parallel* structure, as shown in Figure 1.1(b). A rather appealing, albeit speculative, interpretation of such a model structure, therefore, is a two-layer flow pattern, with the shorter residence time ADZ representing the fast-moving surface layer, and the longer one representing the slower subsurface behaviour. This seems quite a reasonable conjecture in the case of the River Conder, where the physical nature of the channel could create the right conditions for such multi-layer behaviour.

This example emphasises our earlier comments that, while objective statistical criteria such as the YIC can be a great help in ensuring that the model is not over-parametrised, they should not provide the *only* criteria on which the model is judged. If the model has a reasonable physical interpretation, as here, then physical considerations may well be used to override the purely statistical considerations and add to the model's credibility. This seems quite reasonable from a Bayesian statistical standpoint, which would suggest that prior knowledge of this type should be allowed to influence the modelling. It is interesting to note that the recursive approach to time-series analysis used here can be considered directly in Bayesian terms; indeed, the recursive update at each sample from the *a priori* to *a posteriori* estimate of the parameter vector is the very embodiment of Bayesian estimation (Young, 1984).

But we should not allow this influence to be too great without careful thought: in this example, for instance, the reason for the ambiguity between first- and second-order models arises partly from the poor excitation provided by the impulsive input and so, before reaching a final decision on which model is most appropriate, we might think of conducting other experiments to clarify the situation (see Jakeman and Young, 1980). Indeed, in this case, numerous experiments were carried out on the River Conder over a wide range of discharge conditions (see below) and the parallel structure was indicated most of the time. It should be emphasised, however, that either of the two best identified models (first- or second-order) would perform quite well in simulation or predictive applications, since they both explain over 90% of the experimental data (much better than an ADE model) and are reasonably efficient in their parametrisation.

1.4.1 Wider implications of the ADZ model

These ADZ modelling results show the potential value of the time-series approach to modelling. By analysing the experimental tracer data in a careful and largely objective, statistical manner, it is not only possible to model the data efficiently, but also to expose an important flow-invariant parameter; namely the dispersive fraction. It is always satisfying if such an invariant characteristic can be discovered in scientific investigations and it is already helping to stimulate new theoretical (see e.g. Smith, 1987) and practical developments on dispersion modelling. In practical terms, studies carried out with North West Water, Yorkshire Water and the Water Research Centre have shown the potential value of the model in incident prediction and the evaluation of storm-water overflows.

Finally, it should be noted that the kind of ADZ modelling discussed here can be considered as a method of calibrating the river in terms of its transport and dispersive properties. As a result, the model can form a useful basis for the development of more complex water-quality models for interacting, nonconservative, pollutants or biological variables (see e.g. Young, 1990; Riberio da Costa, 1992).

1.5 OTHER PRACTICAL EXAMPLES OF THE ADZ-TYPE MODEL

In this section, we consider two other practical examples where the ADZ type of model has proved valuable in characterising solute transport processes. Both examples are concerned with the modelling of soil-water processes based on breakthrough curve data obtained from tracer experiments on saturated soil columns.

1.5.1 Solute transport in soils: I

This example concerns the transport of solutes through saturated soils and is based on the paper by Beven and Young (1988). The manner in which solutes are transported through soils is clearly an important factor in determining how dissolved pollutants will be transported and dispersed in the natural environment. The conventional approach to modelling the transport and dispersion of solutes in soils has been by recourse to the ADE model (1.4), which is usually termed the convective-dispersion equation (CDE) when applied to soil-water problems. This model has often proved useful in describing the observed dispersion of solutes with adequate accuracy, at least for repacked columns of soil in the laboratory. As in the case of dispersion in rivers and open channels, however, there has been growing awareness that solute transport in more realistic conditions, for example through soils in the field and in undisturbed soil columns, does not often conform to the predictions of CDE theory. In particular, and again in sympathy with the situation of streamflow dispersion, the graph of the concentration of the solute in the water leaving an undisturbed soil column, or the *breakthrough curve* as it termed in the literature, tends to show more rapid initial rise and a longer tail than expected, following the application of an impulsive input of solute at the top of the column (e.g. see Elrick and French, 1966; Smettem, 1984; and many others).

The situation encountered here is very similar to that in the case of riverine pollutant dispersion and it is not surprising that the systems contribution to the subject is closely related to that used in rivers, namely the ADZ model described previously in this chapter. In applying ADZ concepts to soils, it is assumed that there may be inefficient mixing between the faster flow pathways in larger structural pores and slower flows in the matrix, but that an aggregate effective mixing volume again may be able to model the resulting dispersion. However, it is more likely in the soil that there will be part of the soil-water storage that takes no part in the transport process (except by slow molecular diffusion) and so acts as a true dead zone. Consequently, in order to avoid confusion with more traditional concepts, we have used the name *aggregated mixing zone* (AMZ) to describe the model (Beven and Young, 1988).

As in the riverine ADZ case, the main limitation of this AMZ model lies in the fact that it is essentially a linear, small-perturbation model similar to earlier transfer function models introduced by Jury and his co-workers (e.g. Jury, 1982; Jury and Sposito, 1986). The AMZ model, however, has the advantage that it is obtained using a data-based approach based on recursive estimation which, as we have seen, allows for objective identification of the model order and the ability

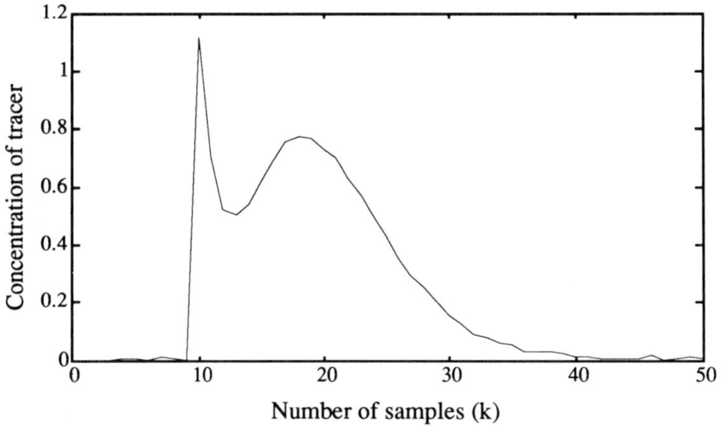

Figure 1.4 Soil tracer experiment: breakthrough curve for impulsive input of tracer

to estimate any changes in the parameters that might be expected as the flow regime changes.

Most of the AMZ models obtained by Beven and Young (1988) are of second or third order and their bimodal response shape again suggests the presence of parallel flow pathways. In the case of Figure 1.4 in the Beven and Young paper, for example, the third-order model can be unambiguously decompsed into the following form,

$$y(k) = \frac{b_1 z^{-1}}{1 + a_1 z^{-1}} u(k-2) + \frac{b_2 z^{-1} + b_3 z^{-2}}{1 + a_2 z^{-1} + a_3 z^{-2}} u(k-2) + \xi(k) \qquad (1.20)$$

where $y(k)$ is the concentration of solute in the water leaving the soil column and $u(k)$ is the input concentration (in this experiment, an impulse input). In other words, the model suggests a parallel connection of first- and second-order subsystems: the relatively fast, first-order system (the first term on the right-hand side of equation (1.20)) has a mean AMZ residence time (time constant) of 599 s; while the slower, second-order system is composed approximately of two first-order systems *in series*, each with AMZ residence times of 943 s.

The estimated residence times in this model imply AMZ volumes of 53.9 cm^3 and 94.4 cm^3 respectively, and the total effective mixing volume of all AMZ elements is found to be 21% of the soil sample, which was close to saturation in the experiment. An estimate of the porosity of the sample was not available in this experiment, but it is expected that this effective mixing volume will be much less than the water-filled pore space in the sample, suggesting significant by-passing behaviour in the solute transport process.

1.5.2 Solute transport in soils: II

In this example, we again consider the modelling of solute transport in an experimental soil column based on data collected some years ago. However, since the full results have not been published previously and represent a more difficult problem of parallel flow model decomposition, we discuss them in somewhat more detail. The breakthrough curve data are shown in Figure 1.4 and Table 1.2 shows the identification results obtained by analysis of these data using recursive least squares, instrumental variable and SRIV methods.

Figure 1.5 compares the best model fits obtained with the three different methods; and Table 1.3 presents the detailed estimation results for some of the more interesting examples listed in Table 1.2. Table 1.2 and Figure 1.5 demonstrate the power of SRIV estimation, which is able to explain the data almost perfectly: in constrast, the least squares (LS, effectively linear regression) model yields a very poor fit; and the ordinary IV model yields an improved result, but

Table 1.2 Soil-core identification results

(a) LS algorithm			(b) IV algorithm			(c) SRIV algorithm		
Model	R_T^2	YIC	Model	R_T^2	YIC	Model	R_T^2	YIC
3,3,0	0.788	−1.847	3,3,0	0.978	−6.699	3,3,0	0.999	−18.49
4,4,0	0.995	3.080	4,4,0	0.994	−4.690	4,4,0	0.999	−11.24
4,3,0	0.830	−1.077	4,3,0	0.985	−5.883	4,3,0	0.998	−6.119
3,4,0	0.883	−1.444	3,4,0	0.980	−5.238	3,4,0	0.998	−6.118

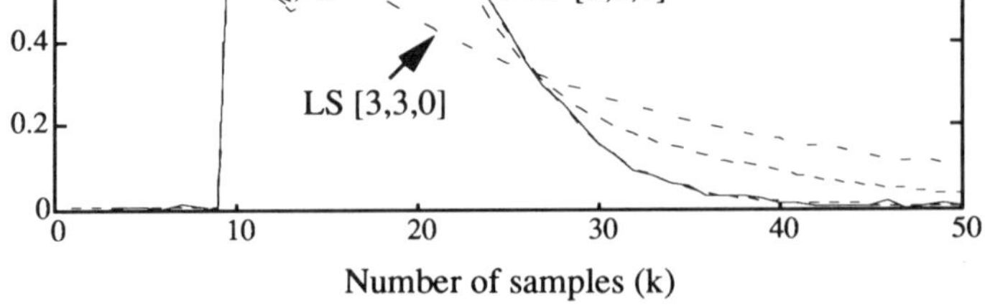

Figure 1.5 Soil tracer experiment: comparison of different microCAPTAIN estimated models

Table 1.3 Selected soil-core estimation results (figures in parentheses are standard errors)

LS estimates

[3,3,0]

$\hat{a}_1 = -2.051(0.825)$; $\hat{a}_2 = 1.444(1.248)$; $\hat{a}_3 = -0.379(0.487)$: roots (den.) 0.953; $0.55 + 0.31j$
$\hat{b}_0 = 1.114(0.140)$; $\hat{b}_1 = -1.585(0.931)$; $\hat{b}_2 = 0.689(0.856)$: roots (num.) 0.953; $0.55 + 0.31j$

IV estimates

[3,3,0]

$\hat{a}_1 = -2.375(0.326)$; $\hat{a}_2 = 1.942(0.486)$; $\hat{a}_3 = -0.554(0.181)$: roots (den.) 0.909; $0.78 + 0.27j$
$\hat{b}_0 = 1.114(0.046)$; $\hat{b}_1 = -1.946(0.366)$; $\hat{b}_2 = 1.017(0.327)$: roots (num.) 0.953; $0.55 + 0.31j$

SRIV estimates

[3,3,0]

$\hat{a}_1 = -2.3975(0.005)$; $\hat{a}_2 = 1.9464(0.009)$; $\hat{a}_3 = -0.5358(0.004)$: root (den) 0.834; $0.78 + 0.18j$
$\hat{b}_0 = 1.115(0.005)$; $\hat{b}_1 = -1.976(0.01)$; $\hat{b}_2 = 1.0224(0.006)$: root (num.) 0.953; $0.55 + 0.31j$

[2,5,0]

$\hat{a}_1 = -1.7453(0.013)$; $\hat{a}_2 = 0.77163(0.012)$: roots (den.) $0.8727 + 0.1j$
$\hat{b}_0 = 1.1133(0.027)$; $\hat{b}_1 = -1.2425(0.055)$; $\hat{b}_2 = 0.1538(0.057)$; $\hat{b}_3 = 0.071(0.055)$; $\hat{b}_4 = 0.2213$ (0.034)

Note almost two real roots at 0.87: i.e. $(1 - 0.87z^{-1})(1 - 0.87z^{-1}) = 1 - 1.74z^{-1} + 0.76z^{-2}$

not nearly so good as the SRIV model. However, Table 1.3 reveals that the LS, IV and [3, 3, 0] SRIV models are all characterised by significantly complex poles (eigenvalues). Since there is no obvious physical mechanism for the oscillatory modes associated with these complex poles, this calls into question their efficacy. However the [2, 5, 0] SRIV model (not listed in Table 1.2) also fits the data almost perfectly and, while it nominally has complex poles, the imaginary part is small so that the model is well approximated by two *real* poles, both at 0.87.

What do these estimation results reveal? From the data, we see that the process is almost certainly exhibiting parallel flow behaviour, with the secondary peak exceptionally significant and strongly redolent of some tracer passing through a slower pathway. But since the impulse response is so complex in this regard, the estimation algorithms are clearly having difficulty in quantifying the nature of the parallel partitioning: it is, yet again, almost certainly an example of poor identifiability caused by the paucity of data (there are, after all, only 40 data points and the input is a single sample impulse).

We can, however, proceed further with the analysis, albeit in a manner which exploits the subjective view that parallel activity is present. To do this, we first observe that the initial peak is probably the result of a first-order mode with zero pure-time delay; while the more rounded second peak is primarily the result of a higher-order, slower mode with 2 or 3 sampling-interval pure delay, which becomes prominent as the effects of the first-order mode decay. Consequently, it makes sense to attempt to exploit the recursive algorithm's ability to handle missing data and model the first-order mode on the basis of only the first few data

Table 1.4 Decomposition estimation results (figures in parentheses are standard errors)

SRIV estimates [1,1,0] (with missing data from sample 13 to 40) $\hat{a}_1 = -0.6678(0.005)$ $\hat{b}_0 = 1.102(0.007)$
[3,1,2] (with first-order [1,1,0] process effect removed) $\hat{a}_1 = -2.5163(0.011)$; $\hat{a}_2 = 2.147(0.020)$; $\hat{a}_3 = -0.622(0.009)$ $\hat{b}_0 = 0.0829(0.002)$ roots (den.) 0.875; $0.821 + 0.19j$: **Note** almost three equal real roots

samples, with samples 13 to 40 completely removed, but retaining samples 40 to 50, since by then the effects of the second-order mode will have decayed away. The results of this estimation are given in Table 1.4, together with those obtained by fitting a [3, 1, 2] model to the *original data* but *with the response of the first-order estimated mode removed by substraction* prior to estimation. The poles of this third-order model are at 0.875 and $0.821 \pm 0.19j$ in the complex z plane and, since the imaginary parts are fairly small, we see that, within the uncertainty bounds on the estimates, the model is closely approximated by three *real* poles, as required from physical considerations.

The final model obtained by this two-step decomposition approach takes the

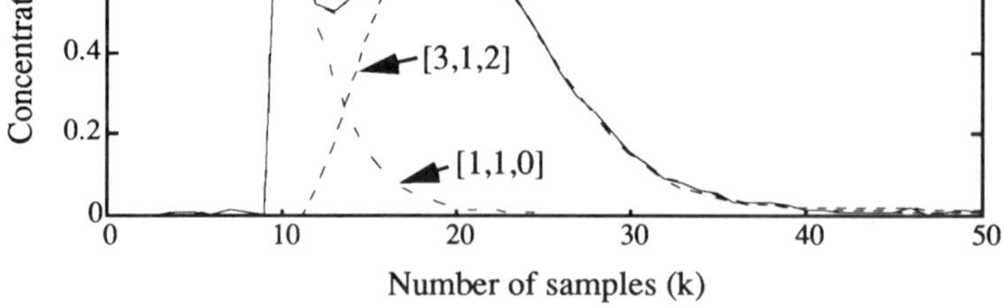

Figure 1.6 Soil tracer experiment: model interpreted as parallel connection of a first-order and second-order system

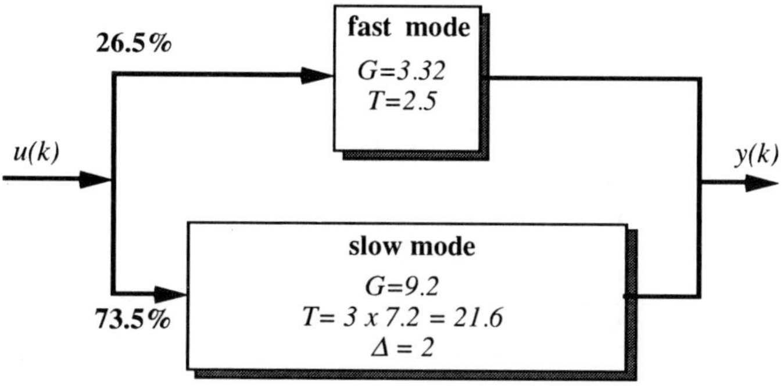

Figure 1.7 Soil tracer experiment: TF block diagram showing partitioning of water along parallel pathways (G denotes steady-state gain, T the time constant or residence time and Δ the pure time delay)

following form,

$$y(k) = \frac{1.102}{1 - 0.668\,z^{-1}}\,u(k) + \frac{0.083\,z^{-2}}{1 - 2.516\,z^{-1} + 2.147\,z^{-2} - 0.622\,z^{-3}}\,u(k) + \xi(k)$$

and the response of the two parts of the model are compared with the data in Figure 1.6. The composite response, as obtained by summing these two individual responses, fits the data very well (similar to the initial SRIV model), confirming the effectiveness of the decomposition. It is clear from these results that the data could well have been generated by the two-parallel-pathway model shown in Figure 1.7, with the fast pathway dominated by a first-order AMZ process having a residence time of 2.5 samples; and the slower pathway consisting of three, similar, first-order AMZ processes, each with a residence time of 7.2 samples (i.e. 21.6 samples in all). This model yields a flow partitioning of 26.5% down the fast pathway and 73.5% down the slower pathway, again with some degree of uncertainty.

Although the conjecture of parallel flow with the partitioning in Figure 1.7 has not been proved by the analysis outlined here, it seems a reasonable supposition if we consider the physical nature of the system. At the very least, it seems a satisfactory result with such a limited data set. Of course, the best approach at this stage would be to carry out more experiments, if possible with better input excitation, in order to confirm the results obtained here.

1.6 A GENERALISATION OF THE ADZ MODELLING CONCEPT: THE ACTIVE MIXING VOLUME (AMV)

The examples in the previous sections demonstrate the wide potential of the modelling procedures discussed in this chapter. However, recent research

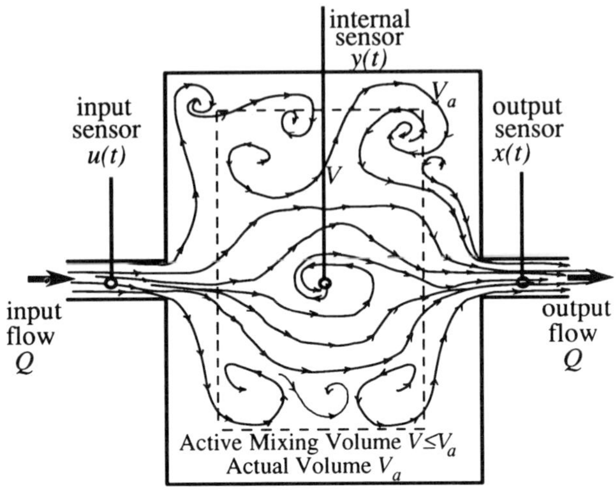

Figure 1.8 The active mixing volume (AMV): an important new concept in environmental systems modelling.

suggests that the concept of an *active mixing volume*, as discussed earlier in relation to the ADZ model, has even wider theoretical and practical significance. In particular, if the idea of an active mixing volume applies in the context of dynamic mass conservation in flowing media, might it not be applicable in the characterisation of other important flow phenomena, such as heat flow modelled via the formulation of energy balance equations?

Figure 1.8 is a diagrammatic representation of the AMV concept in these more general terms. Here, it is assumed that the flow process, whatever its precise form, will normally be rather inefficient, in the sense that the AMV will, in practical situations, be less than the actual volume V_a. Because it provides the main dispersive mechanism, however, it is this AMV which will dictate the observed dynamic behaviour of the system; be it the changes in the concentrations of pollutants in a river; the variations in temperature in soil; or any other flow process of this general type. In this Section, we consider a practical example which demonstrates that the AMV concept does, indeed, appear to apply in this wider context: it concerns the modelling of temperature variations at various depths in a soil column.

1.6.1 An AMV model for heat flow in soils

This example considers the incorporation of a data-based AMV soil temperature model into mechanistic energy-balance equations and shows how the data-based AMV approach to modelling reveals a potentially important problem of

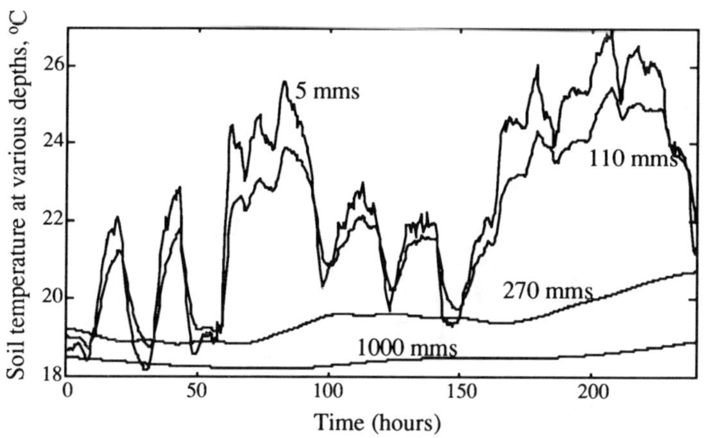

Figure 1.9 Temperature variations at various levels in a soil column: measurements made at 5, 110, 270 and 1000 mm levels in the soil of the Venlo glasshouse at the AFRC Silsoe Research Institute near Bedford, England

identifiability. The various models are evaluated in relation to the set of soil temperature data shown in Figure 1.9, which was recorded at five-minute sampling intervals in the Venlo glasshouse at the AFRC Silsoe Research Institute, over the period 17–27 May 1991 (see *Acknowledgements*). The complete mechanistic soil energy-balance model is part of a much larger, physically based, glasshouse microclimate model which provides a computationally tractable representation of the glasshouse for the purpose of general research on glasshouse systems.

The glasshouse model, which has been developed at Silsoe over the last few years (see Chalabi and Bailey, 1989), simulates the non-steady-state energy and moisture balance inside a Venlo glasshouse. The model is described by a set of coupled, nonlinear, first-order differential equations. The water vapour and energy equations are defined at each internal 'node' of the greenhouse 'network': internal air, crop, roof and ground (6 nodes); and, at the boundary nodes: external air, deep ground layer, heating pipes and the sky. The model is driven by external processes specified at the boundary nodes, for example external air temperature. It is parametrised by a set of constant coefficients relating to the thermal, optical and geometric properties of the glasshouse. Calculations of radiative, conductive and latent heat fluxes are based on well-known physical equations, for example the radiative flux is based on Stefan's law. Figure 1.10 shows the glasshouse network, its associated nodes and the heat fluxes.

In a typically reductionist fashion, the overall model is constructed from appropriate interconnection of the deterministic energy and moisture balance equations specified at each node. Recently, as part of an AFRC-sponsored joint research project between Silsoe and the *Centre for Research on Environmental*

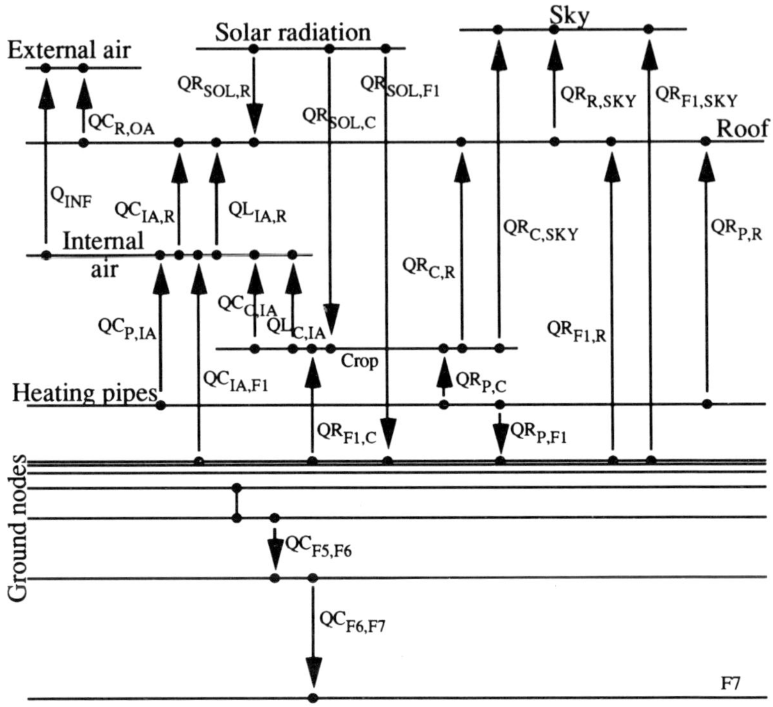

Figure 1.10 The Venlo glasshouse microclimate model 'network'

Systems (*CRES*) of the University of Lancaster *Insitute of Environmental and Biological Sciences.*, the resulting differential equations have been solved using SIMULINK^{TM}, a computer program for simulating dynamic systems. SIMULINK, which is an extension to the MATLAB^{TM} numerical computational package, allows large dynamic models to be simulated using rather novel 'iconographic' programming techniques.

Here, the model is constructed using mouse-driven commands to connect a variety of TF block diagrams contained within SIMULINK's iconographic block libraries. The program then automatically formulates and solves the model equations on the basis of the overall block diagram, utilising user-specified numerical integration routines. This procedure is, of course, particularly appropriate to the present discussion since it involves TF model formulations. Figure 1.11 shows the SIMULINK block diagram of the glasshouse model: note that this diagram is of the overall model and contains several subsystem blocks, many of which are as complicated as the overall block diagram, illustrating the complexity of the complete model.

The soil energy-balance equations, as described below, are used to model the soil temperature as part of the larger energy and moisture balance climate model

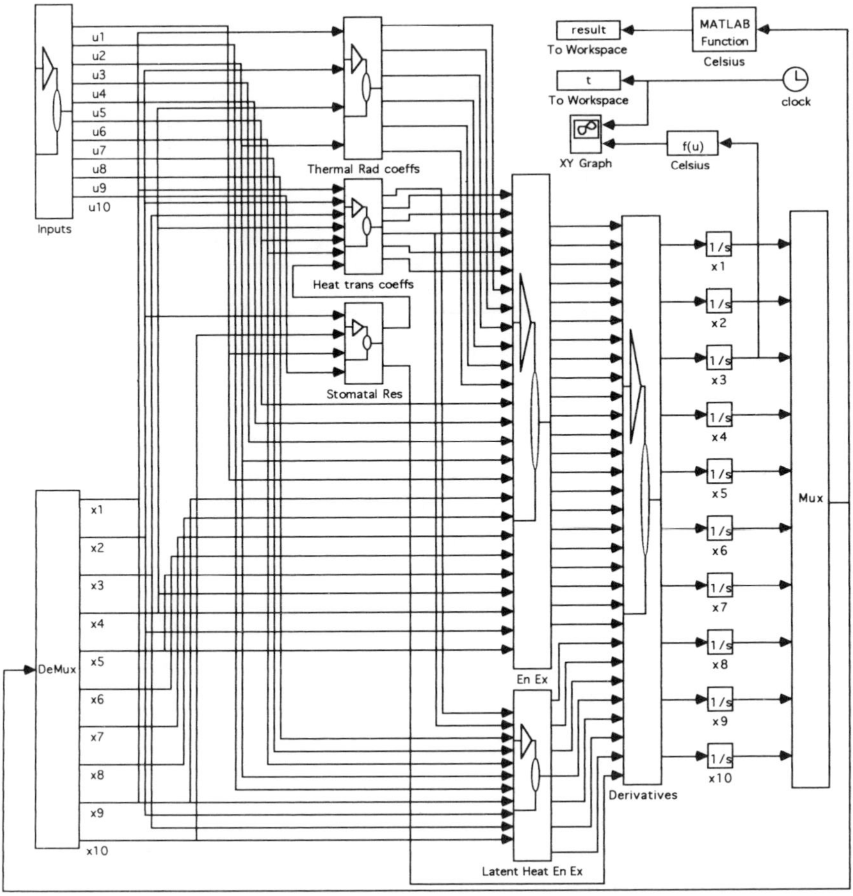

Figure 1.11 SIMULAB block diagram of the mechanistic glasshouse microclimate model

shown in Figure 1.10. This mechanistic subsystem model describes the energy balance for several soil layers: for example, Figure 1.12 shows a three-layer representation: V_U is volume of the upper soil layer; T_M is the middle-layer temperature and $QC_{M,L}$ is the conductive heat flow between the middle and lower layers. The temperature of each layer is dependent on the conductive energy fluxes from the upper and lower layers.

This conductive heat flow between the upper and lower layers can be described in the normal manner as the product of a heat transfer coefficient and the temperature difference, i.e.

$$QC_{U,M} = \frac{2\,Kf}{(V_U + V_M)} \times (T_U - T_M) \qquad (\mathrm{W\,m^{-2}})$$

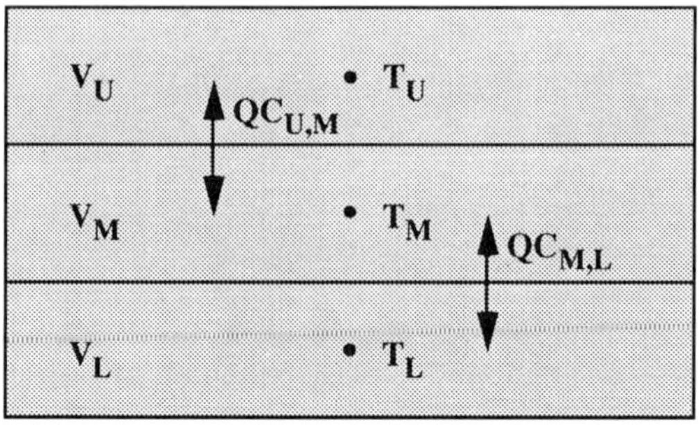

Figure 1.12 Three layer representation of the soil energy-balance model

where Kf is thermal conductivity of the soil. The energy-balance equation for the middle layer is a first-order linear differential equation which describes the temperature change of the soil layer. This is obtained by dividing the energy change by the heat capacity of the layer, i.e.

$$\frac{\mathrm{d}T_{\mathrm{M}}}{\mathrm{d}t}=\frac{QC_{\mathrm{U,M}}-QC_{\mathrm{M,L}}}{MC_{\mathrm{M}}} \qquad (\mathrm{K}). \tag{1.21}$$

The heat capacity MC_{M} is calculated as follows,

$$MC_{\mathrm{M}}=cpf\times \mathrm{d}f\times V_{\mathrm{M}} \qquad (\mathrm{JK}^{-1}\,\mathrm{m}^{-2})$$

where cpf is the specific heat capacity and $\mathrm{d}f$ is the density of the layer.

Combining the above equations, the energy-balance equation (1.21) for the middle layer can be written in the following form,

$$\frac{\mathrm{d}T_{\mathrm{M}}}{\mathrm{d}t}=-\alpha T_{\mathrm{M}}+\beta T_{\mathrm{U}}+\gamma T_{\mathrm{L}}, \tag{1.22}$$

where

$$\alpha=\frac{2\,Kf}{MC_{\mathrm{M}}(V_{\mathrm{U}}+V_{\mathrm{M}})}+\frac{2\,Kf}{MC_{\mathrm{M}}(V_{\mathrm{M}}+V_{\mathrm{L}})}, \tag{1.23}$$

$$\beta=\frac{2\,Kf}{MC_{\mathrm{M}}(V_{\mathrm{U}}+V_{\mathrm{M}})}, \tag{1.24}$$

$$\gamma=\frac{2\,Kf}{MC_{\mathrm{M}}(V_{\mathrm{M}}+V_{\mathrm{L}})}. \tag{1.25}$$

The parameters used to define the relations in (1.22) to (1.25) relate to the dimensions and thermal properties of the soil layers. The layers are assumed to be homogeneous with the same values of thermal conductivity, specific heat capacity and density used for all the layers.

This form of the energy-balance equation suggests that the temperatures at the different layers are related by a first-order differential equation such as (1.22). One way of testing this hypothesis is to estimate the parameters in an equation similar to (1.22) on the basis of the measured temperature data in Figure 1.9; in other words, we can consider the data-based estimation of the equation

$$\frac{dT_M}{dt} = -AT_M + BT_U + CT_L \tag{1.26}$$

or, in continuous-time TF form,

$$T_M = \frac{B}{s+A}T_U + \frac{C}{s+A}T_L.$$

This represents a standard time-series estimation problem; so that estimates of the A, B and C parameters in (1.26) can be obtained using the identification and estimation procedures outlined in previous sections of this chapter.

As in the previous examples, such identification and estimation analysis was carried out using the microCAPTAIN time-series analysis modelling package. Figure 1.13 compares the simulated output of the model obtained in this manner with the measured temperature data at 110 mm depth. It is clear that, whereas the simulated output of the mechanistic model (1.22), also shown in Figure 1.13, produces a very bad explanation of the temperature variations, the

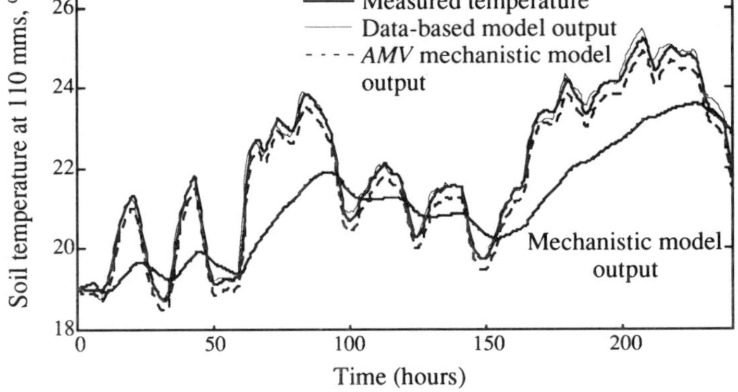

Figure 1.13 Results of data-based and mechanistic models for the 110 mm level, compared with the measured temperature

data-based model (1.26) produces an excellent fit. Here, the parameters in the mechanistic model were set to values which related directly to the physical nature of the soil in the glasshouse, with the soil volumes set to their measured values and the Kf, cpf and df parameters based on the (assumed) known properties of the soil.

Under the modelling philosophy presented in this chapter, we should now try to relate the data-based model parameters obtained in the above manner to those 'physically meaningful' parameters of the mechanistic model: in other words, we should equate the estimates of the parameters A, B and C in equation (1.26) to the parameters α, β and γ, respectively, in equations (1.23–1.25), and then solve for the unknown meachanistic model parameters. Unfortunately, this presents some serious problems in this case. The three equations (1.23–1.25) have six unknwon parameters: Kf, cpf and df and the three volumes V_U, V_M and V_L. Without prior assumptions about certain of the parameter values, therefore, the mechanistic model parameters are not *identifiable* (see the previous section on the subject) from the data-based model parameters.

The conventional approach to this problem would probably to be assume that the soil volumes are known *a priori*, since they can be 'measured', and to then attempt the estimation of the other, 'more uncertain' heat transfer parameters Kf, cpf and df. But the AMV concept justifies a move against this convention: under the AMV hypothesis, the 'effective' AMV volumes are likely to be some fraction of the measured values and cannot be assumed equal *a priori* to these measurements; so we might do better to assume knowledge of Kf, cpf and df. Even with this assumption, however, there remains a problem. At first sight, there are three equations (1.23–1.25) relating the three remaining unknown parameters V_U, V_M and V_L, so that the algebraic relationship between the data-based parameters and the soil-layer volumes appears to be solvable. Unfortunately this is not the case: on inspection, it becomes clear that $\alpha = \beta + \gamma$, so that we are still, in fact, confronted by an *unidentifiable* system of equations with three unknowns and only two independent equations (1.23 and 1.25).

But it is possible to proceed futher: although the AMV's are not all uniquely *identifiable* the following interrelationships can be obtained quite straight-forwardly;

$$V_U = \frac{2\,Kf}{\alpha \times cpf \times df \times V_M} - V_M$$

$$V_L = \frac{2\,Kf}{\gamma \times cpf \times df \times V_M} - V_M.$$

Now by assuming a value for V_M, V_U and V_L can be computed: e.g. assuming the AMV of V_M to be 30% of the actual volume, and taking $\alpha = b$ and $\gamma = c$, gives AMVs of 21% and 82% for V_U and V_L, respectively, which do not seem physically unreasonable. The resulting AMV model output is compared with

Table 1.5 Soil model estimation results (number in parenthesis are standard errors)

Model	Estimates of:	α or A	β or B	γ or C
Mechanistic		0.031	0.0187	0.0123
Data-based		0.578(0.035)	0.422(0.025)	0.164(0.0125)
AMV		0.578	0.422	0.164

the data-based and mechanistic model outputs in Figure 1.13. The improvement in the explanation of the temperature data when compared with the traditional mechanistic model output is quite obvious. A comparison of the parameter values in the various models is given in Table 1.5.

Of course, we can make the above identifiability problem seem rather simpler if we are prepared to be a little more radical and define some less conventional physical parameters. For example, suppose that the following definitions are introduced:

$$Q_T = \frac{2Kf}{MC_M} \mathrm{m^3h^{-1}}; \quad V_{UM} = V_U + V_M \mathrm{m^3}; \quad V_{ML} = V_M + V_L \mathrm{m^3} \qquad (1.27)$$

where the letters Q and V are used to faciliate comparison with the previous ADZ model equations and help to emphasise that Q_T can be considered as a 'flow' measure, while V_{UM} and V_{ML} are volumes, and

$$\beta = \frac{Q_T}{V_{UM}}; \quad \gamma = \frac{Q_T}{V_{ML}}; \quad \alpha = \beta + \gamma. \qquad (1.28)$$

Also by analogy with the ADZ model, the 'residence time' T and 'steady-state gains' (SSGs) G_{UM} and G_{ML} of the model are defined as

$$T = \frac{1}{\alpha} = 1.73 \text{ hours}; \quad G_{UM} = 0.73; \quad G_{ML} = 0.28$$

where G_{UM} is the SSG between the upper and middle layers and G_{ML} is the SSG between the lower and middle layers.

Considering the relationships in (1.28), there are now effectively two equations in three unknowns and the system is still not identfiable. However, if we can assume some value (measure?) for the 'flow' parameter Q_T, then the two AMV values can be computed from the estimates of the data-based model parameters A, B and C. For example, if $Q_T = 7.9\,\mathrm{m^3h^{-1}}$, then $V_{UM} = 1.86\,\mathrm{m^3}$ and $V_{ML} = 4.82\,\mathrm{m^3}$. Finally, note from Table 1.5 that, within the uncertainty on the estimates of A, B and C,

$$\hat{A} = \hat{B} + \hat{C}; \quad \text{i.e. } 0.578 \approx 0.422 + 0.164$$

as required by the relationship $\alpha = \beta + \gamma$ in the mechanistic model and the associated need for heat conservation. This is also reflected in the SSG values which sum approximately to unity and suggest that, in an equilibrium situation, 0.73 of temperature level in the upper layer affects the middle layer, while it is only affected by 0.28 of the level in the lower layer; implying a predominant heat flow in the downward direction, as might be expected. By analogy with the ADZ model, the associated percentage 'flows' are 72.3% and 27.7%, respectively.

What can we conclude about this example? Because of the identifiability problems, it clearly does not prove that the AMV concept is applicable within the energy balance context. But it does show that, within the unavoidable constraints of identifiability, the AMV assumptions yield *resonable* results. After all, we have obtained a model which explains the measured data very well; which has a structure with a reasonable physical interpretation; and which is characterised by parameter values that are also physically meaningful under the AMV hypothesis. Moreover, it is a model that is clearly superior to the fully mechanistic model, which has to be rejected as a theory of dynamic behaviour because it provides a totally inadequate explanation of the measured data. And, finally, if we are prepated to live with the ambiguity about the mechanistic model parameters, it is a model can be used quite confidently as a subsystem in the glasshouse model.

1.7 CONCLUSIONS

This chapter has outlined one particular approach to data-based mechanistic modelling that has proved very useful in many environmental applications over the last few years. By means of several practical example, it has demonstrated how the model builder can exploit this approach to rapidly extract important, physically meaningful, dynamic information from limited sets of time-series data. The chapter has also attempted to show that this general approach to model building has much wider implications. In particular, it reveals, first through the ADZ and related AMZ models for solute transport in rivers and soils, and then from the AMV interpretation of a soil temperature model, how the AMV concept may be applicable to a wide variety of flow processes.

The ADZ model is now widely accepted in the hydrological literature and has been evaluated in many different river systems. In contrast, the soil temperature example represents only the first attempt to apply the AMV concept in a wider context. However, it clearly demonstrates that the concept can be translated to the problem of heat flow in soils and, in doing so, it reveals certain important identifiability problems that may occur when the data-based mechanistic approach to modelling is applied in this context.

The implications of these practical examples are profound. Indeed, if the AMV concept does prove as widely applicable as these examples suggests, it could change the manner in which environmental systems are modelled via the formulation of dynamic conservation equations. No longer could it be assumed that the volumes which are conventionally associated with the medium under consideration, and which appear in the conservation equations, are the actual, measured volumes and are, therefore, known *a priori*: rather they would simply become other, unknown parameters in the model. Such AMV parameters would then be seen as important properties of the medium; properties which, within the natural environment, might well vary considerably, depending upon the nature of the medium at different geographical locations. And, as such, they would need to be determined empirically, using either a procedure similar to that discussed in this chapter, or some other, alternative estimation techniques.

ACKNOWLEDGMENTS

Parts of the research described in this chapter has been supported by grants from three Research Councils, for which the authors are most grateful. The first author is grateful to Natural Environment Research Council (NERC) for supporting the research on ADZ modelling, under grant number GR3/5171; and to the Science and Engineering Research Council (SERC) for supporting the research on model identification and estimation under grant number GR3/E81562. Both authors are benefiting from a recent grant from the Agriculture and Food Research Council (AFRC) under grant number LR89/314. The authors also wish to thank the AFRC Silsoe Research Institute for providing the data on soil temperature variations as part of the latter research programme. Particular thanks are due to our colleagues Dr Z. Chalabi and Dr Bernard Bailey at Silsoe who have both developed the glasshouse model over the last few years and designed the monitoring systems within the Venlo glasshouse.

APPENDIX 1.1 TRANSFER FUNCTION MODELS

Continuous-time, differential equations have dominated the modelling of most physical dynamic systems since the time of Newton, since they often provide the most convenient and physically meaningful starting-point for the modelling of physical environmental systems. Any single-input, single-output linear dynamic system can be described by the following, general ordinary differential equation,

$$\frac{\mathrm{d}^n x(t)}{\mathrm{d}t^n} + A_1 \frac{\mathrm{d}^{n-1} x(t)}{\mathrm{d}t^{n-1}} + \cdots + A_n x(t)$$

$$= B_0 \frac{\mathrm{d}^m u(t-\tau)}{\mathrm{d}t^m} + B_1 \frac{\mathrm{d}^{m-1} u(t-\tau)}{\mathrm{d}t^{m-1}} + \cdots + B_m u)(t-\tau) \quad \text{(A1.1)}$$

where t represents time in appropriate time units; $u(t)$ is the input signal; $x(t)$ is the output signal; τ is a pure-time delay of the kind encountered in many environmental systems; and A_i, $i = 1, 2, \ldots, n$; B_j, $j = 0, 1, 2, \ldots, m$ are real coefficients or parameters that characterise the differential equation and define its dynamic characteristics. In dynamic systems analysis it is convenient to consider such a differential equation description in terms of a *transfer function* (TF) representation, formulated in either continuous or discrete-time.

The time-derivative (s operator) TF model

The most direct TF equivalent of the linear differential equation is obtained by introducing the time-derivative operator s, where

$$s = \frac{\mathrm{d}}{\mathrm{d}t}$$

to yield the following TF relationship between $x(t)$ and $u(t - \tau)$:

$$x(t) = \frac{B(s)}{A(s)} u(t - \tau). \qquad \text{(A1.2)}$$

The operator s is utilised here as the time-derivative operator because of its relationship with the Laplace operator, which is probably the most established icon of control and systems theory: e.g. the Laplace operator transfer function for a zero time delay differential equation, with zero initial conditions on $x(t)$, $u(t)$, and all their derivatives, is identical to the s operator TF in equation (A1.2).

Here the polynomials $A(s)$, $B(s)$ are defined as

$$A(s) = s^n + A_1 s^{n-1} + \cdots + A_n,$$
$$B(s) = B_0 s^m + B_1 s^{m-1} + \cdots + B_m$$

and it is normally assumed that these polynomials do not have any common factors; i.e. that the model is controllable and observable. Otherwise, there are no limitations on the transfer function $B(s)/A(s)$, which may be marginally stable, unstable, or possess nonminimum phase characteristics.

This continuous-time TF holds a revered position in the history of control and systems theory and has many attractions as a vehicle for applied systems analysis. However, most experimental and monitored time-series are now obtained in discrete-time, sampled-data form and are analysed in the digital computer. As a result, it is necessary to introduce other, equivalent, discrete-time TF representations which provide logical replacements for (A1.2) in this digital context.

The backward-shift (z^{-1} operator) TF model

If the $u(t)$ and $x(t)$ are sampled regularly at a sampling interval of Δt time units, then the discrete-time, sampled-data equivalent of equation (A1.1) takes the following form,

$$x(k) + a_1 x(k-1) + \cdots + a_p x(k-p) = b_0 u(k-\Delta) + \cdots + b_p u(k-p-\Delta)$$

where $u(k)$ and $x(k)$ are the sampled values of $u(t)$ and $x(t)$, respectively, at the kth sampling instant (i.e. the values sampled after $k\Delta t$ time units); $p = \max(n, m)$; and a_i, $i = 1, 2, \ldots, p$; b_j, $j = 0, 1, \ldots, p$ are suitably defined parameters; and $\Delta = \tau/\Delta t$. Introducing the backward shift operator, z^{-i}, where

$$z^{-i}x(k) = x(k-i),$$

we then obtain the z^{-1} operator TF representation,

$$x(k) = \frac{B(z^{-1})}{A(z^{-1})} u(k-\Delta), \tag{A1.3}$$

where $A(z^{-1})$ and $B(z^{-1})$ are the following polynomials in z^{-1},

$$A(z^{-1}) = 1 + a_1 z^{-1} + \cdots + a_n z^{-p}$$
$$B(z^{-1}) = b_0 + b_1 z^{-1} + \cdots + b_m z^{-p}.$$

Again, except for controllability and observability conditions, no prior assumptions are made about the nature of the transfer function $B(z^{-1})/A(z^{-1})$, which may be marginally stable, unstable or possess nonminimum phase characteristics.

The discrete differential (δ) operator TF model

An interesting alternative to the z^{-1} operator TF model is the following 'discrete differential operator' model, which was revived recently under the title 'δ operator' by Goodwin and his co-workers (see Middleton and Goodwin, 1990),

$$x(k) = \frac{B(\delta)}{A(\delta)} u(k-\Delta), \tag{A1.4}$$

where $A(\delta)$ and $B(\delta)$ are polynomials of the following form,

$$A(\delta) = \delta^p + \alpha_1 \delta^{p-1} + \cdots + \alpha_p,$$
$$B(\delta) = \beta_0 \delta^p + \beta_1 \delta^{p-1} + \cdots + \beta_p$$

again with the index $p = \max(n, m)$. The δ operator, for the sampling interval Δt, is the simplest discrete-time differential operation, which is defined as follows it terms of the backward-shift operator z^{-1},

$$\delta = \frac{1 - z^{-1}}{z^{-1}\Delta t} = \frac{z-1}{\Delta t}, \quad \text{i.e. } \delta x(k) = \frac{x(k+1) - x(k)}{\Delta t}.$$

Remarks

(1) As $\Delta t \to 0$, the δ operator reduces to the derivative operator $s = d/dt$ in continuous time (i.e $\delta \to s$ as $\Delta t \to 0$); indeed, the definition of δ is, of course, the usual starting point for the derivation of the differential coefficient in calculus.
(2) A given polynomial of any order p in the forward-shift z operator will be exactly equivalent to some polynomial in δ, also of order p. As a consequence of this, we can easily move between the z and δ operator domains. Also, the δ operator model coefficients (α, β) are related to the forward z^{-1} coefficients (a, b) by simple vector matrix equations (see Chotai *et al*, 1990).
(3) One attraction of the δ operator model to those who prefer to think in continuous-time terms is that it can be considered as a direct approximation to a continuous-time system. For example, it is easy to see that the unit stability circle in the complex z plane maps to a circle with centre $-1/\Delta t$ and radius $1/\Delta t$ in the complex δ plane; so that, as $\Delta t \to 0$, this circular stability region is transformed to the familiar, left-half stability region of the complex s plane (the 'left-half-plane').

An illustrative example

Clearly, while there is only one continuous-time TF which represents the differential equation (A1.1), there are an infinite number of z^{-1} operator and δ operator TF models, depending upon the sampling interval Δt. For very rapidly sampled systems, however, the δ operator model can be considered in almost continuous-time terms, with the parameters α_i and β_j tending towards the parameters A_i and B_j of the continuous-time model; and the eigenvalue (pole) positions in the δ plane close to those of the 'equivalent' continuous-time model in the s plane.

In order to illustrate these remarks, let us consider a simple example of a river channel represented by four first-order ADZ models in series, each with a pure time delay of 0.5 time units and a residence time of 2 time units. The overall TF in the s operator then takes the form

$$x(t) = \frac{0.0625}{s^4 + 2s^3 + 1.5s^2 + 0.5s + 0.0625} u(t-2). \tag{A1.5}$$

The response of this model to a unit impulse (Dirac delta function)input, i.e.

$$u(t) = \begin{cases} 1 \text{ for } t = 0 \\ 0 \text{ for } 0 > t > 0 \end{cases}$$

is shown in Figure 1.3, as discussed earlier in the section on the ADZ model.

For very small $\mathbf{\Delta}t$, the resulting δ operator TF is identical to (A1.5) with δ replacing s. As $\mathbf{\Delta}t$ increases, however, small differences start to appear: e.g. at $\mathbf{\Delta}t = 0.005$, the TF becomes

$$x(k) = \frac{0.0005\,\delta + 0.0622}{\delta^4 + 1.9975\,\delta^3 + 1.4963\,\delta^2 + 0.4981\,\delta + 0.0622}\,u(k - 400), \quad \text{(A1.6)}$$

while at $\mathbf{\Delta}t = 0.1$,

$$z(k) = \frac{0.0003\,\delta^2 + 0.0086\,\delta + 0.0566}{\delta^4 + 1.9508\,\delta^3 + 1.4271\,\delta^2 + 0.464\,\delta + 0.0566}\,u(k - 20). \quad \text{(A1.7)}$$

In this latter case, although the TF parameters are still recognisably similar to those of the s operator form, they have noticeably different numeric values.

In contrast to the δ operator TF, the parameters of the backward-shift-operator transfer function are, of course, always very different from those of the s operator TF since they are obtained by substituting for δ in terms of z^{-1}: e.g. at $\mathbf{\Delta}t = 0.005$, this yields

$$x(k) = \frac{10^{-10}(0.0163\,z^{-1} + 0.1783\,z^{-2} + 0.178\,z^{-3} + 0.0161\,z^{-4})}{1 - 3.99\,z^{-1} + 5.97\,z^{-2} - 3.97\,z^{-3} + 0.99\,z^{-4}}\,u(k - 400) \quad \text{(A1.8)}$$

while at $\mathbf{\Delta}t = 0.1$, the TF becomes

$$x(k) = \frac{10^{-5}(0.025\,z^{-1} + 0.2645\,z^{-2} + 0.2541\,z^{-3} + 0.0222\,z^{-4})}{1 - 3.8049\,z^{-1} + 5.4290\,z^{-2} - 3.4428\,z^{-3} + 0.8187\,z^{-4}}\,u(k - 20). \quad \text{(A1.9)}$$

Finally, it is important to emphasise that, although the transfer functions in (A1.5) to (A1.9) are, to a greater or lesser degree, different both in form and parametrically, they all generate exactly the same response characteristics at the sampling instants defined by the chosen sampling interval $\mathbf{\Delta}t$. In other words, for a specified input signal $u(t)$, the transfer functions (A1.6) to (A1.9) generate the same sampled output response $x(k)$ to the sampled input $u(k)$, obtained by sampling $u(t)$ at the sampling interval $\mathbf{\Delta}t$; moreover, these responses are identical at these same sampling instants to the response of the s operator TF in (A1.5). More specifically in the above example, the impulse response in Figure 1.3 is the impulse response of *all* the TF models (A1.5) to (A1.9).

2

Encounter Sampling and Modelling in Ecological and Environmental Studies Using Weighted Distribution Methods

G. P. Patil, C. Taillie and S. Talwalker
Pennsylvania State University, USA

2.1 BACKGROUND AND INTRODUCTION

In environmental sampling and monitoring work, situations arise where seemingly unusual, but locally ingenious, encounter sampling and modelling protocols become relevant and fruitful. Appropriate statistical modelling approaches help accomplish unbiased inference in spite of the biased data and, at times, even provide a more informative and economical set-up.

First of all, we must recognise that there is no substitute for good data, and in no way should we be complacent about it. As we know, statistical thinking is an aid to the interpretation of data. It may help clarify seeming confusion. It may confuse seeming clarity! What is important is that the statistical approach is expected to contribute to the overall insight and perspective of the substantive issue and its resolution in the light of the evidence on hand, be it in the nature of empirical data, of literature-assembled data, of expert opinion data, or a combination.

What makes the problem of ecological and environmental investigations

Statistics for the Environment. Edited by Vic Barnett and K. Feridun Turkman

different from studies in the physical sciences and engineering is that, unlike in the hard sciences, we have a longer span of investigations depending on life stages and their age lengths. Also, the instrumentation changes are there in response to the advancing technology. That puts us in a rather difficult cycle of: no information, new information and non-information. When a question is asked of us, we promptly say, 'Well, we don't have information. We need to collect new information.' And, after the new information is collected, we are ready to answer by saying that the information collected is already noninformation. However, noninformation is no information, and thus we see that we go through this cycle involving data that does not produce information. This unaffordable cycle we need to break.

At the least, we need to be aware of this fundamental issue in ecological and environmental work. It may be described by the concept of soft data. Not only that there is biological or ecological or environmental variability contributing to large variability usually present in these data sets but also there are measurement errors arising from a variety of sources—some known, some unknown, and some unknowable. Thus the data are soft data and they call not merely for hard analysis, for which we have software and software packages, but they need very hard looks if we are to make prudent decisions. This is particularly true when we recognise that, while there are questionable statistical routines, there are no routine statistical questions.

Traditional statistical theory and practice have been occupied largely with statistics involving randomisation and replication. But in ecological and environmental work, observations most often fall in the non-experimental, non-replicated and non-random categories. Additionally, the problems of model-specification and data-interpretation acquire special importance and great concern. In several situations of diverse nature in statistical ecology and environmental statistics, the theory of weighted distributions provides a perceptive and unifying approach for the problems of model-specification and data-interpretation (Patil and Rao, 1977, 1978; Patil *et al.*, 1988; Patil and Taillie, 1989; Patil, 1991).

In this chapter, we discuss some real-life examples in the context of line-transect estimation of ecological abundance, and in the context of teratological binary data collected in developmental toxicity studies. It is both interesting and encouraging that appropriate statistical modelling approaches can help accomplish unbiased inference in spite of the biased data and, at times, even provide a more informative and economical set-up.

2.2 EXTRA-DISTANCE SOURCES OF ENCOUNTER BIAS IN TRANSECT SAMPLING

The method of line-transect sampling has been used to estimate the abundance of plants or animals of a particular species in a given region. The line-transect method consists of drawing a baseline across the region to be surveyed and then drawing a line transect through a randomly selected point on the baseline.

The surveyor looks around while walking along the line transect and includes the sighted objects of interest in the sample.

It is obvious that the nearer the object or the larger its size, the higher is the probability of sighting the object. Similarly, when the individuals cluster in groups, such as schools or herds, then it is appropriate to regard clusters as the basic sampling units with their encounter probabilities being affected by cluster size. Estimates of cluster abundance can be adjusted to individual abundance using the recorded cluster sizes.

Encounter probabilities in transect sampling can be influenced by numerous other factors such as varying terrain and vegetation cover, weather conditions, time of day, and systematic responsive movement toward or away from the transect. Some of these factors are characteristics of the objects themselves and will vary from object to object. 'Size' is an example of such a factor. Other factors, notably environmental features, are survey characteristics but can vary from segment to segment in a multi-segmented survey. We refer to members of these two classes as *object-factors* and *survey-factors*, respectively.

It is clearly desirable to account for as many of these factors as possible, and much of the recent transect sampling literature has been concerned with this issue. Caughley (1972, 1974, 1976) has documented a number of extra-distance factors in aerial surveys. Object size, and especially cluster size, have been extensively studied by Drummer and McDonald (1987), Otto and Pollock (1990), Quang (1991), Quinn (1979, 1985), and Rao *et al.* (1981). Scott *et al.* (1986) have accounted for habitat differences in surveys of Hawaiian birds. Turnock and Quinn (1991) investigate responsive animal movement. Ramsey *et al.* (1987) present a regression-like method for incorporating general covariate information into the sighting function.

Section 2.2.1 summarises some of our own work involving a seabed transect survey for red crabs in which monotonicity of the sighting function was flagrantly violated. The cause was eventually identified and accounted for by putting an additional multiplicative factor in the sighting function. Section 2.2.2 surveys the work of Ramsey *et al.* (1987) and Drummer and McDonald (1987). Each of these papers incorporates the extra-distance factors into the scale parameter of the sighting function. Ramsey *et al.* are concerned with the effects of survey factors, in which case a purely conditional, regression-like, analysis is adequate. Drummer and McDonald examine an object factor, specifically size, and need to consider the probability distribution of that factor as well as the visibility bias that occurs in the recorded sizes. Finally, Section 2.2.3 points out that a failure to account for object factors does not destroy the validity of the abundance estimates or of their estimated standard errors.

2.2.1 Deep-sea red crab

Patil *et al.* (1979, 1980) describe a photographic survey in which features of the optical geometry partially masked the sighting distance bias by inflating the recorded counts at larger distances from the transect. The survey's purpose was to determine the abundance of the deep-sea red crab, *Geryon quinquedens* Smith, in continental slope waters off the north-eastern United States. Water depth at 33 sampling stations ranged from about 200 to 1500 metres. The sampling device was an underwater camera system mounted on a 1200 kg steel sled that was lowered to the ocean floor at each station and towed for 30 to 75 minutes depending upon local conditions. Several hundred non-overlapping photographs were obtained for each station; about half, representing the best quality, were selected for quantitative analysis.

Roughly one crab was sighted for every four frames analysed. Even so, determination of the perpendicular sighting distance to each crab proved excessively labour-intensive, and an alternative was devised in which every photograph was divided into five zones or strips running parallel to the transect and representing widths of 1.22 metres on the ocean floor. (For technical reasons, the zone closest to the camera was only half as wide, requiring appropriate adjustments; see Patil, *et al.*, 1979.) The zones were delineated on an overlay and the number of crabs in each zone was counted.

The sighting frequency had been expected to fall off rapidly with distance because of factors such as turbidity, increasingly diffused lighting, and roughness of the seabed. (The camera was angled in a way that exposed the bottoms of seabed depressions provided they were close to the sled.) But, in fact, the histograms of sighting-frequency versus distance showed only a very gradual decline and, at small distances, suggested that frequency might even increase with distance. The explanation was eventually found when it was noted that what appeared in the photograph as a rectangular zone actually represented a trapezoid on the ocean floor with the shorter side closest to the camera. Thus, at sighting-distance x, the photograph exposes a length $a + bx$ where the constants a and b could be determined from the geometry. The data were analysed using a composite weight function of form

$$w(x) = (a + bx) \cdot v(x; \theta),$$

where the sighting function $v(x; \theta)$ represents the pure sighting-distance bias and would typically involve unknown parameters θ. For the red crab study, the exponential-power sighting function gave a reasonable fit to the data and yielded abundance estimates that were consistent with the results of other survey methods (Wigley *et al.*, 1975).

2.2.2 Bivariate sighting-functions

We let L be the length of the transect and Ω the width of the sighting strip on either side of the transect. Sometimes Ω is established by the sampling design. More often, Ω is taken to be the visible (or audible) horizon, in which case it is convenient to formally let Ω go to infinity and to estimate abundance as a density instead of a count (see below).

With $w(x)$ as the sighting function, the mean detection probability is

$$P = \frac{1}{\Omega}\int_0^{\Omega} w(x)\mathrm{d}x,$$

or

$$\omega = P\Omega, \qquad (2.1)$$

where

$$\omega = \int_0^{\Omega} w(x)\mathrm{d}x \qquad (2.2)$$

is known as the *effective half-width*. Writing N and n for the population and sample counts, one has $E[n] = PN$ so that abundance can be estimated as

$$\hat{N} = n/\hat{P} = n\Omega/\hat{\omega}, \qquad (2.3)$$

provided an estimator $\hat{\omega}$ is available for ω. The awkward quantity Ω can be eliminated if (2.3) is divided by the survey area $2L\Omega$ to obtain the density estimator

$$\hat{N}/2L\Omega = n/2L\hat{\omega}. \qquad (2.4)$$

The recorded right-angle distances x have their probability density function given by $w(x)/\omega$, $0 < x < \Omega$, so that ω can, in principle, be estimated from these distances. See Burnham *et al.* (1980) or Seber (1976) for further details on line-transect sampling in general.

When Ω can be taken to be infinite, Ramsey (1979) has suggested a general technique for constructing parametric visibility functions in which the effective half-width ω appears explicitly as a scale parameter. Starting with a *kernel* $h(t;\theta)$, $0 < t < \infty$, which is monotone decreasing in t and satisfies $h(0;\theta) = 1$ and $\int_0^\infty h(t;\theta)\mathrm{d}t = 1$, Ramsey's family of sighting functions is defined by

$$w(x;\omega,\theta) = h(x/\omega;\theta). \qquad (2.5)$$

Here, θ is a vector of nuisance parameters that regulate the shape of the sighting function. A common choice of kernel is the exponential-power form

$$h(t;\gamma) = \exp(-\Gamma(1+\gamma^{-1})t^{\gamma}), \qquad (2.6)$$

which includes the negative exponential and the half-normal as special cases.

Ramsey *et al.* (1987) propose that covariate information $\mathbf{y} = (y_1, \ldots, y_p)$ be incorporated into the sighting function by letting ω in (2.5) be a parametric function of $\mathbf{y}$. Notice that this effectively yields a multivariate sighting function $w(x, \mathbf{y})$. Ramsey *et al.* suggest the specific form

$$\ln(\omega(\mathbf{y})) = \beta_0 + \sum_{j=1}^{p} \beta_j y_j, \tag{2.7}$$

where the covariates may need to be transformed before inclusion in (2.7). Ramsey *et al.* develop the maximum likelihood estimators conditional upon the recorded covariate values. This conditional approach is appropriate for what we have referred to as survey factor in a multi-segmented survey. Here one wants to pool data from the various segments in order to obtain precise estimates of the visibility functions. However, density is estimated separately for each segment using the segment-specific estimate of ω.

Ramsey *et al.* give an example involving tropical birds with time of day as the covariate. The raw counts suggested a declining abundance through the course of the day; this apparent effect could be accounted for as due to declining visibility (audibility) over time.

Drummer and McDonald (1987) have considered the problem of estimating minke whale abundance where the group size, an object-specific factor, becomes important. They take the group as the basic sighting object and include the group size y as a scaling factor in the exponential-power sighting function

$$w(x, y) = w(x, y; \lambda, \alpha, \gamma) = \exp\{\Gamma(1 + \gamma^{-1})(x/\lambda y^{\alpha})^{\gamma}\}. \tag{2.8}$$

Drummer and McDonald employ a slightly different parametrisation. Notice that (2.8) falls into the framework of (2.7) with the log of group size as covariate since

$$\ln(\omega) = \ln(\lambda) + \alpha \ln(y). \tag{2.9}$$

Since y is an object-factor, its random variation from object to object needs to be taken into account, and the conditional approach of Ramsey *et al.* is no longer sufficient. Let $f(y)$ be the 'natural' distribution of group size. The joint distribution of recorded (x, y) is then proportional to

$$w(x, y)f(y), \tag{2.10}$$

with normalising constant given by

$$\iint w(x, y)f(y)\,\mathrm{d}x\,\mathrm{d}y = \int \omega(y)f(y)\,\mathrm{d}y$$

$$= \lambda \int y^{\alpha} f(y) \mathrm{d}y$$

$$= \lambda \mu_{\alpha} = \bar{\omega}. \tag{2.11}$$

This argument also shows that recorded y follow the size-biased distribution of order α, $f^{*}_{\alpha}(y) = y^{\alpha} f(y)/\mu_{\alpha}$. It is readily seen that equations (2.1)–(2.4) remain valid when ω is replaced by $\bar{\omega}$.

If one were prepared to assume a parametric form for $f(y)$ then estimation could proceed via maximum likelihood in conjuction with (2.10) and (2.11). Drummer and McDonald prefer a two-stage approach in which λ and α (as well as the nuisance parameter γ) are estimated using the conditional procedure of Ramsey *et al.* The moment μ_{α} in (2.11) is then estimated nonparametrically using $\mu_{\alpha} = 1/E_{f^{*}_{\alpha}}[y^{-\alpha}]$, a variant of the Cox (1969) identity for size-biased distributions. Similarly, $\mu_1/\mu_{\alpha} = E_{f^{*}_{\alpha}}[y^{1-\alpha}]$, so that the mean group size, μ_1, can be estimated for the purpose of adjusting group abundance to individual whale abundance.

2.2.3 Random sighting-functions do not cause overdispersion

Otto and Pollock (1990) have evaluated the Drummer and McDonald technique using data collected earlier by Otto in a field trial of line-transect sampling. Clusters consisting of 1, 2, 4, or 8 beer cans were placed at random in a field near Raleigh, North Carolina, and nine observers were asked to walk a transect and note down the clusters detected.

Since the experiment was replicated nine times corresponding to the nine observers, it was possible to obtain a direct estimate of the detection probability for each cluster. These probabilities were superimposed on the plots of fitted sighting-functions and, as noted by Otto and Pollock, the scatter about the fitted curves exceeded what would be expected in a binomial experiment. For example, a given cluster might be seen by most or even all observers, while another cluster of the same size and similar distance might be seen by none. This would indicate that there are additional local factors that significantly affect visibility.

Accounting for these factors is ordinarily impossible, with the result that the sighting function is itself a random variable. In this connection, we wish to address the following question: if we ignore randomness in the sighting function and, instead, use a deterministic (i.e. average) sighting function, are the resulting inferences, especially variance estimates, still valid?

The question is not specific to transect sampling but can arise for a general weighted distribution

$$w(x)f(x)/\omega, \tag{2.12}$$

where x can be multivariate. However, we suppose that $w(x)$ is not completely determined by x but comes from a distribution having x as one of its parameters. For example, w might arise from a two-parameter beta distribution whose mean is determined by x and whose second parameter controls the dispersion of w about its mean. In general, suppose w is realised from a density $g(w; x, \theta)$, where θ represents any additional parameters. Now, the joint distribution of selected (w, x) is proportional to

$$wg(w; x, \theta)f(x). \tag{2.13}$$

Since w is not observable, we consider the marginal distribution of selected x which is proportional to

$$\left[\int wg(w; x, \theta)\mathrm{d}w\right]f(x) = E[w; x, \theta]f(x) = w(x, \theta)f(x). \tag{2.14}$$

Thus a randomly weighted distribution is exactly the same as a deterministically weighted distribution whose deterministic weight function is the average of the random weight function.

2.3 INFORMATIVENESS OF WEIGHTED DISTRIBUTIONS

Bayarri and DeGroot (1986) posed the question of whether 'weighted' observations could be more informative than the corresponding 'natural' observations especially from the point of view of Blackwell's (1951, 1953) comparison of experiments. Intrigued by the idea that encountered, and biased, data might sometimes be more scientifically revealing than experimental data, we have begun to investigate the issue further.

Making precise the notion of informativeness, especially in a multiparameter setting, is neither simple nor unique. Our own work in this area has focused upon Fisher information and estimation efficiency; but even here there can be considerable latitude in how informativeness is defined. Let $f(x; \boldsymbol{\theta})$ be the natural distribution and $f_w(x; \boldsymbol{\theta}) \propto w(x) \cdot f(x; \boldsymbol{\theta})$ be the corresponding weighted distribution, where $\boldsymbol{\theta}$ is a vector of parameters. Also let $J(\boldsymbol{\theta})$ and $J_w(\boldsymbol{\theta})$ be the respective Fisher information matrices for $\boldsymbol{\theta}$. Intrinsic comparison between the weighted and natural observations is possible when the difference,

$$J_w(\boldsymbol{\theta}) - J(\boldsymbol{\theta}),$$

is either positive definite or negative definite. For example, positive definiteness would mean that *every* scalar-valued function of $\boldsymbol{\theta}$ could be estimated with smaller asymptotic standard error under the weighted observation. The natural observations are similarly favoured when the difference is negative definite. However, definiteness of $J_w(\boldsymbol{\theta}) - J(\boldsymbol{\theta})$ appears to be a rare occurrence (unless $\boldsymbol{\theta}$

is a scalar), and one usually has to conclude that relative informativeness depends upon the question being asked.

When $J_w(\boldsymbol{\theta}) - J(\boldsymbol{\theta})$ is indefinite, comparison of weighted and natural observations can be made in at least two ways:

- in terms of a suitable scalar-valued measure of joint information for $\boldsymbol{\theta}$. Here the reciprocal of the generalised variance, $\det J(\boldsymbol{\theta})$, naturally suggests itself.
- in terms of the standard error of estimation for the scalar-valued function of $\boldsymbol{\theta}$ that is considered most relevant to the scientific problem at hand.

The first possibility has been studied by Patil and Taillie (1989) for the two-parameter gamma, negative binomial, and log-normal distributions and with known weight function $w(x) = x^{\alpha}$. In all these cases, $J_w(\boldsymbol{\theta}) - J(\boldsymbol{\theta})$ is indefinite. For the gamma and negative binomial distributions, the weighted observations turn out to be jointly less informative for $\boldsymbol{\theta}$ when the generalised variance is the criterion. However, the relative efficiency is quite close to unity unless the shape parameter is small (less than 0.5, say). The log-normal distribution is an interesting case because $\det J_w(\boldsymbol{\theta}) = \det J(\boldsymbol{\theta})$ for all $\boldsymbol{\theta}$ so that weighted and unweighted observations are equally informative in terms of the generalised variance.

The next section applies the second possibility to the truncated log-normal distribution with parameters μ and σ^2. The example is motivated by Preston's (1948, 1962) log-normal distribution of species abundance. Scientific interest resides primarily in the parameter σ^2 which is related to the evenness component of species diversity (Patil and Taillie, 1979).

2.3.1 Species abundance

Preston (1948, 1962) argues that in large ecological communities, the various species have their abundance X' distributed according to a log-normal curve with parameters μ' and σ^2. If an investigator samples a fraction t of the individuals in the community, then a species with natural abundance X' has recorded abundance $X = t \cdot X'$. (As Preston points out, the analysis is only approximate since it ignores sampling fluctuation. With Poisson sampling, for example, the Poisson-log-normal would be a more appropriate sampling distribution.) Now $X = tX'$ follows a log-normal distribution with parameters μ and σ^2, where $\mu = \mu' + \ln t$. The parameter of primary interest is $\varphi = 1/\sigma^2$ since it is a measure of species evenness.

Preston also notes that species with $X < 1$ will go unrecorded so that the observed data actually follow a truncated log-normal distribution with truncation point located at $X = 1$ (or at $\ln X = 0$). Preston refers to the truncation point as the *veil line*: when the sampling intensity increases, the frequency curve moves to the right relative to the veil line and additional species become 'unveiled'.

With the preceding a motivation, we investigate the question: for the purpose of estimating σ^2, what is the gain or loss of information resulting from truncation of the log-normal curve, and how does that gain/loss change as the sampling intensity increases (i.e. as μ increases)? The case of practical importance has μ as unknown nuisance parameter, but we also consider the case in which μ is known.

The analysis is most conveniently done in terms of the truncated normal distribution, for which Cohen (1949, 1950) has given a detailed discussion of maximum likelihood estimation. We follow Cohen's notation:

$$\xi = -\mu/\sigma$$

$$Z = Z(\xi) = \frac{\exp(-\xi^2/2)/\sqrt{2}\pi}{1-\Phi(\xi)}.$$

Note that ξ is the truncation point expressed in standardised normal units, while Z is the reciprocal of Mill's ratio and, despite the notation, is not to be confused with the ordinate of the standard normal curve. The following result will be used several times and is a simple consequence of Gordon's (1941) inequality (see Johnson and Kotz, 1970, p. 279):

Lemma 2.1. *For all real* ξ, (a) $Z-\xi>0$, (b) $1-\xi(Z-\xi)>0$, (c) $1-Z(Z-\xi)>0$.

Cohen parametrises the truncated normal(μ, σ^2) distribution by ξ, σ and obtains the following for the information matrix per observation (1949, eq. 24–6; 1950, eq. 25–6):

$$J_w = J_w(\xi, \sigma) = \begin{bmatrix} 1-Z(Z-\xi) & -\dfrac{Z-\xi}{\sigma} \\ -\dfrac{Z-\xi}{\sigma} & \dfrac{2-\xi(Z-\xi)}{\sigma^2} \end{bmatrix}. \tag{2.15}$$

When sampling from the full (untruncated) normal cuvc, the information matrix is $J(\mu, \sigma) = \text{diag}(1, 2)/\sigma^2$. In terms of ξ and σ, this becomes

$$J = J(\xi, \sigma) = \begin{bmatrix} 1 & \dfrac{\xi}{\sigma} \\ \dfrac{\xi}{\sigma} & \dfrac{2+\xi^2}{\sigma^2} \end{bmatrix}. \tag{2.16}$$

We can now show that $J_w - J$ is indefinite, implying that no intrinsic comparisons can be made about the relative informativeness of samples from truncated and untruncated normal populations. First recall that a symmetric, nonsingular

2×2 matrix is indefinite if and only if it has negative determinant. But

$$J_w - J = \begin{bmatrix} -Z(Z-\xi) & -Z/\sigma \\ -Z/\sigma & -Z\xi/\sigma^2 \end{bmatrix}$$

and

$$\det(J_w - J) = \frac{Z^2}{\sigma^2}[\xi(Z-\xi) - 1]$$

which is negative according to part (b) of the lemma.

Next consider comparisons on the basis of the generalised variance. We have

$$\det J_w = \frac{2}{\sigma^2} - \frac{Z(Z-\xi)[2+1-\xi(Z-\xi)]}{\sigma^2} \tag{2.17}$$

$$= \frac{2}{\sigma^2}[1 - Z(Z-\xi)] - \frac{Z(Z-\xi)[1-\xi(Z-\xi)]}{\sigma^2}, \tag{2.18}$$

$$\det J = \frac{2}{\sigma^2}. \tag{2.19}$$

In view of the lemma, we see that $\det J_w < \det J$, so that observations from the truncated normal are less informative for joint estimation of (μ, σ^2) in terms of the generalised variance. It is interesting to note that the relative joint efficiency,

$$\left[\frac{\det J_w}{\det J}\right]^{\frac{1}{2}} = \{1 - Z(Z-\xi)[2+1-\xi(Z-\xi)]/2\}^{\frac{1}{2}},$$

depends only on ξ and not on σ^2.

Finally, we consider estimation of σ with ξ (or μ) as a nuisance parameter. Let $\mathrm{var}(\hat{\sigma})$ and $\mathrm{var}_w(\hat{\sigma})$ designate the asymptotic variance of the MLE for samples from the full and truncated distributions. Recall that the asymptotic relative efficiency, $\mathrm{var}(\hat{\sigma})/\mathrm{var}_w(\hat{\sigma})$, is unchanged when the parameter σ is replaced by a smooth monotonic transformation such as $\varphi = 1/\sigma^2$. Our next theorem shows that samples from the truncated curve are less informative and have a very low efficiency relative to samples from the full curve.

Theorem 2.1. *For all values of the parameters, it is the case that* $\mathrm{var}_w(\hat{\sigma}) > \mathrm{var}(\hat{\sigma})$. *Further, the relative efficiency is*

$$e = \mathrm{var}(\hat{\sigma})/\mathrm{var}_w(\hat{\sigma}) = 1 - \frac{Z(Z-\xi)[1-\xi(Z-\xi)]}{2[1-Z(Z-\xi)]}. \tag{2.20}$$

Proof: Equations (2.15) and (2.16) give

$$n \cdot \mathrm{var}(\hat{\sigma}) = 1/\det J$$

$$n \cdot \mathrm{var}_w(\hat{\sigma}) = [1 - Z(Z - \xi)]/\det J_w,$$

so that

$$e = \frac{1}{1 - Z(Z - \xi)} \frac{\det J_w}{\det J}.$$

Formula (2.20) is now an easy consequence of (2.18) and (2.19). Also from the Lemma, we can conclude that $e < 1$.

The following tabulation shows that the efficiency is quite low, being only 12% when the mode is first unveiled and remaining below 70% when as much as 98% of the curve has been unveiled.

	Mode veiled				Mode unveiled			
$-\xi = \mu/\sigma$:	-3	-2	-1	0	1	2	3	4
efficiency :	0.7%	2%	4%	12%	33%	67%	93%	99.5%

The low efficiency is not a direct result of truncation but instead is caused by the presence of the nuisance parameter μ, which prevents one from accurately locating the mode relative to the veil line. (Along these lines Cohen (1950) has shown that estimation of σ improves markedly if one knows the number of observations that were truncated from the normal curve, i.e. the number of veiled species. In effect, such knowledge helps in locating the mode.) To see the effect more clearly, we take up the problem of estimating σ when μ is known.

Theorem 2.2. *Consider samples of equal size from the truncated and from the full normal* (μ, σ^2) *population where* μ *is known. Let* $\mathrm{var}_w(\hat{\sigma})$ *and* $\mathrm{var}(\hat{\sigma})$ *be the respective asymptotic variances for estimating* σ*. Then* $\mathrm{var}_w(\hat{\sigma})$ *if and only if* $\mu < 0$*. Further, the relative efficiency is*

$$e = \mathrm{var}(\hat{\sigma})/\mathrm{var}_w(\hat{\sigma}) = 1 + \tfrac{1}{2}\xi Z[1 - \xi(Z - \xi)]. \tag{2.21}$$

Proof: We first transform the information matrix (2.15) to the μ, σ parametrisation to obtain

$$J_w(\mu, \sigma) = \begin{bmatrix} \dfrac{A}{\sigma^2} & -\dfrac{A\mu + B\sigma}{\sigma^3} \\ -\dfrac{A\mu + B\sigma}{\sigma^3} & -\dfrac{A\mu^2 + 2B\mu\sigma + C\sigma^2}{\sigma^4} \end{bmatrix} \tag{2.22}$$

where

$$A = 1 - Z(Z - \xi),$$
$$B = -(Z - \xi),$$
$$C = 2 - \xi(Z - \xi).$$

Also $J(\mu, \sigma) = \text{diag}(1, 2)/\sigma^2$. Because μ is known, $n \cdot \text{var}_w(\hat{\sigma})$ and $n \cdot \text{var}(\hat{\sigma})$ are the reciprocals of the $(2, 2)$-elements of $J_w(\mu, \sigma)$ and $J(\mu, \sigma)$. This gives

$$e = (A\xi^2 - 2B\xi + C)/2,$$

from which (2.21) follows upon substitution for A, B, C. Now we have to show that $e > 1$ if and only if $\mu < 0$, i.e. if and only if $\xi > 0$. But this is clear from (2.21) and part (b) of lemma 2.1.

The relative efficiency given by (2.21) is tabulated below:

	Model veiled					Mode unveiled			
$-\xi = \mu/\sigma$:	-4	-3	-2	-1	0	1	2	3	4
efficiency :	182%	174%	160%	136%	100%	67%	72%	93%	99.5%

The tabulation reveals several interesting features. First, the efficiency is never very small. A numerical search indicates that the minimum efficiency of 63.6% occurs at $\mu/\sigma = 1.38$. Second, the efficiency is high when only the extreme right tail is unveiled. This seems to be saying that large observations are more informative than small observations for estimating the variance of a normal curve. In fact one can show that the information for estimating σ^2 is an increasing function of T when the interval $[-T, T]$ is removed from the Normal$(0, \sigma^2)$ curve.

We conclude by stating the analogous results that apply when μ is the object of inference. Part (b) of the next theorem was obtained by Bayarri and DeGroot (1986).

Theorem 2.3. *Consider samples of equal size from the truncated and from the full Normal* (μ, σ^2) *population. Let* $\text{var}_w(\hat{\mu})$ *and* $\text{var}(\hat{\mu})$ *be the respective asymptotic variances for estimating* μ. *Then*

(a) *when* σ *is unknown,* $\text{var}_w(\hat{\mu}) > \text{var}(\hat{\mu})$ *and the relative efficiency is*

$$e = \text{var}(\hat{\mu})/\text{var}_w(\hat{\mu}) = 1 - \frac{2Z(Z - \xi) + Z^2[1 - \xi(Z - \xi)]}{2 + \xi Z[1 - \xi(Z - \xi)]} \tag{2.23}$$

(b) *When* σ *is known,* $\text{var}_w(\hat{\mu}) > \text{var}(\hat{\mu})$ *and the relative efficiency is*

$$e = \text{var}(\hat{\mu})/\text{var}_w(\hat{\mu}) = 1 - Z(Z - \xi). \tag{2.24}$$

The relative efficiencies (2.23) and (2.24) for estimating μ are tabulated below.

	Mode veiled					Mode unveiled			
$-\xi = \mu/\sigma$:	-4	-3	-2	-1	0	1	2	3	4
efficiency :									
σ unknown:	0%	0%	0%	1%	5%	30%	83%	99%	99.9%
σ known :	5%	7%	11%	20%	36%	63%	89%	99%	99.9%

2.4 OVERDISPERSION AND HETEROGENEITY IN TERATOLOGICAL BINARY DATA

The problem of overdispersion and heterogeneity in binary data arises quite naturally in developmental toxicity studies. A typical toxicity study consists of a certain number t of dosages of a toxic chemical. A certain number m_i of pregnant females are exposed to the ith dose during the period of organogenesis, or during the pregnancy and lactation period, depending upon the purpose of the experiment. At the end of the experiment, the proportions of foetuses showing toxic responses are considered as indicators of the toxicity of the dose. In such experiments, one invariably finds that there is extra variation which cannot be explained by the ordinary. binomial distribution.

Since a pregnant female is exposed to the chemical dose, litter becomes the primary unit. The random effect of the litter, i.e. biological response of the mother to the chemical dose, affects the toxic responses of the foetuses. This potential random litter effect causes heterogeneity, extravariation and also intra-litter correlation between the responses of two foetuses within the litter.

In order to incorporate this random litter effect in the analysis of such binary data, research workers have introduced the beta-binomial model (BTB) which is a binomial mixture model. The binomial parameter p (proportion of toxic responses in a litter) which varies from one litter to another is assumed to follow a beta distribution on $[0, 1]$ with parameters $\alpha, \beta > 0$. Therefore,

$$\Pr(Y_{ij} = y) = \binom{n_{ij}}{y} \cdot \frac{\Gamma(\alpha + y)}{\Gamma(\alpha)} \cdot \frac{\Gamma(\beta + n_{ij} - y)}{\Gamma(\beta)} \cdot \frac{\Gamma(\alpha + \beta)}{\Gamma(\alpha + \beta + n_{ij})} \tag{2.25}$$

where n_{ij} denotes the size of the jth litter in the ith dose group and Y_{ij} denotes the total number of toxic responses in the (ij)th litter.

Therefore

$$E(Y_{ij}) = n_{ij} \cdot \frac{\alpha}{\alpha + \beta} = n_{ij}P = \mu$$

$$\text{var}(Y_{ij}) = n_{ij}P(1 - P)[1 + (n_{ij} - 1)\varphi] = V(\mu)[1 + (n_{ij} - 1)\varphi] \tag{2.26}$$

where

$$\varphi = \frac{1}{\alpha + \beta + 1}.$$

Williams (1975), who used this mixture model to analyse developmental toxicity data, expressed the variance as

$$\text{var}(Y_{ij}) = V(\mu)\left[1 + (n_{ij} - 1)\cdot\frac{\theta_w}{1 + \theta_w}\right], \tag{2.27}$$

where

$$\theta_w = \frac{1}{\alpha + \beta}.$$

It is obvious that when $\theta_w = 0$ or $\varphi = 0$, $\text{var}(Y_{ij}) = V(\mu)$, where $V(\mu)$ denotes the variance of the binomial random variable with mean μ. Thus when $\theta_w = 0$, there is no overdispersion and the binomial mixture model reduces to the ordinary binomial model. Therefore φ or θ_w can be looked upon as the index of overdispersion. Further, φ is also the intra-litter correlation coefficient. Therefore it can also be looked upon as the measure of the proportion of total variation attributable to the random litter effect. Thus the beta-binomial model, regarded as the binomial mixture model, takes care of overdispersion, heterogeneity, and also the clumping of observations within the litter. This clumping of observations occurs because of the tendency of the foetuses within the litter to 'behave' alike.

Efron (1986) and Diaconis and Efron (1985) have introduced a very general family, called the *double exponential family* (DEF), in order to incorporate extra variation into the exponential family. Let $f_\mu(y)$ denote the p.d.f. of the original exponential family, with $E(Y) = \mu$ and $\text{var}(Y) = V(\mu)$ where $V(\cdot)$ is a known function. The corresponding DEF is obtained as

$$\bar{f}_{\mu,\theta}(y) = c(\mu, \theta)\sqrt{\theta}\, f_\mu(y)\left[\frac{f_y(y)}{f_\mu(y)}\right]^{1-\theta}, \tag{2.28}$$

where θ is referred to as the parameter of overdispersion and $c(\mu, \theta)$ is the normalising constant. This family is further discussed in Section 2.5 below. Efron (1986) has shown that

$$c(\mu, \theta) \simeq 1, \quad E(Y) \simeq \mu \quad \text{and} \quad \text{var}(Y) \simeq \frac{V(\mu)}{\theta}, \tag{2.29}$$

provided θ is close to unity.

Efron (1986) has used the double binomial model to analyse toxoplasmosis data collected in 34 cities in Latin America. He has used the model to accommodate the extra-binomial variation occurring due to the clumping of observations. The situation is similar to that of foetuses showing toxic responses in each of the litters at a given dose. Therefore we decided to investigate the use of the double binomial (DBN) model as an alternative model in the analysis of developmental toxicity data. Let the original binomial model with mean μ be written as

$$f_{\mu}(y) = \binom{n}{y}\left(\frac{\mu}{n}\right)^{y}\left(1 - \frac{\mu}{n}\right)^{n-y}$$

with $E(Y) = \mu$ and $\text{var}(Y) = \mu(1 - \mu/n)$ where n is a positive integer. Therefore, according to (2.28)

$$\bar{f}_{\mu,\theta}(y) = c(\mu, \theta)\sqrt{\theta}\binom{n}{y}\left(\frac{\mu}{n}\right)^{y}\left(1 - \frac{\mu}{n}\right)^{n-y}\left[\frac{(y/n)^{y}(1 - y/n)^{n-y}}{(\mu/n)^{y}(1 - \mu/n)^{n-y}}\right]^{1-\theta} \quad (2.30)$$

Using this DBN model, we have analysed the data in a developmental toxicity study used by Williams (1975) to illustrate the use of BTB. There were two dose groups: control and treated with sixteen pregnant females at each dose. Pregnant females were treated during pregnancy and also during the 21-day lactation period. The following tabulation presents the parameter estimates for the two models; P denotes the proportion of foetuses that survived until day 21, out of those which were alive on day 4.

	BTB control group	BTB treated group	DBN treated group	DBN control group
$\hat{P}$:	0.898	0.740	$\hat{P} = 0.772$	0.899
$\hat{\phi}$:	0.0206	0.32		
$\hat{\theta}_w$:	0.0210	0.465	$\hat{\theta}_{\text{DBN}} = 0.247$	0.747
$L_{\max} =$	-51.7	-64.99	$L_{\max} = -64.57$	-51.4

The BTB estimates are those obtained by Williams (1975).

It can be noticed from the above tabulation that both the models fit equally well to the data in both the treatment groups. The maximum values of log-likelihood function denoted as $L_{\max}$ in the table are very much comparable. The estimates of θ_w and θ_{DBN} are not comparable, because the two parameters do not correspond to each other. In the beta-binomial model, as $\theta_w \to 0$, overdispersion $\to 0$; but in the double binomial model, as $\theta_{\text{DBN}} \to 1$, overdispersion $\to 0$. The beta-binomial model provides an estimate of the index of overdispersion and also an estimate of intra-litter correlation coefficient, while the double-binomial model provides an estimate only of the overdispersion parameter.

2.4.1 Comparison of beta-binomial and double binomial as models for overdispersed binary data

The problem of analysis of overdispersed binary data coming from different litters at a given dose or coming from different cities with given rainfall can also be looked upon as the problem of encounter data where one is trying to combine the observational data coming from different sources. Therefore one can look upon both the beta-binomial and the double binomial as weighted binomial distributions where each model has its own separate weight function. We have plotted the weight functions for a broad range of parameters and generally find a close match between the two. In making comparisons, the parameters are fixed so that the two weighted distributions have the same mean and the same variance.

The near coincidence of the two weighted distributions is displayed in Figure 2.1 where we have plotted their probability functions, as well as the probability function of the binomial distribution having the same mean (and, of course, having a smaller variance). For extremely small values of θ, i.e. for large overdispersion, there is a tendency for the double binomial to have a mode at $x = 0$ as well as a more central mode. This tendency is hinted at in Figure 2.1 by the comparatively low value at $x = 1$ of the double binomial probability function.

Our general conclusion is that the two distributions are quite comparable in their capability for describing overdispersed binary data. It appears that the choice between them may be made on such grounds as parameter interpretation and inferential convenience. Since maximum likelihood estimation is simplified for the double binomial distribution when $c(\mu, \theta) \simeq 1$, we next inquire into the accuracy of this approximation.

2.4.2 Accuracy of Efron's approximation formulae

Efron (1986) has stated that $c(\mu, \theta) \simeq 1$ and $E(Y) \simeq \mu$, where μ denotes the mean of the original exponential family model. The second of these relations would imply that $E(Y) \simeq np$ for the double binomial, where p is the underlying binomial response probability. As a check on this approximation, we have plotted $E(Y)/n$ versus p in Figure 2.2 for two values of θ, namely $\theta = 0.5$ and $\theta = 0.1$. The curve for $\theta = 0.5$ follows the 45-degree line rather closely, but marked departures emerge when $\theta = 0.1$. The effect seems to be closely related to the tendency of the double binomial to produce a mode at $x = 0$ for small θ, thereby substantially affecting the mean value. One consequence of the breakdown in the approximation is that the fitted value of p is no longer interpretable as the response probability; one must use instead $E(Y)/n$ for this probability and accept that the individual response probability depends upon the litter size n. In the beta-binomial model, the response probability is not litter-size-dependent.

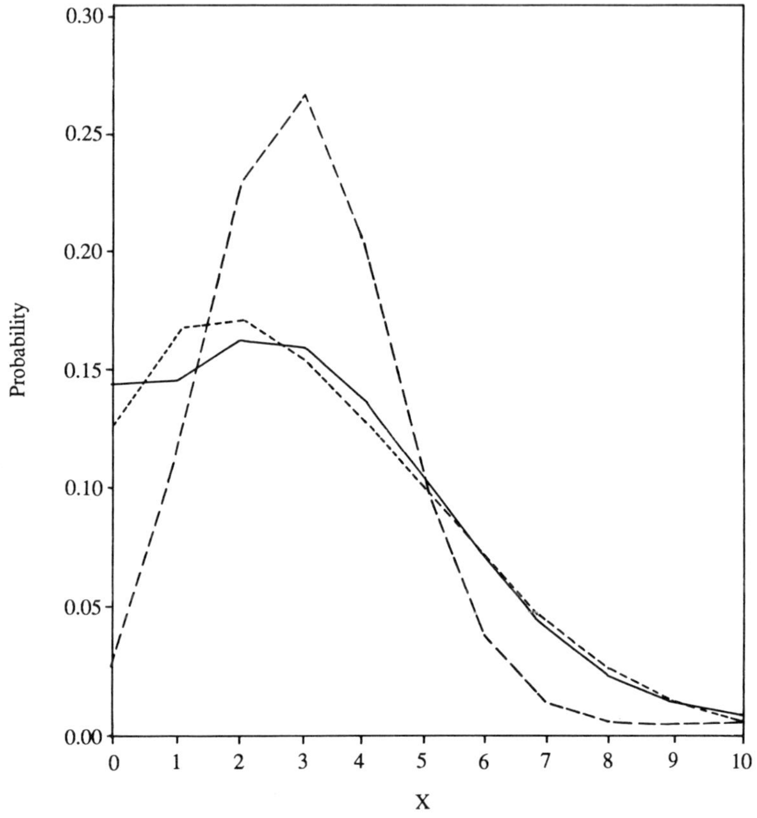

Figure 2.1 Probability functions for the double binomial distribution (solid), the beta-binomial distribution (short dashes), and the binomial distribution (long dashes). The double binomial parameters are $n = 10$, $p = 0.3$ and $\theta = 0.4$. The plotted beta-binomial has the same mean and variance as the double binomial; the plotted binomial has the same mean as the double binomial

Figure 2.3 plots the normaliser $c(\mu, \theta)$ versus θ for $p = 0.15$ and $p = 0.5$. The value of the normaliser is only modestly affected by p and is quite close to unity unless θ is very small, say $\theta < 0.1$. If one has reason to think that one's data exhibit extreme overdispersion, then it is advisable to expressly calculate $c(\mu, \theta)$ when maximising the likelihood.

2.5 MODELLING BIVARIATE OVERDISPERSION

In teratology, one sometimes has more than two possible responses. For example, the responses might be *death*, *malformation* and *normal*. Here, the simplest model

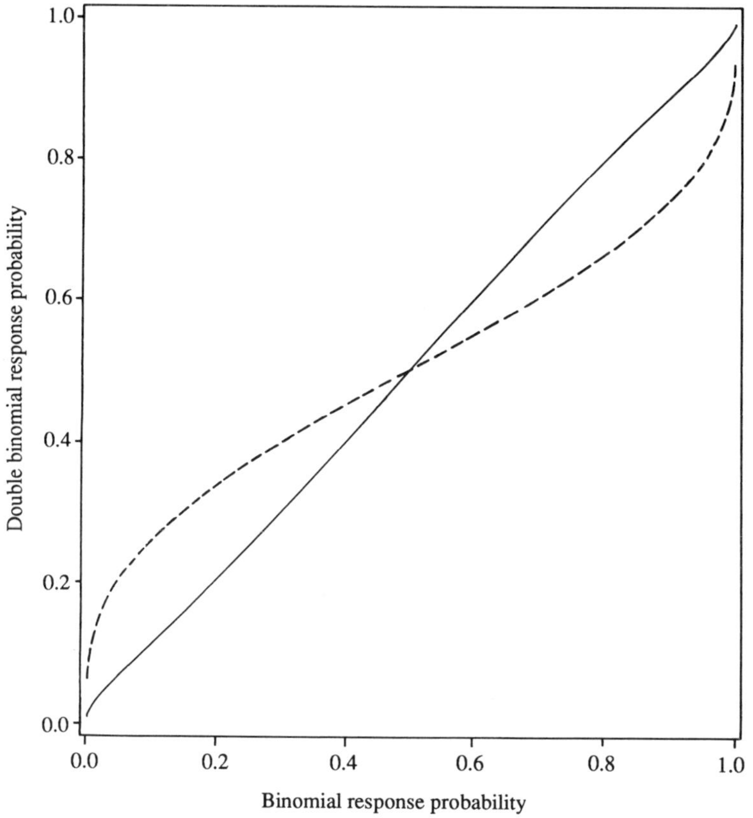

Figure 2.2 Relation between the binomial response probability p and the double binomial response probability $E(Y)/n$, with $n = 10$. Curves are drawn for $\theta = 0.5$ (solid) and $\theta = 0.1$ (dashes). The latter curve departs more markedly from the 45-degree line

would be the trinomial, but this would probably have to be generalised to account for overdispersion. Two such generalisations suggest themselves immediately: (i) the Dirichlet-trinomial and (ii) a multivariate DEF given by (2.28) with $f_\mu(x)$ as the trinomial distribution and μ and x multivariate. Each possibility has the serious defect of having only one overdispersion parameter, whereas, in multivariate data, one would expect the degree of overdispersion to vary with the direction. For example, trinary data can be analysed by fitting one of the marginals and, separately, the conditional distributions. Not infrequently, the two dispersion parameters are found to be significantly different.

Section 2.5.1 attempts to develop a bivariate form of the Diaconis–Efron DEF with multiple overdispersion parameters. The solution that we offer appears promising but, unfortunately, is not completely general since the datum $\mathbf{x}$ is required to have a rectangular support, thereby ruling out the trinomial example

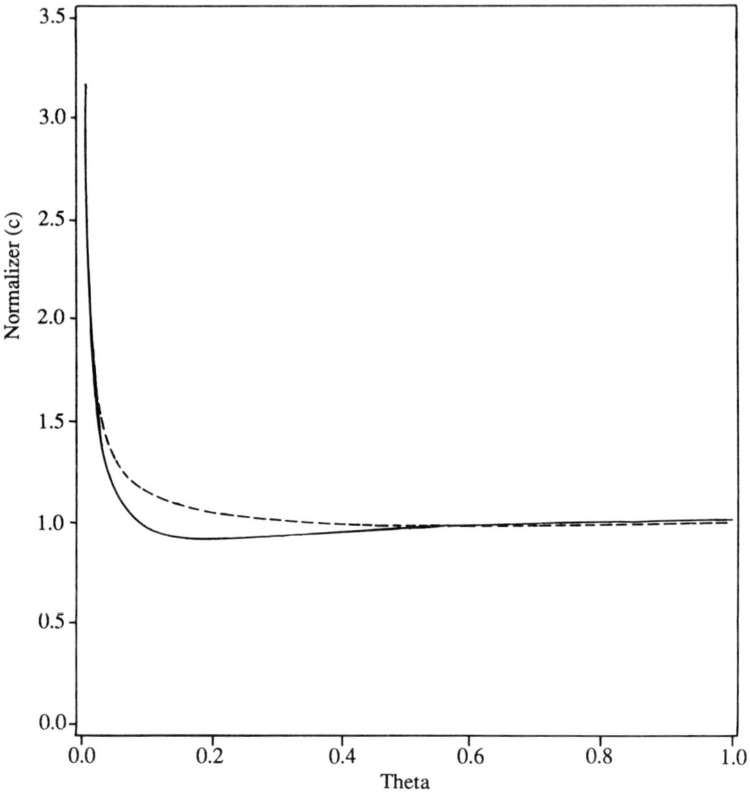

Figure 2.3 Normalising constant $c(\mu, \theta)$ versus θ for the double binomial distribution with $n = 10$. Curves are drawn for $p = 0.5$ (solid) and $p = 0.15$ (dashes)

discussed above. Sections 2.5.2 and 2.5.3 examine a special case: the double bivariate normal distribution. In the univariate case, the double normal is again normal with an altered variance, making it rather uninteresting. Similarly, the double bivariate normal remains bivariate normal but with the interesting feature that the correlation is affected by the overdispersion. In particular, the recorded observations can exhibit correlation even when the natural observations are independent. Section 2.5.4 studies the informativeness of observations from the double bivariate normal as compared with observations from the original bivariate normal. We find that, in a certain overall sense, observations from the double bivariate normal are the more informative in spite of overdispersion.

2.5.1 A bivariate form of the Diaconis–Efron DEF

Starting from an exponential family of univariate distributions indexed by the mean μ, Diaconis and Efron (1985) and Efron (1986) have attempted to account for overdispersion in observed x by introducing the family of densities

$$[f_\mu(x)]^\theta [f_x(x)]^{1-\theta}. \tag{2.30}$$

Here, and throughout this section, we omit normalising constants. This family can be regarded as a sort of geometric mixture expressing the loss of information due to overdispersion: a fraction θ of each observation contributes information for inference on μ; the remaining fraction $1-\theta$ is associated with the parameterless factor $f_x(x)$ and carries no information regarding μ. This interpretation implicitly assumes that $0 \leqslant \theta \leqslant 1$. But, in fact, (2.30) is often a valid probability distribution for values of θ outside the unit interval. For $\theta > 1$, the exponent $1-\theta$ is negative and (2.30) represents a situation of underdispersion or overinformativeness for μ (relative to the basic distribution $f_\mu(x)$). Similarly, as θ approaches zero, we sometimes find that the limiting form of (2.30) does not represent the most extreme overdispersion imaginable. Letting θ become negative allows for further overdispersion. Thus the range of permissible values for θ has to be determined in each instance according to the requirement of normalisability of (2.30).

As indicated in (2.28), the DEF can also be expressed as a weighted form of the basic density,

$$\left[\frac{f_x(x)}{f_\mu(x)}\right]^{1-\theta} \cdot f_\mu(x), \tag{2.31}$$

with weight function

$$w(x;\mu,\theta) = \left[\frac{f_x(x)}{f_\mu(x)}\right]^{1-\theta}. \tag{2.32}$$

While (2.30) highlights the information loss associated with overdispersion, (2.31) and (2.32) emphasise the spreading out about the mean that accompanies overdispersion. The representation (2.31) also demonstrates that overdispersion can be regarded as the result of selection bias where the weight function has the very unusual feature of depending on the parameter μ of the natural distribution.

Our goal is to extend the Diaconis–Efron family to bivariate distributions. A straightforward approach simply replaces $f_\mu(x)$ and $f_x(x)$ in (2.30) by their multivariate counterparts. But this involves only a single overdispersion parameter θ and implies that each component of the bivariate observation $\mathbf{x} = (x_1, x_2)$ is overdispersed to the same degree, which is very limiting. We can overcome the limitation with a bivariate density of the form

$$[f_{\mu_1\mu_2}(\mathbf{x})]^\theta \cdot [f_{\mu_1 x_2}(\mathbf{x})]^{\theta_1} \cdot [f_{x_1\mu_2}(\mathbf{x})]^{\theta_2} \cdot [f_{x_1 x_2}(\mathbf{x})]^{1-\theta-\theta_1-\theta_2}. \tag{2.33}$$

The interpretation is similar to that of (2.30): a fraction θ of each observation carries information for (μ_1, μ_2) jointly, a fraction θ_1 carries information for μ_1 alone, etc.

As discussed in the univariate case above, it is possible for one or some of $\theta, \theta_1, \theta_2$ to fall outside the unit interval. But here the possibilities are even more interesting since one can have all three θ's in the unit interval and yet find that $\theta + \theta_1 + \theta_2 > 1$, suggesting some kind of overall 'super' information. We return to this point in the discussion of the normal distribution below.

Analogous to (2.31), the weighted representation of (2.33) is

$$\left[\frac{f_{\mu_1 x_2}}{f_{\mu_1 \mu_2}}\right]^{\theta_1}\left[\frac{f_{x_1 \mu_2}}{f_{\mu_1 \mu_2}}\right]^{\theta_2}\left[\frac{f_{x_1 x_2}}{f_{\mu_1 \mu_2}}\right]^{1-\theta-\theta_1-\theta_2} \cdot f_{\mu_1 \mu_2}. \tag{2.34}$$

The bivariate DEF considered here requires that the mean vector (μ_1, μ_2) vary over a Cartesian product. Otherwise, it cannot be ensured that the subscripts in $f_{\mu_1 x_2}$ and $f_{x_1 \mu_2}$ represent valid parameter values. For example, (2.33) cannot be applied to overdisperse the trinomial distribution since $\mu_1 + \mu_2 \leqslant 1$.

2.5.2 Double bivariate normal distribution

We take the basic distribution to be bivariate normal with mean (μ_1, μ_2) and variance matrix

$$\Sigma = \begin{bmatrix} \sigma_1^2 & \rho\sigma_1\sigma_2 \\ \rho\sigma_1\sigma_2 & \sigma_2^2 \end{bmatrix}. \tag{2.35}$$

An easy calculation shows that the corresponding bivariate DEF (2.33) is bivariate normal with the same mean vector (μ_1, μ_2) and with variance matrix

$$\Sigma_w = \begin{bmatrix} \sigma_{1w}^2 & \rho_w\sigma_{1w}\sigma_{2w} \\ \rho_w\sigma_{1w}\sigma_{2w} & \sigma_{2w}^2 \end{bmatrix}, \tag{2.36}$$

where

$$\rho_w = \rho \frac{\theta}{\sqrt{(\theta + \theta_1)(\theta + \theta_2)}}, \tag{2.37}$$

$$\sigma_{1w}^2 = \varphi^2 \frac{\sigma_1^2}{\theta + \theta_1}, \tag{2.38}$$

$$\sigma_{2w}^2 = \varphi^2 \frac{\sigma_2^2}{\theta + \theta_2}, \tag{2.39}$$

$$\varphi^2 = \frac{1 - \rho^2}{1 - \rho_w^2}.$$

Positive definiteness of Σ_w imposes the restrictions

$$|\rho\theta| < \sqrt{(\theta+\theta_1)(\theta+\theta_2)}, \tag{2.40}$$

$$\theta+\theta_1 > 0, \tag{2.41}$$

$$\theta+\theta_2 > 0. \tag{2.42}$$

A curious case occurs when $\theta = 0$: then $\rho_w = 0$ and the weighted distribution has independent components irrespective of any correlation in the original distribution.

It is also useful to note that

$$\Sigma_w^{-1} = \frac{1}{1-\rho^2}\begin{bmatrix} \dfrac{\theta+\theta_1}{\sigma_1^2} & -\dfrac{\rho\theta}{\sigma_1\sigma_2} \\ -\dfrac{\rho\theta}{\sigma_1\sigma_2} & \dfrac{\theta+\theta_2}{\sigma_2^2} \end{bmatrix}, \tag{2.43}$$

whereas

$$\Sigma^{-1} = \frac{1}{1-\rho^2}\begin{bmatrix} \dfrac{1}{\sigma_1^2} & -\dfrac{\rho}{\sigma_1\sigma_2} \\ -\dfrac{\rho}{\sigma_1\sigma_2} & \dfrac{1}{\sigma_2^2} \end{bmatrix}. \tag{2.44}$$

It follows that the weight function is given by

$$\exp(-\tfrac{1}{2}(x-\mu)'(\Sigma_w^{-1}-\Sigma^{-1})(x-\mu))$$

which simplifies to

$$\exp\left(-\frac{1}{2(1-\rho^2)}\mathbf{z}'\begin{bmatrix} \theta+\theta_1-1 & \rho(1-\theta) \\ \rho(1-\theta) & \theta+\theta_2-1 \end{bmatrix}\mathbf{z}\right) =$$

$$\exp\left(-\frac{(\theta+\theta_1-1)z_1^2+(\theta+\theta_2-1)z_2^2+2\rho(1-\theta)z_1z_2}{2(1-\rho^2)}\right), \tag{2.45}$$

where $z_i = (x_i - \mu_i)/\sigma_i$ is the standardised variate.

2.5.3 A weighted bivariate normal correlation model

As a special case of the preceding section, we study the effect of the bivariate weight function,

$$w(z_1, z_2; \lambda) = \exp(\lambda z_1 z_2), \tag{2.46}$$

on the bivariate normal distribution. Cox (1991) points out that the weight function (2.46) provides an interesting way of generating association in bivariate distributions; see also Goodman (1981; 1991, p. 1136). As before, the z_i are the standardised variates, while λ is a fixed parameter not depending upon the parameters of the normal distribution. The contours of constant weight are rectangular hyperbolae. Large weights occur in the first and third quadrants for positive λ, and in the second and fourth quadrants for negative λ. Accordingly, we would expect positive λ to increase the (signed) correlation and negative λ to decrease it.

The weighted distribution of this section is a special case of the bivariate normal DEF discussed in the last section. Matching coefficients in (2.45) and (2.46), we nced

$$\theta + \theta_1 = 1, \tag{2.47}$$

$$\theta + \theta_2 = 1, \tag{2.48}$$

$$\lambda = -\frac{\rho}{1-\rho^2}(1-\theta). \tag{2.49}$$

The first two of these relations imply that $\theta + \theta_1 + \theta_2 > 1$ so that there may be some overall overinformativeness in spite of possible overdispersion in the marginals. We make this precise in the next section.

Since we take λ to be independent of ρ, (2.49) requires that θ depends upon ρ. Putting (2.47)–(2.49) into (2.37)–(2.39) yields

$$\rho_w = \rho\theta, \tag{2.50}$$

$$\rho_w = \rho + \lambda(1-\rho^2), \tag{2.51}$$

$$\sigma_{1w}^2 = \varphi^2\sigma_1^2, \tag{2.52}$$

$$\sigma_{2w}^2 = \varphi^2\sigma_2^2. \tag{2.53}$$

Requiring that $-1 < \rho_w < 1$ in (2.51) gives the restriction

$$-\frac{1}{1-\rho} < \lambda < \frac{1}{1+\rho} \tag{2.54}$$

on the parameter λ. Also, when $\rho = 0$, (2.51) gives $\rho_w = \lambda$ so that the weighting can introduce correlation into the normal model with independent components. This does not contradict (2.50) since θ becomes infinite as ρ goes to zero.

2.5.4 Relative informativeness of the weighted observations

We investigate the relative informativeness of observations from the weighted bivariate normal distribution of Section 2.5.3. Throughout, the parameter λ of the weight function (2.46) is regarded as fixed and known. As in Section 2.3, our comparisons are on the basis of the Fisher information matrices J_w and J. Since J_w and J depends upon which of the parameters $\mu_1, \mu_2, \sigma_1^2, \sigma_2^2, \rho$ are unknown, we distinguish three cases. Proofs of the following results are complicated and are omitted.

Case 1: Let all five parmameters $\mu_1, \mu_2, \sigma_1^2, \sigma_2^2, \rho$ be unknown. Then $J_w - J$ is indefinite unless $\lambda = 0$ so intrinsic comparison of weighted and unweighted observations is not possible. On the other hand,

$$\det J_w/\det J = \left[1 + \frac{(\rho_w - \rho)^2}{1 - \rho_w^2}\right]^2 \geqslant 1. \tag{2.55}$$

Thus weighted observations are always more efficient in the overall sense of the generalised variance. This is true even in the presence of overdispersion and confirms the conjecture of the previous section.

Case 2: Let $\sigma_1^2, \sigma_2^2, \rho$ be unknown and μ_1, μ_2 be known. Definiteness of $J_w - J$ is still an open question for this case. But

$$\det J_w/\det J = \varphi^2\left[1 + \frac{(\rho_w - \rho)^2}{1 - \rho_w^2}\right]^2, \tag{2.56}$$

which can be greater than or less than unity depending upon ρ and ρ_w.

Case 3: Let μ_1, μ_2 be known and $\sigma_1^2, \sigma_2^2, \rho$ be known. Then $J_w - J$ is indefinite unless $\lambda = 0$. Also,

$$\det J_w/\det J = \frac{1}{\varphi^2}, \tag{2.57}$$

so that weighted observations are more informative for inferences on the means exactly as in the case of underdispersion.

3

Comparing Sampling Designs for Monitoring Ecological Status and Trends: Impact of Temporal Patterns

N. Scott Urquhart, W. Scott Overton and David S. Birkes,
Oregon State University, USA

3.1 INTRODUCTION

Environmental monitoring is beginning. This chapter presents a comparison of two classes of sampling designs suggested for long-term environmental monitoring. One class is based on a rotating panel design that prescribes visiting a panel of sites for several consecutive years with partial replacement of sites each year; the other class is based on a serially alternating design which prescribes returning to sampling sites every four years. A random-effects linear model is used to investigate and compare the proposed sampling designs because the sites to be visited and the years of visits may manifest substantial random contributions.

Ecological monitoring can provide a mechanism for detecting gradual, but sustained, changes in our environment. Gradual environmental changes could reflect a positive impact of legislation or regulation, or could result from unrecognised environmental insults.

Statistics for the Environment. Edited by Vic Barnett and K. Feridun Turkman

Regulators concerned with enforcing legislative mandates for environmental improvement have had to adopt broad spatial perspectives. But they have located their monitoring resources where problems were perceived to be, and at those points only as long as seemed necessary. Virtually all monitoring for enforcement has been accomplished without the benefit of a sampling plan based on probability. Consequently, past environmental monitoring efforts provide a very limited basis for making statements about ecological health in a region. Yet many regulatory or legislative mandates, for example those related to acid rain, air quality, or water quality, have an impact at least at the regional level; some have an influence even at national and international levels.

Skalski (1990) offered a design for environmental monitoring. Although he described his offering as sampling with partial replacement, Duncan and Kalton (1987), in general considerations of issues of design and analysis of surveys across time, called the design Skalski advanced a **rotating panel**. The *top part* of Table 3.1 displays an extension of the temporal pattern of this design. To see the basic design, ignore the column labelled '*Set size*' and the row labelled '*Repeated*'. Once in full operation, this design prescribes that a series of sites will be visited for several consecutive years, four in this example, and then removed from future consideration; as each set of sites is removed, another set is started.

The United States Environmental Protection Agency (USEPA) has embarked on the Ecological Monitoring and Assessment Program (EMAP) as a successor to several successful one-time probability surveys (Linthurst, *et al.*, 1986; Landers, *et al.*, 1987; Kaufmann, *et al.*, 1988). Overton *et al.* (1991) prescribe a sampling design for this programme that is both serially alternating and spatially interpenetrating. The present consideration focuses on temporal patterns only; the *bottom part* of Table 3.1 displays the temporal pattern of the **serially alternating** design. (Again ignore the column labelled '*Set size*' and the row labelled '*Repeated*' in Table 3.1). General features of this design have been recognised since at least Rao and Graham (1964), but it appears not to have been advanced for ecological monitoring before its prescription for EMAP.

Both of these designs, as shown in Table 3.1, have been augmented with an additional set of sites to be visited annually for the duration of the monitoring programme. Although Kish (1987) describes such augmentation as *split panel designs*, his attempt at standardising terminology does not foresee designs of the temporal complexity shown in Table 3.1. Our use of the word 'augmentation' is consistent with the usage of Federer and Searle (1961). Kish (1987) suggests that the two parts (base and augmented) of such designs should be optimised separately for different objectives. Because both the base and augmented parts of these designs contain information about status and trends, our approach will be to ask what portion of the resources should be expended on each part to achieve the best precision for several interesting criteria. The '*Set size*' column in Table 3.1 illustrates one such allocation of 48 units of sampling per year effort between the augmented and base parts of the designs. By varying the sizes of the

Table 3.1 Temporal sampling pattern and an example of the distribution of sampling effort for the augmented rotating panel and the augmented serially alternating designs

Augmented Rotating Panel Design

Site set	**Set size (n_i)**	**Years** 1	2	3	4	5	6	7	8	9	10	11	12
Repeated	8	X	X	X	X	X	X	X	X	X	X	X	X
1	10	X											
2	10	X	X	X									
3	10	X	X	X	X								
4	10	X	X	X	X	X							
5	10		X	X	X	X	X						
6	10				X	X	X	X					
7	10					X	X	X	X				
8	10						X	X	X	X			
9	10							X	X	X	X		
10	10								X	X	X	X	
11	10									X	X	X	X
12	10										X	X	X
13	10											X	X
14	10												X
15	10												

Augmented Serially Alternating Design

Site set	**Set size (n_i)**	**Years** 1	2	3	4	5	6	7	8	9	10	11	12
Repeated	8	X	X	X	X	X	X	X	X	X	X	X	X
1	40	X				X				X			
2	40		X				X				X		
3	40			X				X				X	
4	40				X				X				X

augmentation set of sites and the sets visited less than annually, these augmented designs vary from pure rotating panel or pure serially alternating to complete annual visits.

Although it may seem attractive to execute a context-free comparison of sampling designs, a biometrician usually uses a specific context to motivate assumptions. EMAP will provide a context for motivating assumptions here, but the alternatives considered will not be limited to those currently assumed by that programme. EMAP focuses on national coverage of all ecological resources, with sufficient sampling to allow regional inferences. The programme is directed to make statements, with known confidence, about the status of, and trends in, the health of the nation's ecological resources. The programme is evaluating various responses as indicators of ecological condition; in keeping with EMAP terminology, we discuss 'indicators' rather than 'responses', the term more familiar to biometricians. The indicators will be evaluated during a single visit to each site; visits will occur no more frequently than annually. Of course visits can be timed to match the best possible time of year for the suite of indicators appropriate for the ecological resource being studied.

For the purpose of defining status and trend, suppose that the expected value of an indicator is of the form $\beta_0 + \beta_1 j$, where $j = 1, 2, \ldots, t$ indexes years and β_1 provides a measure of trend. In the absence of trend ($\beta_1 = 0$), status will be interpreted as the mean response of an indicator over all years. In the presence of trend ($\beta_1 \neq 0$), status will be interpreted as the indicator mean, projected along the trend line (curve) to the most recent time of sampling ($\beta_0 + \beta_1 t$). Trend concerns a continuing and directional change in an indicator across years, like β_1 above. For some number of years, it is anticipated that only linear trend will be detectable; thereafter, simple nonlinear trends will be detectable. Even if the actual trend is not linear, a continuing, directional change will manifest a large and detectable linear component.

Most conventional sampling perspectives recognise the random contribution of units as a consequence of the sampling process, analogous to the randomly selected sites here. The EMAP context has at least two further random aspects: the error in the evaluation of biological and chemical variables, and the potential contribution of the year (hot or cold, wet or dry, etc.) to all sites visited that year. In contrast, many surveys with a temporal component, such as the consumer price survey, seek to estimate the effect of each time increment and remove its effect from the results. The EMAP context needs to estimate the magnitude of the (random) year component, but identify a trend if such exists.

3.2 MODELLING APPROACH

This examination utilises a flexible linear model and the tools of experimental design to make approximate comparisons of two kinds of sampling designs. In doing so, we do not suggest that the approach exactly depicts every detail of the sampling situation. Nevertheless, the flexibility of the model allows comparison of the two basic sampling designs and exploration of the importance of various assumptions and characteristics of the situation.

Consider an indicator of ecological condition (Y) evaluated at a site and in a particular year. Because the sites have been randomly selected, we will assume their effects are random in the sense of a random effect in a linear model. This random contribution of each individual site may remain constant across years. but in other situations a part of a site's effect may decay across years. For example, consider a deep lake in which only 5% of the lake's water volume is replaced annually. If such a lake is above the regional average one year, it is very likely to remain above it the next year and for several following years. Likewise, because the years selected are of no intrinsic interest, we also will assume this factor has a random effect, but years are consecutive. Part of the effect of years will be a consequence of atmospheric conditions, which may have some (auto)correlation through time. For both sites and years, the correlation between responses will diminish the farther apart in time the responses are. We will model their covariance structure as a stationary Markov process. Machin (1975) used this covariance structure; Morrison (1970) investigated more general aspects of repeated measurements using the same covariance structure; and Rao and Graham (1964) even used this structure in a study of sampling designs.

If sites are revisited in some organised manner, as anticipated here, sets of sites will be visited in the same pattern of years. A reasonable plan, and the one assumed here, is that sites to be visited will be partitioned into disjoint sets for which the visit pattern will be the same. Thus if any site in a site set is visited in a year, all sites in that set will be visited that year. Let i index these sets of sites, $i = 1, 2, \ldots, s$; let j index years, $j = 1, 2, \ldots, t$; and let k index sites within a site set, $k = 1, 2, \ldots, n_i$. If the site effects remain constant across years, then the random contributions to an indicator of ecological condition (Y_{ijk}) at a particular site (i, k) in a particular year (j) can be modelled as:

$$Y_{ijk} = S_{ik} + T_j + E_{ijk}. \tag{3.1}$$

The three terms on the right of (3.1) represent the site, time (year), and residual effects, respectively. The residual term includes all effects not otherwise modelled, including any site–year interaction; more generally, the residual effect includes all random contributions specific to a site during a particular year, including measurement error. If the effect of sites decays with years, the site term above must be indexed also by j; this subscript will be enclosed in parentheses, however, to preserve the identification of sites by their two subscripts, i and k, but allow a correlation of site effects across years of less than the unity implied by (3.1):

$$Y_{ijk} = S_{i(j)k} + T_j + E_{ijk}. \tag{3.2}$$

Suppose the random components in (3.1) and (3.2) have variances σ_S^2, σ_T^2 and σ_E^2, and are mutually uncorrelated except for those correlations between the $S_{i(j)k}$ and between the T_j explicitly discussed below. For now we will assume the site components are uncorrelated between sites; possible effects of an anticipated spatial correlation between sites will be the topic of a subsequent consideration. Define the matrix function $\mathbf{\Gamma}_t(\rho)$ by

$$\mathbf{\Gamma}_t(\rho) = \begin{bmatrix} 1 & \rho & \rho^2 & \cdots & \rho^{t-1} \\ \rho & 1 & \rho & \cdots & \rho^{t-2} \\ \rho^2 & \rho & 1 & \cdots & \rho^{t-3} \\ \rho^3 & \rho^2 & \rho & \cdots & \rho^{t-4} \\ \cdot & \cdot & \cdot & \cdots & \cdot \\ \cdot & \cdot & \cdot & \cdots & \cdot \\ \cdot & \cdot & \cdot & \cdots & \cdot \\ \rho^{t-1} & \rho^{t-2} & \rho^{t-3} & \cdots & 1 \end{bmatrix}.$$

If we denote the autocorrelation between year effects by ρ_T, then the covariance among the year contributions $(T_1, T_2, \ldots, T_t)$ can be modelled as $\mathbf{\Sigma}_T = \sigma_T^2\mathbf{\Gamma}_t(\rho_T)$; likewise, the covariance among the responses at the same site in consecutive years (namely $S_{i(1)k}, S_{i(2)k}, \ldots, S_{i(t)k}$), can be modelled as $\mathbf{\Sigma}_S = \sigma_S^2\mathbf{\Gamma}_t(\rho_S)$. This assumed structure is equivalent to requiring that

$$S_{i(j)k} = \rho_S S_{i(j-1)k} + \sqrt{1-\rho_S^2}F_{ijk}, \tag{3.3}$$

where the F_{ijk} represent new independent parts of yearly random contributions to the site effects satisfying $\text{var}(S_{i(j)k}) = \text{var}(F_{ijk}) = \sigma_S^2$, but the F_{ijk} are otherwise uncorrelated. This correlation structure provides a way to investigate impacts of possible correlations under our expectation that correlations between the $S_{i(j)k}$ (or the T_j), if present, will decline as the $S_{i(j)k}$ (or T_j) become farther apart in time. Our utilisation of this structure does not imply we believe it exactly models the correlation structure that exists in reality; it provides a suitable approximation that will support an investigation of possible effects of correlation. The case $\rho_T = 0$ and $\rho_S = 1$ deserves special note because it corresponds to a randomised complete block experimental design with additive treatment (S_{ik}) and block (or year, T_j) effects. Thus, in practice, we expect ρ_T to be fairly small, say less than 0,25, and ρ_S to be fairly close to 1.

If all sites in a site set are visited in the same year, the overall behaviour of status and trend can be evaluated from the means in a table indexed by site set and year. The averaging described here could be taken over the entire population, or over a relevant subpopulation. Thus consider the vector of site set-year means organised by site set within year:

$$\bar{\mathbf{Y}}' = (\bar{Y}_{11\cdot}, \bar{Y}_{21\cdot}, \ldots, \bar{Y}_{s1\cdot}, \bar{Y}_{12\cdot}, \bar{Y}_{22\cdot}, \ldots, \bar{Y}_{s2\cdot}, \ldots, \bar{Y}_{st\cdot})\,.$$

The general variance–covariance among these means can be represented in scalar form, using the Kronecker δ, as

$$\text{cov}(\bar{Y}_{ij\cdot}, \bar{Y}_{i'j'\cdot}) = \delta_{ii'}\sigma_S^2\rho_S^{|j-j'|}/n_i + \sigma_T^2\rho_T^{|j-j'|} + \delta_{ii'}\delta_{jj'}\sigma_E^2/n_i, \tag{3.4}$$

or in matrix form, using the Kronecker product of matrices, as

$$\mathrm{cov}(\bar{\mathbf{Y}}) = \mathbf{\Phi} = \mathbf{\Sigma}_S \otimes \mathbf{D}^{-1}(n_i) + \mathbf{\Sigma}_T \otimes \mathbf{1}_s \mathbf{1}'_s + \sigma_E^2 \mathbf{I}_t \otimes \mathbf{D}^{-1}(n_i), \tag{3.5}$$

where $\mathbf{1}_n$ denotes a column of n ones and $\mathbf{D}(\cdot)$ denotes a diagonal matrix having the indicated value on its diagonal. This variance–covariance matrix clearly depends on all of $\sigma_S^2, \sigma_T^2, \sigma_E^2, \rho_S, \rho_T$, and $n_1, n_2, \ldots, n_t$, and the pattern of cells containing means.

Now consider the estimation of status and trend as if the trend were the same at all sites and use generalised least squares incorporating the variance–co-variance matrix given by (3.5). If we assume that the $E(Y_{ijk})$ are finite, then $\bar{\mathbf{Y}}$ are linearly sufficient in the sense of Barnard (1963a). Consequently, the generalised least squares estimator, based on all the observations Y_{ijk}, actually can be expressed as a linear combination of the site-set-year means $\bar{Y}_{ijk}$. Let $\mathbf{X}$ denote a (regression) matrix having as many rows as $\bar{\mathbf{Y}}$ has means, a column of ones, and a column of the year index associated with each mean. Estimates of the regression coefficients are given by

$$\hat{\boldsymbol{\beta}} = (\mathbf{X}'\mathbf{\Phi}^{-1}\mathbf{X})^{-1}\mathbf{X}'\mathbf{\Phi}'^{-1}\bar{\mathbf{Y}}, \tag{3.6}$$

and their variances and covariances are given by

$$\mathrm{cov}(\hat{\boldsymbol{\beta}}) = (\mathbf{X}'\mathbf{\Phi}^{-1}\mathbf{X})^{-1}. \tag{3.7}$$

The second element of $\hat{\boldsymbol{\beta}}$, the slope, clearly gives a measure of the trend. Status can be estimated by the value of the fitted line at the most recent time, namely, at time t. If the 2×2 matrix $\mathbf{C}_0$ has a first row containing 1 and t, and a second row consisting of 0 and 1, then $\mathbf{C}_0\hat{\boldsymbol{\beta}}$ estimates status and trend and has a variance–covariance matrix given by

$$\mathbf{C}_0(\mathbf{X}'\mathbf{\Phi}^{-1}\mathbf{X})^{-1}\mathbf{C}'_0. \tag{3.8}$$

When a sampling plan is imposed on this context, it specifies the n_i and the years in which specific site sets are visited. Specifically, n_{ij}, the number of observations taken in each cell of the conceptual mean table, is either 0 or n_i, as specified by the sampling design. Consequently, a proposed sampling design can be defined in terms of the number of site sets, the n_i, and the years in which each site set is to be visited. If only some of the $n_{ij} > 0$, all of $\bar{\mathbf{Y}}, \mathbf{X}$ and $\mathbf{\Phi}$ need to be replaced by $\bar{\mathbf{Y}}^*, \mathbf{X}^*$ and $\mathbf{\Phi}^*$, where these represent merely that subset of rows (and columns, if appropriate) of the respective vector and matrices corresponding to cell s where $n_{ij} > 0$.

A fair comparison between variations on the two sampling designs, described in Section 3.1 and displayed in Table 3.1, must be made under some contextual constraints. Specifically, suppose the resources available for data acquisition would be the same, independent of the sampling plan. For the sake of illustration, suppose available resources allow acquiring indicators of interest at 48 sites in a region of interest each year. (The specific numerical results presented later do depend on this number of observations, but the results for other numbers of observations can be related to the results developed using 48 sites. The number of observations per year must be a multiple of 4 for the rotating panel design. The choice of 48 allows 13 splits of the number of observations dedicated to repeated (annual) visits and to the rest of the design.)

The number of sites in the repeated (annual) visit set, depicted in Table 3.1 as 8, will be denoted generally as n_r, and the number in each of the other site sets will be denoted by n_s. These relate to the n_i given earlier in the following way: n_r is *any one* of the n_i, and the other n_i are *all equal* to n_s. Constant resources for annual sampling would imply for the serially alternating design that $n_s = 48 - n_r$, i.e. all site sets other than the augmentation one have to have the same size. Even though all site sets have the same size for the rotating panel design depicted in Table 3.1, other configurations are possible; for example, 10, 10, 10, 10, ... could be replaced site-set sizes of by 34, 2, 2, 2, 34, 2, ... In fact any four-year cycle of sizes totalling n_s would work. Although it will not be discussed further here, site sets of unequal size never performed better than the equal distribution plan; this fact was established using the approaches applied below.

Comparisons of the designs require some criteria. One objective could be to pick the design that would given the most precise estimates of relevant functions of parameters. Equations (3.7) and (3.8) and the context suggest several possible objective functions: (i) the variance of the intercept or (ii) the variance of the slope, of the fitted line, (iii) the mean response value, or (iv) the projected current status. Several of these are correlated, but only two are linearly independent. A fifth measure of precision is the generalised variance, $|\mathrm{cov}(\hat{\boldsymbol{\beta}})|$, a simultaneous evaluation of the joint precision of correlated estimators; an experimental design minimising this determinant is sometimes called D-optimal (Raktoe *et al.* 1981). We will investigate all five of these measures of precision.

Equations (3.7) and (3.8) give variances of estimators in terms of components of variance. If we divide (3.4) through by the year component of variance, it can be rewritten as

$$\mathrm{cov}(\bar{Y}_{ij\cdot}, \bar{Y}_{i'j'\cdot})/\sigma_T^2 = \delta_{ii'}(\{\sigma_S^2/n_i\}/\sigma_T^2)\rho_S^{|j-j'|} + \rho_T^{|j-j'|} + \delta_{ii'}\delta_{jj'}(\{\sigma_E^2/n_i\}/\pi_T^2). \quad (3.9)$$

The year component of variance can be viewed as noise; the ratios of variance components on the right of (3.9) can be viewed as signal-to-noise ratios. Further, (3.9) shows that sample sizes can be incorporated with the site and measurement components of variance. The effect of a range of values of the ratios of the site and measurement components of variance to year component of variance can be interpreted equivalently as a change of sample size for fixed ratios of the components of variance.

3.3 RESULTS

Table 3.2 displays three metrics with results for the two designs in adjacent columns for $\sigma_S = 5$, $\sigma_T = 1$, $\sigma_E = 0.1$, $\rho_S = 0.95$, $\rho_T = 0.25$ and $n_r = 0(4)\ 48$; this set of values represents a realistic possibility. The minimum in each column is underlined and recorded in the last row. If the parameter values used were the correct values in a particular case, and interest lay in minimising the variance of current status, the rotating panel design should be used with $n_r = 0$, whereas if the serially alernating design were used, $n_r = 4$ would be optimal. The serially alternating design would have a slightly lower variance (0.7566 rather than 0.8503). The ratio of these provides a measure of relative efficiency: 0.8503/ 0.7566 = 1.1238.

Equations (3.4) to (3.9) have evaded exact algebraic simplification in all but a few special cases. Consequently, the comparisons of the previous paragraph were made numerically over a wide range of conditions. An examination of (3.9) makes it clear that variation in the variance components can be limited to the ratios σ_S/σ_T and σ_E/σ_T; both were varied across 0.1, 0.5, 1.0, 1.5, 2.0,

Table 3.2 Comparing measures of the precision of the serially alternating and rotating panel designs at the end of 8 years for trend, current Status and generalised variance: $\sigma_S = 5$, $\sigma_T = 1$, $\sigma_E = 0.1$, $\rho_S = 0.95$, and $\rho_T = 0.25$

	Variance of estimates of					
Number in repeated set (n_r)	Trend (slope)		Current status		Generalised variance	
	Rotate	Ser. Alt.	Rotate	Ser. Alt.	Rotate	Ser. Alt.
0	0.0389	0.0376	0.8503*	0.7711	0.0145	0.0117
4	0.0388	0.0358	0.8588	0.7566	0.0149	0.0114
8	0.0387	0.0353	0.8685	0.7597	0.0153	0.0115
12	0.0387	0.0352	0.8794	0.7680	0.0157	0.0119
16	0.0386	0.0352	0.8919	0.7797	0.0162	0.0123
20	0.0385	0.0353	0.9061	0.7945	0.0167	0.0128
24	0.0385	0.0354	0.9225	0.8129	0.0174	0.0134
28	0.0384	0.0356	0.9416	0.8357	0.0181	0.0142
32	0.0383	0.0359	0.9641	0.8642	0.0190	0.0152
36	0.0383	0.0363	0.9908	0.9007	0.0200	0.0166
40	0.0382	0.0367	1.0230	0.9489	0.0212	0.0183
44	0.0381	0.0373	1.0626	1.0153	0.0227	0.0208
48	0.0381	0.0381	1.1123	1.1123	0.0246	0.0246
Minimum	0.0381	0.0352	0.8503	0.7566	0.0145	0.0114
ratios	1.08		1.12		1.27	

*The minimum value in each column is underlined.

2.5, 3, 4, 5 and 6. The correlations ρ_S and ρ_T were varied across 0, 0.05, 0.10, 0.25, 0.50, 0.75, 0.90, 0.95 and 1.00. Years of the study ranged across 2, 3, 4, 8 and 12. For each of these 40 500 combinations of values, each metric for each design was evaluated and minimised across $n_r = 0(4)48$. Figure 3.1(a–f) displays contour plots of the minimised variance of current status for each design and their ratio in relation to ρ_S and ρ_T for $\sigma_S = 5$, $\sigma_T = 1$, $\sigma_E = 0.1$, and in relation to σ_S/σ_T and σ_E/σ_T for $\rho_S = 0.95$ and $\rho_T = 0.25$. The results presented in Table 3.2 are displayed in the appropriate part of Figure 3.1. Figure 3.2 displays the same kind of results, and has the same relation to Table 3.2 as Figure 3.1 has, but Figure 3.2 displays the generalised variance rather than the variance of current status. Observe that both of these measures of relative efficiency change substantially with σ_S/σ_T, but show little influence from σ_E/σ_T. The cases displayed here typify the many cases not explicitly discussed.

The serially alternating design had a smaller (or equal) variance for the estimation of trend under all conditions examined. For the other four metrics, the serially alternating design generally has a smaller value, sometimes substantially smaller. However, in a few cases where neither design is particularly good, the rotating panel design is slightly better; nevertheless, ratios of the metrics (rotating panel/serially alternating) never fell below 0.99.

The considerations so far have held the cycle length at four years. Limited numerical investigations confirm what seems intuitively reasonable: longer cycles, other things constant, lead to modestly improved estimates and gradual gains in the advantages of the serially alternating design over the rotating panel design. The primary disadvantage of longer cycles lies in delayed opportunities to investigate assumptions. Once the serially alternating design has two complete cycles, estimates of change, and thereby trend, can be based on differences of indicators between successive visits. Because the validity of such estimates rests on the design rather than on a model such as that used here, their availability would provide an important and very defensible check on assumptions. Realistically, checks on such assumptions cannot be deferred much more than 10 years; after factoring in time for field work, laboratory analysis and statistical analysis, a cycle of four years appears to be the longest feasible cycle.

The effect of augmentation depends on parameters indexing other features of the situation. The size of the augmentation set that minimised the various metrics was recorded and subjected to a factorial analysis of variance ($5 \times 10 \times 10 \times 9 \times 9$). This particular discussion relates to the generalised variance and the serially alternating sampling design, but similar comments apply to the other metrics and to both designs. The analysis of variance showed that variation in the size of the augmentation set changed in a major way in response to years of sampling and the two correlations; the variance components had substantially less effect. Table 3.3 shows that the size of the optimal augmentation set decreases dramatically with years of sampling, regardless of the correlations, but generally, the higher the correlation, the larger the size of the optimal augmentation set. The analysis of variance indicated a high degree of interaction from both

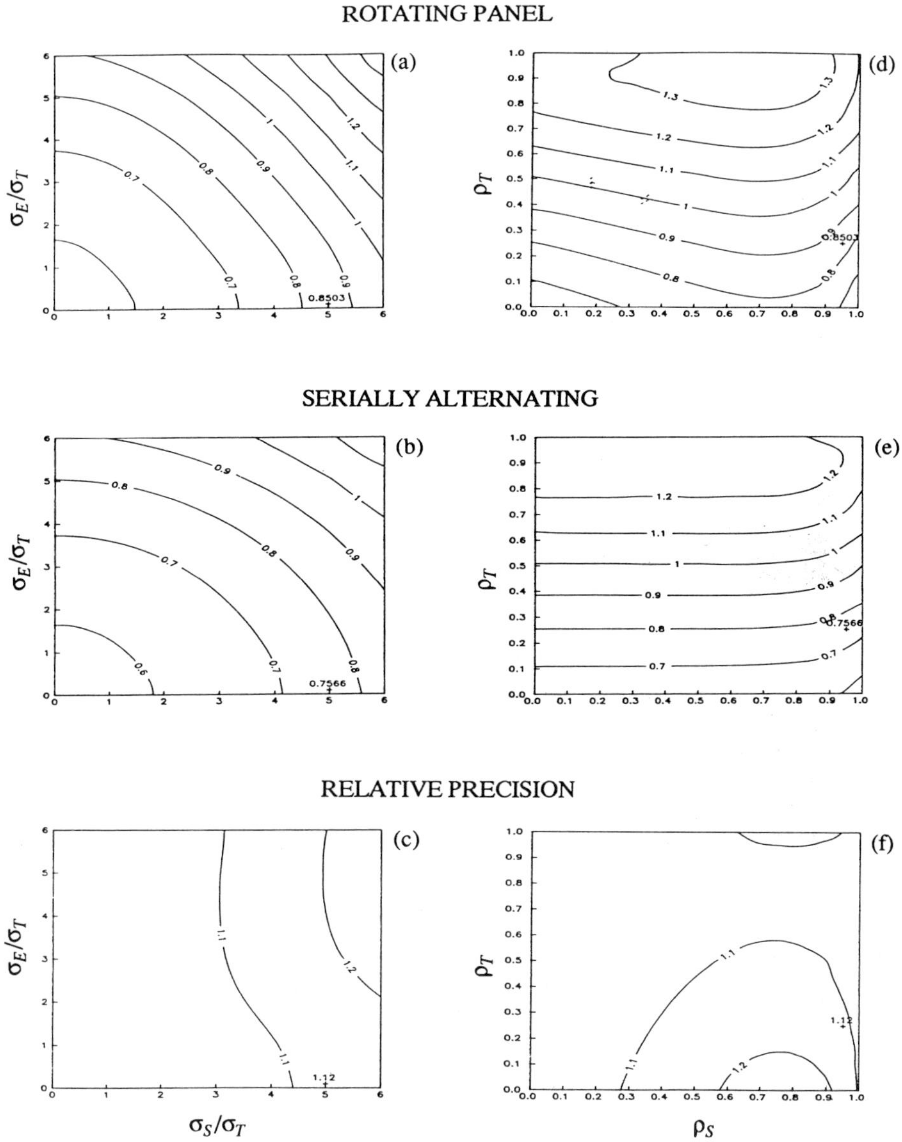

Figure 3.1 Contour plots of the variance of the estimate of current status at the end of 8 years relative to site and year autocorrelations, and ratios of components of standard deviation (σ_S/σ_T and σ_E/σ_T): $\rho_S = 0.95$, and $\rho_T = 0.25$, for Parts a, b and c; $\sigma_S = 5$, $\sigma_T = 1$, $\sigma_E = 0.1$ for Parts d, e and f. Underlined values from Table 3.2 are indicated where they occur on these graphs

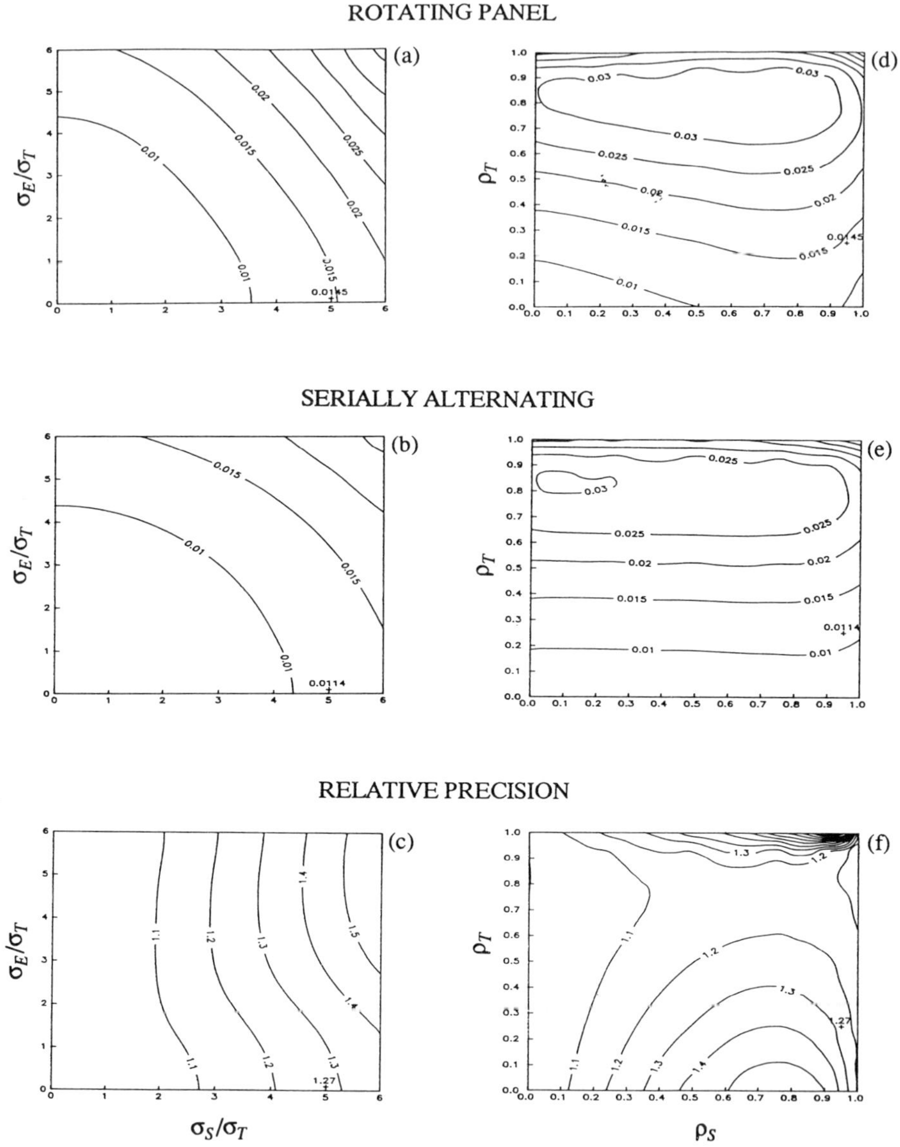

Figure 3.2 Contour plots of the generalised variance at the end of 8 years relative to site and year autocorrelations, and ratios of components of standard deviation (σ_S/σ_T and σ_E/σ_T): $\rho_S = 0.95$, and $\rho_T = 0.25$, for Parts a, b and c; $\sigma_S = 5$, $\sigma_T = 1$, $\sigma_E = 0.1$ for Parts d, e and f. Underlined values from Table 3.2 are indicated where they occur on these graphs

correlations. The bottom part of Table 3.3 provides the basis for completing these results. Earlier we explained why we expect the year-effect correlation (ρ_T) to be near 0.0 and the site correlation (ρ_S) to be near 1.0. After four years, the size of the optimal augmentation set is small, generally less than 15% of sampling resources, and after eight years the size of the optimal augmentation set is nearly zero. This led us to investigate the value of augmentation only during years 2, 3 and 4. That design performs nearly as well as the augmented design previously discussed. Of course the size of the optimal augmentation set depends on the parameters of the situation, but parameter values in the range anticipated usually fell into the range of 10% to 20% of sampling resources.

Table 3.3 Average number of sites in the repeated set n relation to years of sampling, year autocorrelation (ρ_T) and site autocorrelation (ρ_S) for the serially alternating sampling design and the generalised variance measure of precision

Years of sampling	Autocorrelation of year effects (ρ_T)								
	0.00	0.05	0.10	0.25	0.50	0.75	0.90	0.95	1.00
2	23	32	34	37	40	41	42	42	43
3	3	3	4	5	11	22	29	31	34
4	2	2	2	3	4	11	18	20	22
8	0	0	0	0	1	2	3	4	4
12	0	0	0	0	0	1	2	2	3

Years of sampling	Autocorrelation of site effects (ρ_S)								
	0.00	0.05	0.10	0.25	0.50	0.75	0.90	0.95	1.00
2	0	46	45	44	43	41	40	39	37
3	0	2	5	11	20	25	27	27	25
4	0	0	0	0	3	15	21	22	21
8	0	0	0	0	0	1	2	3	9
12	0	0	0	0	0	0	1	1	5

Corr. sites (ρ_S)	After 4 years					After 8 years				
	Autocorrelation of year effects (ρ_T)									
	0.00	0.05	0.10	0.25	0.50	0.00	0.05	0.10	0.25	0.50
0.50	0	0	0	0	0	0	0	0	0	0
0.75	0	0	0	1	5	0	0	0	0	0
0.90	4	5	5	7	11	0	0	0	0	1
0.95	6	6	7	8	12	1	1	1	1	2
1.00	6	6	6	7	11	1	1	1	2	3

3.4 DISCUSSION

In practice, we expect the site component of variance to be substantially larger than the year component, and for the year component to be substantially larger than the residual component (see, for example, Paulsen *et al.*, 1991); the temporal correlation between year effects should be modest to relatively low and the site autocorrelation should be fairly close to unity. In such cases, the interpenetrating design always outperforms the rotating panel design.

What would happen if the number of observations were increased substantially? Let m denote a positive integer greater than unity. Suppose the n_i were increased to mn_i. Equations (3.3), (3.4) and (3.7) would change simply: n_i would be replaced by mn_i, or equivalently, σ_S^2 and σ_E^2 would be replaced respectively by σ_S^2/m and σ_E^2/m. The right-hand sides of Figures 3.1 and 3.2 have axes with values scaled by σ_T; these would change in a corresponding manner by the factor $1/\sqrt{m}$. A particular configuration of variance components, with an annual number of observations of 48, fixes points on the horizontal and vertical axes, leading to results that could be read from the contour plots. If m were taken to be some value greater than unity, but nothing else were changed, the new points on the axes would be located at $1/\sqrt{m}$ times the original values. Because the two sampling designs are nearly equivalent near the origin of Figures 3.1 and 3.2, increasing m leads to progressively less difference between the performance of the two designs.

The principal difference between the two sampling designs lies in how they perform with modest numbers of sites. A monitoring programme seeking to identify emerging trends needs to identify subpopulations of sites, either by geographic boundaries or by values of critical indicators, for further study. The results developed earlier apply to such a subpopulation, as well as to whole populations. Consequently, the serially alternating design will be at least as good, relative to all of the metrics used, as the rotating panel design, provided both of the designs place the same number of sample sites in the subpopulation of interest.

The serially alternating design, however, has a distinct advantage in the total number of different sites it is likely to place in a subpopulation of interest during the early to intermediate years of a monitoring effort. For the sake of illustration, again consider the n_i given in Table 3.1. At the end of the first year of sampling, both designs anticipate visits to 48 sites, but after 2, 3 and 4 years, the serially alternating design plans visits to 88, 128 and 168 sites whereas the the rotating panel design would plan visits to only 58, 68 and 78 sites. In subsequent years, the serially alternating design would add no new sites because sites would be revisited on a 4-year cycle; the rotating panel design would plan visits to 10 new sites each year. It would take 13 years for the rotating panel design to catch up to the number of sites planned for recurring visits by the serially alternating design. Because we believe these differences in the number of different sites visited are very important, the serially alternating design is substantially better than the rotating panel design for the study of subpopulations over realistic periods of time.

This comparison of sampling designs has proceeded with minimal attention to issues of estimation because these issues involve many considerations beyond those already investigated here. One important aspect, however, could affect the measures of precision we have examined. To use generalised least squares in practice, components of variance would have to be known. If they were not known, ordinary least squares might be used in place of generalised least squares, in which case $\hat{\boldsymbol{\beta}} = (\mathbf{X}^{*\prime}\mathbf{X}^{*})^{-1}\mathbf{X}^{*\prime}\bar{\mathbf{Y}}$ and $\text{cov}(\hat{\boldsymbol{\beta}}) = (\mathbf{X}^{*\prime}\mathbf{X}^{*})^{-1}\mathbf{X}^{*}\boldsymbol{\Phi}^{*}\mathbf{X}^{*\prime}(\mathbf{X}^{*\prime}\mathbf{X}^{*})^{-1}$. Once a programme such as EMAP becomes fully operational, σ^2 and σ_E^2 will established in a few years; precise estimates could replace parameter values in generalised least squares without materially affecting the comparison of sampling designs. The potential effect of σ_T^2 cannot be dismissed quite so easily because many years would have to elapse before that component of variance would become practically known. Nevertheless, this fact has little effect on the comparison of the sampling designs. If σ_T^2 is large compared to σ_S^2, the year component will dominate the variances of the estimates, so there is no basis for choice of sampling design. If σ_T^2 is small relative to σ_S^2, the serially alternating design clearly is superior to the rotating panel design. (See the right-hand sides of Figures 3.1 and 3.2) Consequently, conclusions about the choice between sampling designs does not change because we have to estimate components of variance.

The performance of both designs that have been advanced for environmental monitoring, the rotating panel sampling design and the serially alternating sampling design, can be improved by augmentation. The two designs perform similarly for a wide range of conditions that probably do not occur in practice. When the performance of the designs differs, the augmented serially alternating design always does better, for all of the measures of performance studied. However, the main value of augmentation occurs during the early years of sampling; augmentation has substantial value for the serially alternating sampling design during the first four years, and a modest value during the next four years, but virtually no value thereafter.

ACKNOWLEDGMENTS

The work reported here was conducted primarily under the auspices of cooperative agreement CR 816721 between the Environmental Research Laboratory, USEPA, Corvallis, Oregon, and the Department of Statistics at Oregon State University. Cliff Pereira of that department suggested evaluating certain properties of the serially alternating design using a linear model equivalent to that given by (3.1). D. L. Stevens, Jr., of ManTech Environmental Technology, Inc., of Corvallis summarised the initial findings of Birkes and Overton using that model. We acknowledge helpful reviews by D. S. Robson, L. L. McDonald, J. A. Flueck and S. J. Christie.

PART II

Measuring Levels and Consequences of Pollution and Contamination

4

Point Process Modelling in Environmental Epidemiology

Peter J. Diggle,
Lancaster University, UK

4.1 INTRODUCTION

We define environmental epidemiology as the study of the spatial or spatio-temporal distribution of disease in relation to possible environmental factors. Perhaps the earliest and most celebrated example is John Snow's detection of the Broad Street water-pump as the source of the cholera outbreak in London in the 1840s. More recently, interest in environmental epidemiology has grown with the recognition that some industrial processes may constitute health hazards for the population at large. In the United Kingdom, one of the best-publicised, and most controversial, examples concerns the possible elevation of childhood leukaemia rates near nuclear installations. Following anecdotal evidence reported on a television documentary programme, the UK government asked an independent advisory group to investigate whether the apparently high incidence of childhood leukaemia cases in the village of Seascale might be linked to the nearby nuclear installation at Sellafield. The group's report (Black, 1984) concluded that although the incidence of childhood leukaemia in Seascale *was* unusually high, there was no evidence of a causal relationship between increased radioactivity and increased leukaemia. Subsequently, Gardner *et al.* (1990) presented results which suggest that the link may be an indirect one with the father's exposure to radiation at the workplace prior to conception, rather than with direct exposure of the child.

The literature on environmental epidemiology has grown enormously since

Statistics for the Environment. Edited by Vic Barnett and K. Feridun Turkman

1984. Reviews include Marshall (1991), Bithell and Stone (1989), Muirhead and Darby (1989), Elliott (1989) and Williams (1984).

In this chapter we assume that the data available for analysis consist of a reference location $\mathbf{x}_i$ for each of n cases of the disease in question, these cases representing a complete census or, if this is not possible, an effectively random samples of cases within a given geographical region over a given time interval $(0, T)$. The data may be supplemented by an associated reference time t_i for each case, by information on the spatial distribution of the population at risk, or by collateral spatial information on relevant environmental factors.

The appropriate modelling framework for data of this kind is a *spatial point process*: a stochastic process for generating a countable set of points in the plane. Within this framework, we shall view the data $\mathbf{x}_i$: $i = 1, \ldots, n$ as that part of the realisation of the process contained within the region A. An alternative approach is to aggregate the data into *counts*, $N(A_j)$: $j = 1, \ldots, m$, of the numbers of cases in each of m subregions A_j which partition A. This aggregation loses the fine spatial structure of the data, and should be avoided if possible. Also, even when the data are available only in aggregated form, it is preferable to derive a statistical model for the $N(A_j)$ from an underlying point process model, rather than to attempt to model the $N(A_j)$ directly. This is because many commonly used statistical models for count data, such as the negative binomial distribution or other mixed Poisson distributions, lack a proper spatial interpretation (Diggle and Milne, 1983); this interpretational difficulty would not arise if the A_j constituted a natural set of discrete spatial units, but in practice they are almost invariably defined in terms of arbitrary political boundaries.

A possible argument against using point process models concerns the limited spatial accuracy of the data. However, it is important here to distinguish between technical precision and substantive relevance. Modern post-coding systems can define individual locations to a high degree of precision, perhaps 10 metres in either coordinate direction, and this is more than adquate for point process modelling of spatial variation on a geographical scale. Substantive relevance is a more delicate matter. Should the reference location indicate place of residence, or place of birth? Similarly, should the reference time denote diagnosis, or estimated time of onset, or death? A possible way to handle this problem within a point process modelling approach is to regard the observed locations as random perturbations of the underlying 'correct' locations. Analyses of aggregated data tend to conceal, rather than solve, the problem.

A final comment on data quality concerns the earlier reference to 'an effectively random sample' of cases. This is not to suggest that a random sample should deliberately be taken. However, in any epidemiological study, incomplete ascertainment of cases is inevitable, if only because some people die of unrelated causes before their condition is diagnosed. Unrecognised spatial structure in ascertainment rates, for example because of inconsistent reporting policies in different subregions, must invalidate any spatial statistical analysis. However, it does seem reasonable to insist that the interpretation of the analysis should remain valid under *random* non-ascertainment of cases.

The substantive issues arising in environmental epidemiology include the following: spatial clusters and clustering; space–time interaction; raised incidence near known sources of pollution.

The first two of these issues are exploratory in nature. Identification of 'clusters', or locally elevated rates, of a disease may point to hitherto unsuspected risk factors with a strong spatial focus. Incidentally, there is an important methodological distinction between the identification of individual clusters and the detection of clustering as a general spatial phenomenon. Also, it is important to define clusters and clustering relative to known spatial heterogeneity in the population at risk.

Space–time interaction has been studied in detail since the pioneering work of Knox (1963, 1964). Knox argued that whereas diseases should be expected to exhibit both spatial and temporal clustering because of environmental heterogeneity, in the absence of either an infectious agent or changes in environmental risk factors which are both spatially and temporally localised, the spatial and temporal clustering processes should operate independently. Thus, detection of space–time interaction would establish the case for further investigation of the disease in question.

The third issue is more confirmatory in nature. Once it is known that there are spatially or temporally localised instances of high risk, and there is a well-formulated hypothesis concerning the source, or sources, of that risk, it is appropriate to assess the strength of statistical evidence for association between the disease and the putative risk factor. Typically, as with concerns about the health effects of nuclear or other industrial installations, an appropriate model would be one which postulates an elevated risk in the spatial neighbourhoods of one or more point sources.

In Section 4.2 below, we summarise the theory of point processes and their statistical analysis to the extent that it is needed to provide possible solutions to the substantive issues identified above. More detailed accounts can be found in Ripley (1981), Diggle (1983), Stoyan *et al.* (1987) and Cressie (1991). In Sections 4.3 to 4.5 we see how this theory can be applied to the investigation of spatial clustering, space–time interaction and raised incidence near one or more point sources. Section 4.6 describes a recently developed software environment for this type of work. Section 4.7 contains concluding remarks of a general nature, and some speculative thoughts on the detection of clusters.

4.2 POINT PROCESSES AND THEIR STATISTICAL ANALYSIS

4.2.1 First- and second-moment properties

Let $\mathbf{x}_i$: $i = 1, 2, \ldots$ denote the points of a point process in $\mathbb{R}^2$. We call the $\mathbf{x}_i$ the *events* of the process to distinguish them from arbitrary points in $\mathbb{R}^2$. Let A be any planar region, write $|A|$ for the area of A, and $N(A)$ for the number of events in A. Let $d\mathbf{x}$ denote an infinitesimal region containing the point $\mathbf{x}$. Then, a useful partial description of a point process is in terms of its first- and second-moment properties, which we now define. Our treatment is mathematically informal, and assumes, among other things, that the process cannot generate multiple coincident events; for a more rigorous treatment, see Stoyan *et al.* (1987).

The (*first-order*) *intensity function*, $\lambda(\mathbf{x})$, of a spatial point process is defined by the limiting operation,

$$\lambda(\mathbf{x}) = \lim_{|d\mathbf{x}| \to 0} \left\{ \frac{E[N(d\mathbf{x})]}{|d\mathbf{x}|} \right\}.$$

The function $\lambda(\mathbf{x})$ represents the mean number of events per unit area in the neighbourhood of the point $\mathbf{x}$. The *second-order intensity function*, $\lambda_2(\mathbf{x}, \mathbf{y})$, is similarly defined as

$$\lambda_2(\mathbf{x}, \mathbf{y}) = \lim \left\{ \frac{E[N(d\mathbf{x})N(d\mathbf{y})]}{|d\mathbf{x}||d\mathbf{y}|} \right\},$$

where the limit is taken as both $|d\mathbf{x}| \to 0$ and $|d\mathbf{y}| \to 0$. Note that if $N(d\mathbf{x})$ and $N(d\mathbf{y})$ are uncorrelated, $\lambda_2(\mathbf{x}, \mathbf{y}) = \lambda(\mathbf{x})\lambda(\mathbf{y})$. The quantity

$$\gamma(\mathbf{x}, \mathbf{y}) = \lambda_2(\mathbf{x}, \mathbf{y}) - \lambda(\mathbf{x})\lambda(\mathbf{y})$$

is sometimes called the *covariance density* of the process.

A point process is (*second-order*) *stationary* if $\lambda(\mathbf{x}) = \lambda$ and $\lambda_2(\mathbf{x}, \mathbf{y}) = \lambda_2(\mathbf{x} - \mathbf{y})$, and *isotropic* if $\lambda_2(\mathbf{x} - \mathbf{y}) = \lambda_2(\|\mathbf{x} - \mathbf{y}\|)$, where $\|\mathbf{x}\| = \sqrt{(x_1^2 + x_2^2)}$ denotes Euclidean distance. Thus, stationarity and isotropy correspond to invariance of first- and second-moment properties under translation and rotation, respectively. From a theoretical point of view, these assumptions generate the internal replication which is necessary to enable inference from a single realisation of the process. From a practical point of view, we will see that they provide a reasonable basis for the development of useful data-analytic methods, and are less restrictive than might appear to be the case at first sight. In particular, we will see that they do *not* preclude analysis in the presence of spatial heterogeneity.

Under stationarity and isotropy, $\lambda = E[N(A)]/|A|$ is the expected number of events per unit area in *any* region A. It is less obvious how to give a physically tangible interpretation to the second-order intensity function $\lambda_2(s)$. We consider instead the *reduced second-moment measure*, or *K-function*,

$$K(s) = 2\pi\lambda^{-2} \int_0^s \lambda_2(u) u du \tag{4.1}$$

(Ripley, 1976). Note that the *K*-function is obtained by integrating $\lambda_2(u)$ over a disc of radius *s*, and scaling the result by λ^2. Since $\lambda_2(\|\mathbf{x}\|)/\lambda$ represents the conditional first-order intensity of an event at $\mathbf{x}$ given an event at the origin, it follows easily from (4.1) that:

Theorem 4.1. *$\lambda K(s)$ is the expected number of further events within distance s of an arbitrary event.*

This result gives the *K*-function a tangible interpretation as a (scaled) expectation. Further useful results follow. Firstly,

Theorem 4.2. *$K(s)$ is invariant under random thinning.*

This follows from the fact that when the events of an underlying process are, independently, recorded with probability p or lost with probability $1-p$, then the intensity of the process of recorded events is $p\lambda$, the expected number of *recorded* events within distance *s* of a recorded event is also multiplied by *p*, and the invariance of $K(s)$ follows from Theorem 4.1.

Secondly, suppose that events are generated by a stationary Poisson process of intensity λ. By this, we mean that $N(A)$ follows a Poisson distribution with mean $\lambda|A|$, and for any two disjoint regions *A* and *B* the random variables $N(A)$ and $N(B)$ are independent. Then, the expected number of events in a disc of radius *s* is $\lambda\pi s^2$. But by the independence property, this is also the expected number of further events at the centre of the disc (think of a disc with a tiny hole in the middle, and let the hole shrink to a point). Then, again from Theorem 4.1,

Theorem 4.3. *For a stationary Poisson process,*

$$K(s) = \pi s^2.$$

Theorem 4.3 is not of direct interest to us, since the stationary Poisson process is not a useful model in environmental epidemiology. However, it does suggest using πs^2 as a benchmark value for the *K*-function, in the same way that zero is used as a benchmark for the correlogram of a time series. For this reason, we use $K(s) - \pi s^2$, rather than $K(s)$ itself, as a measure of the degree of spatial aggregation in a point process. For example, consider a *cluster point process* in which cluster centres form a stationary Poisson process of intensity ρ, events are randomly allocated amongst cluster centres and, within a cluster, the distribution function of the distance between two randomly chosen events is $H(s)$. Then,

$$K(s) = \pi s^2 + \rho^{-1} H(s). \qquad (4.2)$$

Formula (4.2) shows clearly how the K-function reflects both the expected number of events per cluster since, for given intensity λ, the expected number of events per cluster is $\lambda\rho^{-1}$, and the tightness of the spatial distribution within a cluster, as described by the rate at which $H(s) \to 1$.

Now, consider a *bivariate* point process in which events are of two qualitatively distinguishable types. Let $N_j(A)$ denote the number of type-j events in a region A. Then, we define first-order and second-order intensity functions

$$\lambda_j(\mathbf{x}) = \lim_{|d\mathbf{x}| \to 0} \left\{ \frac{E[N_j(d\mathbf{x})]}{|d\mathbf{x}|} \right\}: j = 1, 2$$

and

$$\lambda_{ij}(\mathbf{x}, \mathbf{y}) = \lim \left\{ \frac{E[N_i(d\mathbf{x}) N_j(d\mathbf{y})]}{|d\mathbf{x}||d\mathbf{y}|} \right\}: i = 1, 2; j = 1, 2.$$

In the stationary, isotropic case, $\lambda_j(\mathbf{x}) = \lambda_j$, the expected number of type-j events per unit area, $\lambda_{ij}(\mathbf{x}, \mathbf{y}) = \lambda_{ij}(\|\mathbf{x} - \mathbf{y}\|)$ and the bivariate K-functions are

$$K_{ij}(s) = 2\pi(\lambda_i \lambda_j)^{-1} \int_0^s \lambda_{ij}(u) u \, du. \tag{4.3}$$

Note that (4.3) reduces to (4.1) if $i = j$, and that $K_{21}(s) = K_{12}(s)$. Also, we have the following analogoue of Theorem 4.1:

Theorem 4.4. *$\lambda_2 K_{12}(s)$ is the expected number of type-2 events within distance s of an arbitrary type-1 event.*

Can we establish a useful benchmark for $K_{12}(s)$, analogous to Theorem 4.3? There are at least two possibilities.

Theorem 4.5. *Suppose that type-1 and type-2 events are generated by any two* independent *stationary, isotropic point processes. Then, $K_{12}(s) = \pi s^2$.*

Theorem 4.6. *Suppose that type-1 and type-2 events are generated by* random labelling *of a stationary, isotropic point process, i.e. events are designated as type-1 or type-2 in a sequence of independent Bernouilli trials. Then,*

$$K_{11}(s) = K_{12}(s) = K_{22}(s).$$

Theorem 4.5 follows immediately from Theorem 4.4 and the fact that the expected number of type-2 events in a disc of radius s whose centre is positioned independently of the process of type-2 events is $\lambda_2 \pi s^2$. Theorem 4.6 is essentially a restatement of Theorem 4.2.

4.2.2 Estimation of the first- and second-moment properties of a bivariate stationary, isotropic point process

Suppose that the data consist of two types of event in a region A, and that there are n events in total, of which the first n_1 are of type-1 whilst the remaining $n_2 = n - n_1$ are of type-2. Obvious, unbiased estimators for the intensities λ_j are

$$\hat{\lambda}_j = n_j/|A|: j = 1, 2. \tag{4.4}$$

Similarly, if we consider Theorem 4.1 and replace the theoretical expectation by the corresponding observed average value, an obvious estimator for $\lambda_1 K_{11}(s)$ would appear to be

$$n_1^{-1} \sum_{i=1}^{n_1} \left\{ \sum_{j \neq i} I(d_{ij} \leqslant s) \right\},$$

where $d_{ij} = \| \mathbf{x}_i - \mathbf{x}_j \|$ and $I(\cdot)$ is the indicator function which takes the value 1 if its logical argument is true, 0 if its argument is false. In fact, this estimator has a substantial negative bias because of edge-effects: events outside A are not recorded. Several methods for correcting this bias have been proposed. The most widely known, due to Ripley (1976), uses weights w_{ij} such that $w_{ii} = 0$ and, for $j \neq i$, w_{ij}^{-1} is the proportion of the circumference of the disc with centre $\mathbf{x}_i$ and radius d_{ij} which is contained in A. Then, Ripley's estimator of $\lambda_1 K_{11}(s)$ is

$$n_1^{-1} \sum_{i=1}^{n_1} \sum_{j=1}^{n_1} w_{ij} I(d_{ij} \leqslant s).$$

To convert this into an estimator for $K_{11}(s)$, we divide by $\tilde{\lambda}_1 = (n_1 - 1)/|A|$ to give

$$\hat{K}_{11}(s) = |A| \{n_1(n_1 - 1)\}^{-1} \sum_{i=1}^{n_1} \sum_{j=1}^{n_1} w_{ij} I(d_{ij} \leqslant s). \tag{4.5}$$

The reason for using $\tilde{\lambda}_1$ rather than $\hat{\lambda}_1$ will be made clear in Section 4.3. Similarly, our estimators for $K_{22}(s)$ and $K_{12}(s)$ are

$$\hat{K}_{22}(s) = |A| \{n_2(n_2 - 1)\}^{-1} \sum_{i=n_1+1}^{n} \sum_{j=n_1+1}^{n} w_{ij} I(d_{ij} \leq s) \tag{4.6}$$

and

$$\hat{K}_{12}(s) = \{n_1 \tilde{K}_{12}(s) + n_2 \tilde{K}_{21}(s)\}/n \tag{4.7}$$

where

$$\tilde{K}_{12}(s) = |A| \{(n_1 - 1)(n_2 - 1)\}^{-1} \sum_{i=1}^{n_1} \sum_{j=n_1+1}^{n} w_{ij} I(d_{ij} \leqslant s)$$

and

$$\tilde{K}_{21}(s) = |A|\{(n_1 - 1)(n_2 - 1)\}^{-1} \sum_{i=n_2+1}^{n} \sum_{j=1}^{n_1} w_{ij} I(d_{ij} \leqslant s).$$

An alternative edge-correction, due to Ohser and Stoyan (1981), uses weights defined in terms of areal rather than circumferential proportions. Stein (1991) uses two sets of weights to give an estimator for $\lambda_1 K_{11}(s)$ proportional to

$$\sum_{i=1}^{n_1} w_i^* \sum_{j=1}^{n_1} w_{ij} I(d_{ij} \leqslant s),$$

choosing the new weights w_i^* to be small when the corresponding event $\mathbf{x}_i$ is close to the boundary of A. The results of comparisons between Ripley's and Ohser and Stoyan's edge-corrections are somewhat equivocal; see, for example, Ripley (1982). Stein's estimator, on the other hand, appears to be generally more efficient, but at the expense of an element of subjective choice for the weights w_i^*, whose theoretically optimal values depend on the underlying point process.

4.2.3 Models

Spatial heterogeneity is a universal feature of the environment in which we live. If we want our models and methods to be useful, they must therefore be able to accommodate such heterogeneity.

Mathematically, the simplest model for a point process in a heterogeneous environment is the *inhomogeneous Poisson process.* This has a general, *non-stationary* intensity function $\lambda(\mathbf{x})$. Its definition is completed by the following two postulates:

(1) for any region A, the random variable $N(A)$ has a Poisson distribution with mean $\int_A \lambda(\mathbf{x})d\mathbf{x}$,
(2) for any two disjoint regions A and B, the random variables $N(A)$ and $N(B)$ are independent.

Given data $\{\mathbf{x}_i \in A: i = 1, \ldots, n\}$, the likelihood function for the inhomogeneous Poisson process is

$$L = \prod_{i=1}^{n} \lambda(\mathbf{x}_i) \exp\left\{ - \int_A \lambda(\mathbf{x})d\mathbf{x} \right\}. \tag{4.8}$$

To be useful, this needs a parametric specification for $\lambda(\mathbf{x})$. Cox (1972a) considers the analogous model for a temporal point process, and introduces the term *modulated Poisson process* for a model in which $\lambda(\mathbf{x})$ is determined by one or more explanatory variables with associated regression parameters.

In the absence of relevant explanatory variables, we could describe $\lambda(\mathbf{x})$ by a parametric model involving latitude and longitude. However, this point process analogue of trend-surface modelling seems unnatural in the present context. An alternative way forward is to assume that $\lambda(\mathbf{x})$ is itself a realisation of a stationary random process $\Lambda(\mathbf{x})$. Then, the resulting point process is a stationary *Cox process* (Cox, 1955). Furthermore, if $E[\Lambda(\mathbf{x})] = \lambda$ and $\text{cov}\{\Lambda(\mathbf{x}), \Lambda(\mathbf{y})\} = \gamma(\|\mathbf{x} - \mathbf{y}\|)$, then λ and $\gamma(\cdot)$ are also the first-order and covariance intensities of the point process whose K-function is

$$K(s) = \pi s^2 + 2\pi\lambda^{-2} \int_0^s \gamma(u)u du. \tag{4.9}$$

Note that if $\gamma(\cdot)$ is non-negative-valued, then (4.9) and (4.2) are qualitatively similar in that both take the form πs^2 plus a monotone nondecreasing function of s. In fact, Bartlett (1964) showed that these two models are indistinguishable in the following sense. In the cluster process leading to (4.2), let $h(\mathbf{x})$ denote the bivariate p.d.f. of the displacement of any event from its corresponding cluster centre. Then the process is also interpretable as a Cox process with

$$\lambda(\mathbf{x}) = \lambda\rho^{-1} \sum_{i=1}^{\infty} h(\mathbf{x} - \mathbf{Y}_i),$$

where the $\mathbf{Y}_i$ are the events of a homogeneous Poisson process of intensity ρ. This duality explains why environmental heterogeneity typically gives a spurious impression of clustering in point process data. It follows that we can only hope to give a valid analysis of the disease clustering problem if we can somehow make an appropriate adjustment for environmental heterogeneity.

4.2.4 Data-perturbation

A possible way of modelling the discrepancy between the recorded position of an event and its substantively correct position is to assume that the recorded events are obtained by independent random perturbations of their original positions. It is well known that such perturbations generally make the observed point process more like a homogeneous Poisson process than was the original process. In particular, the Poisson process is obtained in the limit as the perturbation distribution becomes stochastically large. See, for example, Cox and Isham (1980). The effect of random perturbation on the K-function can be quantified as follows.

Let $f(\mathbf{x})$ denote the bivariate p.d.f. of the convolution of two independent random perturbations, and assume that $f(\mathbf{x})$ is radially symmetric so as to preserve isotropy. Let λ, $\lambda_2(\mathbf{x})$ and $K(s)$ be the intensity, second-order intensity and K-function of the underlying process, and $\lambda^0, \lambda_2^0(\mathbf{x})$ and $K^0(s)$ the corresponding quantities for the observed process. Of course, $\lambda_2(\mathbf{x}) = \lambda_2(\|\mathbf{x}\|)$, but in what follows it helps to use the argument $\mathbf{x}$ rather than $s = \|\mathbf{x}\|$.

Clearly, $\lambda^0 = \lambda$ since events are neither added nor deleted in the perturbation process. Also, by arguing conditionally on the perturbations of a pair of events and then integrating out over the perturbation distribution,

$$\lambda_2^0(\mathbf{x}) = \int_{\mathbb{R}^2} \lambda_2(\mathbf{x}-\mathbf{y}) f(\mathbf{y}) d\mathbf{y}$$
$$= \int_{\mathbb{R}^2} \lambda(\mathbf{z}) f(\mathbf{x}-\mathbf{z}) d\mathbf{z}.$$

Now, write $\lambda_2(\mathbf{z}) = \lambda^2 + \gamma(\mathbf{z})$, where $\gamma(\cdot)$ is the covariance density. Then

$$\lambda_2^0(\mathbf{x}) = \lambda^2 \int_{\mathbb{R}^2} f(\mathbf{x}-\mathbf{z}) d\mathbf{z} + \int_{\mathbb{R}^2} \gamma(\mathbf{z}) f(\mathbf{x}-\mathbf{z}) d\mathbf{z}.$$

Finally, using (4.1) but without the implicit transformation to polar coordinates in the integration, we obtain

$$K^0(s) = \int_{\|\mathbf{x}\| \leqslant s} \int_{\mathbb{R}^2} f(\mathbf{x}-\mathbf{z}) d\mathbf{z}\, d\mathbf{x} + \lambda^{-2} \int_{\|\mathbf{x}\| \leqslant s} \int_{\mathbb{R}^2} \gamma(\mathbf{z}) f(\mathbf{x}-\mathbf{z}) d\mathbf{z}\, d\mathbf{x} \qquad (4.10)$$

Now, because $f(\cdot)$ is a radially symmetric p.d.f.,

$$\int_{\mathbb{R}^2} f(\mathbf{x}-\mathbf{z}) d\mathbf{z} = \int_{\mathbb{R}^2} f(\mathbf{z}-\mathbf{x}) d\mathbf{z} = 1$$

for any $\mathbf{x}$, and the first term on the right-hand side of (4.10) reduces to πs^2. For the second term, change the order of integration and write

$$R(s, \mathbf{z}) = \int_{\|\mathbf{x}\| \leqslant s} f(\mathbf{x}-\mathbf{z}) d\mathbf{z}.$$

Then,

$$K^0(s) = \pi s^2 + \lambda^{-2} \int_{\mathbb{R}^2} R(s, \mathbf{z}) \gamma(\mathbf{z}) d\mathbf{z}.$$

To interpret this result, note that $R(s, \mathbf{z})$ is the probability that the action of the distribution $f(\cdot)$ will move an event originally at the point $\mathbf{z}$ to a point somewhere in the disc $\|\mathbf{x}\| \le s$. Thus, $R(s, \mathbf{z}) = R(s, \|\mathbf{z}\|)$ and

$$K^0(s) = \pi s^2 + 2\pi\lambda^{-2} \int_0^\infty R(s, u)\gamma(u)udu. \tag{4.11}$$

We compare this with the K-function of the original process,

$$K(s) = \pi s^2 + 2\pi\lambda^{-2} \int_0^s \gamma(u)udu. \tag{4.12}$$

In (4.12) the covariance integral is truncated at s, whereas in (4.11) the integrand is attenuated by the function $R(s, u)$. Now, if the perturbation distribution degenerates to a zero perturbation with probability 1, then $R(s, u) = 1$ if $u \le s$, $R(s, u) = 0$ otherwise, and $K^0(u) = K(u)$, as it must. Conversely, if the perturbation distribution is highly dispersed, then for any fixed s, $R(s, u)$ will be approximately zero for all u, and $K^0(s) \simeq \pi s^2$. In intermediate cases, typically for any value of s, we have that $0 < R(s, 0) < 1$ and $R(s, u)$ is monotone decreasing in u. Since typically $\gamma(u)$ is also positive and monotone decreasing in u, the effect of the attenuation is to *reduce* the value of the covariance integral, so that

$$0 < K^0(s) - \pi s^2 < K(s) - \pi s^2.$$

The practical implication of this result is that second-moment analyses of randomly perturbed data are likely to be conservative, in the sense that they are likely to underestimate the true extent of spatial heterogeneity or clustering. Furthermore, if $f(\cdot)$ is known, or can be estimated independently of the observed pattern, then in principle we can use (2.11) and (2.12) to derive an estimate of $K(s)$ from an estimate of $K^0(s)$. In practice, this is likely to be a delicate business since it would involve the numerical solution of a possibly ill-conditioned integral equation. The problem is the inverse of the one considered in Berman and Diggle (1989), and is not unrelated to the deconvolution problem considered in Diggle *et al.* (1987).

4.2.5 The practical relevance of stationarity

Given the degree of spatial heterogeneity in human settlement patterns, it would seem hard to justify any assumption of stationarity in the context of environmental epidemiology. However, there are at least two counter-arguments to this view.

The first lies in the discussion of inhomogeneous Poisson and Cox processes in Section 4.2.3 above. Both processes are characterised by a variable first-order intensity function $\lambda(\mathbf{x})$. If the known or hypothesised source of the variability in $\lambda(\mathbf{x})$ is directly observable, it is reasonable to reflect this by using a modulated Poisson process as a model for the data, thereby abandoning the stationarity assumption in favour of an economically parametrised nonstationary model. If, on the other hand, the variability in $\lambda(\mathbf{x})$ arises from unknown or unobservable factors, it may be more sensible to regard $\lambda(\mathbf{x})$ as a realisation of a random process, and to use a stationary Cox process as a model. In this case, the K-function reflects the covariance structure of the random process $\Lambda(\mathbf{x})$, which in turn characterises the likely degree of spatial structure in the realised intensity function $\lambda(\mathbf{x})$; slowly decaying covariance functions correspond to smoothly changing $\lambda(\mathbf{x})$ and vice versa. A close analogy is the distinction between fixed and random effects models in the analysis of variance.

The second argument is that we can use the stationarity assumption to construct useful descriptive statistics such as $\hat{K}(s)$, but need not then invoke stationarity for inferential purposes. This idea is illustrated in Sections 4.3 and 4.4.

4.3 SPATIAL CLUSTERING

Suppose that we are given data $\{\mathbf{x}_i \in A: i = 1, \ldots, n_1\}$, representing the locations of cases in a given region A. Our earlier discussion suggests that we must expect such data to exhibit spatial structure, but that it might well be difficult, or even imppossible, to explain such structure unambiguously in terms of either spatial heterogeneity in the environment or clustering of cases. To resolve the ambiguity, we need information on the extent of the environmental heterogeneity.

Cuzick and Edwards (1990) propose a case-control approach to this problem. They suggest supplementing the data on cases by a set of control locations $\{\mathbf{x}_i: i = n_1 + 1, \ldots, n_1 + n_2\}$ constructed as an effectively random sample from the population at risk. They then develop significance tests of an implicit null hypothesis of random labelling, treating cases and controls as events of types 1 and 2 in a bivariate point process.

Diggle and Chetwynd (1991, henceforth DC) suggest using the K-functions to construct suitable test statistics. Since $K_{22}(s)$ reflects environmental heterogeneity only, whereas $K_{11}(s)$ reflects a combination of environmental heterogeneity and clustering, a natural diagnostic for clustering is

$$\hat{D}(s) = \hat{K}_{11}(s) - \hat{K}_{22}(s),$$

where $\hat{K}_{11}(s)$ and $\hat{K}_{22}(s)$ are defined by (4.5) and (4.6), respectively. DC use a combinatorial argument to show that under random labelling, $E[\hat{D}(s)] = 0$ exactly – which incidentally explains the use of $\hat{\lambda}_j$ rather than $\tilde{\lambda}_j$ in our definition of $\hat{K}_{jj}(s)$. DC also give an explicit expression for $\text{cov}\{\hat{D}(s), \hat{D}(u)\}$ for any values of

s and u. The variances of $\hat{D}(s)$ under random labelling are a guide to the interpretation of the empirical function $\hat{D}(s)$, and are used to define a test statistic

$$D = \sum_{j=1}^{m} \hat{D}(s_j)/\surd[\mathrm{var}\{\hat{D}(s_j)\}], \tag{4.13}$$

where s_j: $j = 1, \ldots, m$ span the range of inter-event distances judged to be epidemiologically relevant. Of necessity, s_m must be small relative to the dimensions of the region A; in practice, the epidemiologically relevant range appears often to be *much* smaller than this.

Two approaches to significance testing using (4.13) are:

(1) to construct an approximate standard normal test statistic, $T = D/\sqrt{V}$ where

$$V = m + 2 \sum_{j=1}^{m} \sum_{k=j+1}^{m} \mathrm{corr}\{\hat{D}(s_j), \hat{D}(s_k)\};$$

(2) to implement a Monte Carlo test by ranking the observed value of D amongst values obtained after indepdendent random permutations of the $n = n_1 + n_2$ labels (Barnard, 1963).

The second approach is exact, but may be computationally more expensive – although this depends on the algorithms used for the simulations and for the computation of $\mathrm{cov}\{\hat{D}(s), \hat{D}(u)\}$. If random labelling is rejected, one advantage of $\hat{D}(s)$ over the test statistics suggested by Cuzick and Edwards is that it retains a natural interpretation: $\tilde{\lambda}_1 \hat{D}(s)$ is an estimate of the expected number of excess cases within distance s of a reference case, averaged over all reference cases; alternatively, $\hat{D}(s)/\hat{K}_{22}(s)$ is an estimate of the proportional elevation of risk within a disc of radius s consequent on the observation of a case at the centre of the disc. If partial ascertainment of cases is an issue, the estimate $\hat{\lambda}_1$ would need to be adjusted upwards. However, the interpretation of the ratio $\hat{D}(s)/\hat{K}_{22}(s)$ would be preserved because of the invariance of the K-function under random thinning.

The most important practical questions for the application of this methodology concern the choice and number of controls. For some childhood cancers, a random sample from the birth register for the region and time-period in question is generally assumed to give a reasonable set of controls. However, for many other diseases including adult cancers, it may be desirable to match cases and controls on the basis of age, sex, socio-economic status, smoking habit, or a host of other factors. Clearly, this is difficult. With regard to the number of controls, some simulation experiments led Cuzick and Edwards (1990) to suggest four controls per case as a reasonable working rule. If the approach based on K-functions is used, the variance formulae in DC can provide useful illumination as they identify separate components for $\mathrm{var}\{\hat{K}_{11}(s)\}$, $\mathrm{var}\{\hat{K}_{22}(s)\}$

and $\text{cov}\{\hat{K}_{11}(s), \hat{K}_{22}(s)\}$. Clearly, the last two of these can be made arbitrarily small by letting $n_2 \to \infty$. More realistically, we can use the order-of-magnitude dependence of the variance formulae on n_2 to estimate the improvement in the precision of the estimates $\hat{D}(s)$ as we increase n_2, the number of controls, whilst holding A and n_i fixed. To a first approximation in orders of magnitude of n_2, $\text{var}\{\hat{K}_{11}(s)\} = O(1)$, $\text{var}\{\hat{K}_{22}(s)\} = O(n_2^{-1})$ and $\text{cov}\{\hat{K}_{11}(s), \hat{K}_{22}(s)\} = O(n_2^{-1})$.

DC re-analyse data on childhood leukaemia in North Humberside, UK, presented originally by Cuzick and Edwards (1990) and consisting of 62 cases and 141 controls. For these data, Cuzick and Edwards obtained a wide range of results according to which version of their family of test statistics they used. DC's results, using the K-function, were also somewhat equivocal. However, they did suggest some elevation of risk due to clustering up to a range of about 0.5 km.

4.4 SPACE–TIME INTERACTION

In this section, we suppose that the data $\{(\mathbf{x}_i, t_i): i = 1, \ldots, n\}$ represent the reference locations *and* times of all recorded cases in a given spatial region A and time-period $(0, T)$. The first step in an investigation of space–time interaction is to test the null hypothesis of no interaction, i.e. that the reference locations and times are generated by two *independent* stochastic processes. Note that this null hypothesis makes no reference to spatial or temporal stationarity. As in Section 4.1, we use the stationary paradigm to construct informative descriptive summaries of the data, but require that our inferences do not rely on stationarity. Specifically, we base our inferences on the fact that under the null hypothesis, the joint distribution of the data is invariant under random permutation of the t_i for fixed $\mathbf{x}_i$, or vice versa.

Knox's (1963, 1964) original approach to this problem was to classify each pair of cases as close in space if $\|\mathbf{x}_i - \mathbf{x}_j\| \leqslant s_0$, and close in time if $|t_i - t_j| \leqslant t_0$, where the threshold values s_0 and t_0 are chosen without reference to the data. Under this classification, the $N = \frac{1}{2}n(n-1)$ pairs can be summarised in a 2×2 contingency table according to whether each pair is or is not close in space and/or time. Then, if R denotes the number of pairs close in space, C the number close in time, and X the number close in both space and time, the test statistic is X and its approximate null sampling distribution is Poisson, with mean RC/N. This result does *not* follow from the usual distribution theory of the 2×2 table, since the N entries in the table are not mutually independent, even under the null hypothesis. Its approximate validity for *small* RC/N was established by Barton and David (1966).

One unattractive feature of Knox's test is its dependence on the pair of threshold values (s_0, t_0). In a recent application to childhood leukaemia in the UK, Gilman and Knox (1991) give lengthy tabulations of the statistic and its associated p-value for a range of threshold values. In the absence of a natural choice for (s_0, t_0), the interpretation of the analysis is extremely difficult. One way out of this difficulty is to turn the problem into one of *estimating* an interpretable *function* of (s_0, t_0).

Diggle *et al.* (1991) consider the use of K-functions for the detection of space–time interaction, and in so doing provide a reinterpretation of Knox's statistic as a *biased* estimator for the space–time K-function, the bias arising because Knox made no correction for edge-effects. This reinterpretation is developed as follows.

For a stationary space–time point process, let λ denote the intensity, or expected number of events per unit area per unit time. We could define the second-order intensity, and from it the K-function, by direct analogy with the definitions in Section 4.2.1. However, as these provide no further insights we proceed directly to a definition of the K-function by the analogue of Theorem 4.1, namely:

Definition 4.1. $\lambda K(s, t)$ is the expected number of further events within distance s and time t of an arbitrary event.

An approximately unbiased estimator for $K(s, t)$ follows by analogy with the estimator for the spatial K-function given in Section 4.2.3. Let d_{ij} and w_{ij} be defined as before. Define $u_{ij} = |t_i - t_j|$, $v_{ii} = 0$ and, for $i \neq j$, $v_{ij} = 1$ if $t_i - u_{ij} > 0$ and $t_{ij} + u_{ij} < T$, otherwise $v_{ij} = 2$. Then

$$\hat{K}(s, t) = |A| T \{n(n-1)\}^{-1} \sum_{i=1}^{n} \sum_{j=1}^{n} w_{ij} I(d_{ij} \leqslant s) v_{ij} I(u_{ij} \leqslant t) \tag{4.14}$$

Note that if we put $w_{ij} = v_{ij} = 1$ for all $i \neq j$, then the right-hand side of (4.14) reduces to $|A| T \{n(n-1)\}^{-1} X$, where X is Knox's statistic. Whilst the edge-correction is important, a more fundamental benefit of this reinterpretation of Knox's statistic is its status as an estimator of a tangible property of the underlying space–time process, as given by Definition 4.1. However, before we can reap this benefit, we need to understand the sampling distribution of $\hat{K}(s, t)$ under the null hypothesis of no space–time interaction, and thereby construct a test of significance.

An immediate consequence of the form of (4.14) is that if the underlying spatial and temporal processes are independent, then the expectation of $\hat{K}(s, t)$ factorises into a product of two functions, one of s only and one of t only. Specifically,

$$E[\hat{K}(s, t)] = \hat{K}_1(s) \hat{K}_2(t),$$

where $\hat{K}_1(s)$ and $\hat{K}_2(t)$ are, in effect, estimating the K-functions of the spatial and temporal marginal processes. Furthermore, the same kind of combinatorial arguments used in DC provide explicit expressions for the covariance structure of $\hat{K}(s, t)$ under random permutation of the t_i. This suggests defining a *residual surface*,

$$R(s,t) = \{\hat{K}(s,t) - \hat{K}_1(s)\hat{K}_2(t)\}/\surd[\text{var}\{\hat{K}(s,t)\}]. \quad (4.15)$$

Note that in (4.15) the variance is with respect to the distribution induced by random permutation of the t_i, and the estimated marginal K-functions, $\hat{K}_1(s)$ and $\hat{K}_2(t)$, are fixed quantities.

Possible diagnostic plots derived from (4.15) include a perspective or contour plot of $R(s,t)$, or a 'residuals against fitted values' plot of $R(s,t)$ against $\hat{K}_1(s)\hat{K}_2(t)$. For a formal test of no space–time interaction, a possible test statistic is

$$D = \sum_{k_1=1}^{m_1} \sum_{k_2=1}^{m_2} R(s_{k_1}, t_{k_2}).$$

Possible implementations of a test based on D are a normal approximation using the known null covariance structure of $R(s,t)$, and an exact Monte Carlo test. If the null hypothesis is rejected, a further possible diagnostic is a perspective or contour plot of $\{\hat{K}(s,t) - \hat{K}_1(s)\hat{K}_2(t)\}/\{\hat{K}_1(s)\hat{K}_2(t)\}$, which estimates the proportion by which the risk within a space–time cylinder centred on a case is elevated as a result of the space–time interaction.

Bhopal *et al.* (1992) apply these ideas to data on apparently sporadic cases of legionnaires' disease in Scotland. By combining the information from the space–time distributions of cases in the two largest regions, Glasgow and Edinburgh, they find strong evidence of space–time interaction, suggesting a need for a reassessment of the current working definition of 'sporadic' cases and a search for possible causal factors.

4.5 RAISED INCIDENCE NEAR ONE OR MORE POINT SOURCES

We now move to a more confirmatory type of analysis. Suppose that a particular type of pollutant has been implicated in connection with a particular disease, and that this pollutant emanates from one or more known point sources within a region A. We wish to determine whether there is a significantly raised incidence in the neighbourhood of any point source and, if so, to describe its spatial structure.

Because the point sources are known, we abandon the stationary paradigm and formulate inhomogeneous Poisson process models for the spatial distribution of the disease. What follows is based on Diggle (1989, 1990). Similar ideas have been developed, independently, by Lawson (1989, 1991). See also Bithell and Stone (1989). We consider a single point source, at $\mathbf{x}_0$ say, and assume that cases are observed at points $\mathbf{x}_i \in A$: $i = 1, \ldots, n$.

Diggle (1990) assumes a multiplicative model for the intensity function $\lambda(\mathbf{x})$, of the form

$$\lambda(\mathbf{x}) = \rho\lambda_0(\mathbf{x})f(\mathbf{x} - \mathbf{x}_0; \boldsymbol{\theta}), \tag{4.16}$$

where $\lambda_0(\mathbf{x})$ is the intensity function of the population at risk, ρ is scalling parameter and $f(\mathbf{u}; \boldsymbol{\theta})$ describes the putative raised incidence as a function of location relative to the point source $\mathbf{x}_0$. Assume temporarily that $\lambda_0(\mathbf{x})$ is known, and substitute (4.16) into (4.8). The resulting log-likelihood for ρ and $\boldsymbol{\theta}$ is

$$\begin{aligned} l(\rho, \boldsymbol{\theta}) = n \log \rho + \sum_{i=1}^{n} \log \lambda_0(\mathbf{x}_i) + \sum_{i=1}^{n} \log f(\mathbf{x}_i - \mathbf{x}_0; \boldsymbol{\theta}) \\ - \rho \int_A \lambda_0(\mathbf{x}) f(\mathbf{x} - \mathbf{x}_0; \boldsymbol{\theta}) d\mathbf{x}. \end{aligned} \tag{4.17}$$

Given $\boldsymbol{\theta}$, the maximum likelihood estimator for ρ is

$$\hat{\rho}(\boldsymbol{\theta}) = n \Big/ \int_A \lambda_0(\mathbf{x}) f(\mathbf{x} - \mathbf{x}_0; \boldsymbol{\theta}) d\mathbf{x}.$$

Back-substitution of $\hat{\rho}(\boldsymbol{\theta})$ into (4.17) then gives $\hat{\boldsymbol{\theta}}$ as the value which maximises

$$l^*(\boldsymbol{\theta}) = -n \log \left\{ \int_n \lambda_0(\mathbf{x}) f(\mathbf{x} - \mathbf{x}_0; \boldsymbol{\theta}) d\mathbf{x} \right\} + \sum_{i=1}^{n} \log f(\mathbf{x}_i - \mathbf{x}_0; \boldsymbol{\theta}).$$

Diggle (1990), advocates replacing $\lambda_0(\mathbf{x})$ by a kernel estimator $\hat{\lambda}_0(\mathbf{x})$ constructed from a set of control locations $\mathbf{x}_i$: $i = n + 1, \ldots, n + m$. The kernel estimator is

$$\hat{\lambda}_0(\mathbf{x}) = h^{-1} \sum_{i=n+1}^{n+m} G\{(\mathbf{x} - \mathbf{x}_i)/h\}, \tag{4.18}$$

where

$$G(\mathbf{u}) = (2\pi)^{-1} \exp(-\mathbf{u}'\mathbf{u}/2)$$

and $h > 0$ is chosen to minimise an estimate of the mean squared error of $\hat{\lambda}_0(\mathbf{x})$, as given by Berman and Diggle (1989). Finally, by making the pragmatic choice

$$f(\mathbf{u}; \boldsymbol{\theta}) = 1 + \boldsymbol{\theta}_1 \exp(-\boldsymbol{\theta}_2 \mathbf{u}'\mathbf{u}),$$

Diggle (1990) obtains explicit expressions for $l^*(\boldsymbol{\theta})$ and its derivatives, leading to computationally fast evaluation of $\hat{\boldsymbol{\theta}}$ and its asymptotic variance matrix.

In Diggle (1990) and Diggle *et al.* (1990), this model is applied to the spatial distribution of cancers of the larynx in the neighbourhood of a now-disused industrial incinerator in the Chorley–Ribble area of Lancashire, England. Using lung cancer cases as controls, an admittedly questionable choice, these authors find significant evidence of raised incidence near the incinerator, although the sparsity of the data does not allow for a sensitive validation of the assumed form of $f(\mathbf{u};\boldsymbol{\theta})$. Also in these papers, simulations are presented which suggest that the standard likelihood asymptotics are not reliable in this application, although the substantive conclusions are confirmed by Monte Carlo testing.

The kernel estimator (4.18) may be of interest in its own right. However, it introduces an untidy element to the inferential process, since the inferences for the parameters of interest, $\boldsymbol{\theta}$, are influenced by a somewhat *ad hoc* choice for the method of estimating an infinite set of nuisance parameters, namely the function $\lambda_0(\mathbf{x})$. Diggle and Rowlingson (1992) suggest avoiding this problem by conditioning on the case and control locations $\mathbf{x}_i$: $i = 1, \ldots, n+m$. Then, under the inhomogeneous Poisson formulation (4.16), the *labels* of the $n+m$ events $\mathbf{x}_i$ are determined by a set of independent Bernoulli trials with position-dependent probabilities. Specifically, if $p(\mathbf{x})$ denotes the probability that an event at $\mathbf{x}$ is a case, then it follows from (4.16) that

$$p(\mathbf{x}) = \rho f(\mathbf{x} - \mathbf{x}_0;\boldsymbol{\theta})/\{1 + \rho f(\mathbf{x} - \mathbf{x}_0;\boldsymbol{\theta})\}. \tag{4.19}$$

The binary regression model implied by (4.19) cannot be generalised linear, since $f(\mathbf{u};\boldsymbol{\theta}) \to 1$ as $\|\mathbf{u}\| \to \infty$, so $p(\mathbf{x}) \to \rho/(1+\rho)$, a value strictly between 0 and 1, as $\|\mathbf{x} - \mathbf{x}_0\| \to \infty$. Nevertheless, likelihood inference for ρ and $\boldsymbol{\theta}$ is relatively straightforward, and the 'nuisance function' $\lambda_0(\mathbf{x})$ has been eliminated by the conditioning. There is a connection here with the proportional hazards regression methodology introduced in Cox (1972b).

Diggle and Rowlingson (1992) re-analyse the Chorley–Ribble data using the conditional formulation of the model, and reach essentially the same conclusions as with the unconditional approach, but by a much cleaner inferential route. They also describe a straightforward extension to multiple point sources $\mathbf{x}_{0j}$: $j = 1, \ldots, p$, in which the function $f(\cdot)$ in (4.16) or (4.19) is replaced by a product form

$$\prod_{j=1}^{p} f(\|\mathbf{x} - \mathbf{x}_{0j}\|;\boldsymbol{\theta}).$$

This is, of course, but one of many possible extensions; the limitations on modelling raised incidence near multiple point sources derives more from the amount of data needed to discriminate between models, rather than from any theoretical difficulties.

There is a strong connection between Poisson *point process* modelling, and Poisson *regression* modelling of spatial variation in disease incidence. In the latter, as exemplified by Forman *et al.* (1987) and Cook-Mozaffari *et al.* (1989), the assumption is that counts, $N(A_j)$, of numbers of cases in a set of disjoint subregions A_j can be treated as mutually independent Poisson random variables with means μ_j determined by one or more covariates associated with the A_j. Strictly, the regression model only has a spatial interpretation, in the sense that its meaning is preserved when the partitioning subregions A_j are changed, if *all* spatially referenced covariates are piecewise constant functions whose discontinuities coincide with the boundaries of the original A_j. Since the A_j are almost invariably defined by arbitrary political boundaries, this is clearly a ludicrous proposition. A pragmatic solution is to recognise a degree of over-dispersion in the data, i.e. to allow the variability between subregions with notionally identical means μ to exceed μ. As noted in Section 4.1, this again has *no* genuine spatial interpretation. Nevertheless, it may give useful answers to specific questions, provided that the important covariates show relatively little spatial variation *within* the subregions A_j. This must be more nearly true for small subregions than for large ones, and the ultimate small subregion is, of course, a point!

4.6 SPLANCS: SOFTWARE FOR ENVIRONMENTAL EPIDEMIOLOGY

The statistial methods which we have considered in this chapter are not readily implementable in standard statistical software packages. Also, the formal statistical analysis will typically need to be preceded by a more informal inspection of the basic data. For this, we need relatively sophisticated graphical software which can quickly and easily generate both point-maps of the cases or controls, and perspective, contour, or chloropleth maps of relevant spatial covariates. Geographical Information Systems (GIS) provide the required display facilities, but are typically *very* limited in their capacity for statistical analysis. At Lancaster, the Department of Mathematics and the North West Regional Research Laboratory have been collaborating on an ESRC-funded research project to add statistical analytical capacity to a GIS. Our original approach was to write additional commands into a commercially available GIS, ARC-INFO. However, we subsequently decided that a looser coupling between a GIS-like display environment and an analytic engine might be preferable, particularly as we were in the fortunate position of being able to use the windowing environment on SUN workstations as an almost ready-made loose coupling. Our choice for the analytic engine was S, a statistical programming environment described in Becker *et al.* (1988). The two crucial features of S from our point of view were: firstly, its highly developed existing analytic capacity implemented as a set of high-level statistical *functions*; secondly, the ease with which additional, user-defined functions could be added to the environment, written either in the S language itself or in a more traditional language (Fortran or C). The final stage in the development of our ideas came with the realisation that S could also be used as the GIS-like display environment, particularly in its commercially developed variant known as S-plus, distributed by Statistical Sciences Inc.

Splancs (Rowlingson and Diggle, 1991) consists of an additional set of S-plus functions which implement all of the methodology described in the present paper, together with a number of analytically more basic, but vitally important operations such as mouse-driven definition of polygonal regions, selection of cases within regions and data-mapping. Figure 4.1 shows a typical screen-dump of a Splancs session. It related to the analysis of a set of data on childhood leukaemia as reported in Diggle and Chetwynd (1991) and at the end of Section 4.3 above. The small graphics window at the top left of the screen is a map of the data, which have been extracted from a larger set of data covering the areas of North, South and West Yorkshire and Humberside. The quadrilateral is the one used by Diggle and Chetwynd (1991) to define their region A; their algorithm required a *convex* polygonal A, whereas Splancs allows an arbitrary polygon. The larger graphics window at the bottom left is the plot of $\hat{D}(s)$ with tolerance limits defined as $\pm 2\sqrt{[\mathrm{var}\{\hat{D}(s)\}]}$ under random labelling. The window on the right-hand side of the screen contains all of the Splancs commands used to generate the analysis.

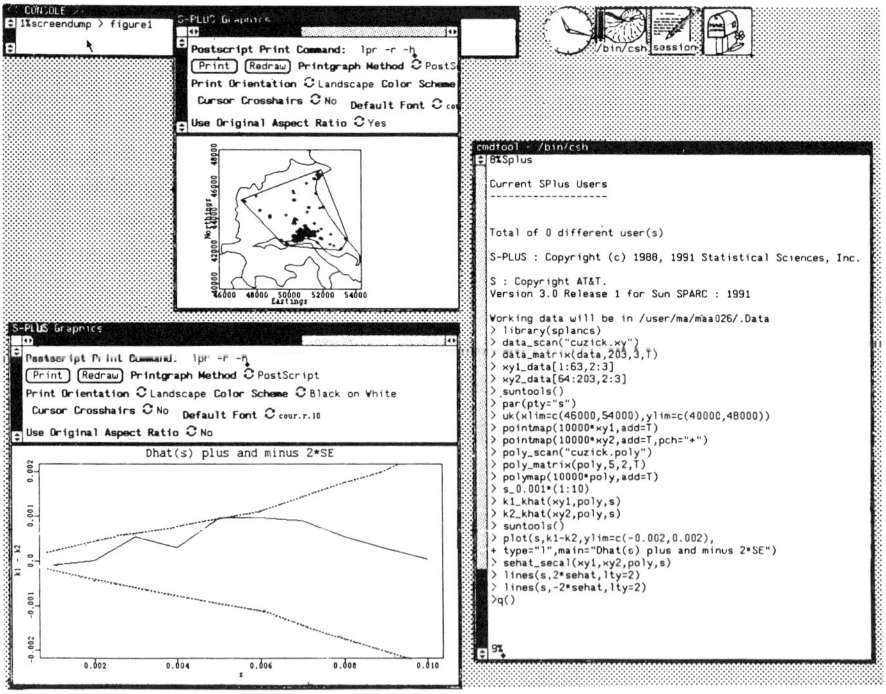

Figure 4.1 A screen-dump of a Splancs session.

Splancs operates on SUN workstations using either the Suntools or Openwindows windowing environments. In principle, it could be modified to run on an appropriately configured 386-based PC, since a DOS version of S-plus is available for this kind of machine. However, we have no first-hand experience of using S-plus in a DOS environment.

4.7 DISCUSSION

The basic methodological framework for point process modelling in environmental epidemiology is well established. But there is still considerable scope for refinement and further development. One general methodological issue concerns the need for a continuing move away from significance testing *per se*, and towards estimation of epidemiologically interpretable quantities, whether nonparametric descriptors of elevated risk such as the function $\hat{D}(s)/\hat{K}_{22}(s)$ suggested at the end of Section 4.3, or parametric models such as the function $f(\mathbf{u}; \boldsymbol{\theta})$ in Section 4.5. A second general issue is to question the heavy reliance on second-moment methods for point pocess data which, of necessity, are highly non-Gaussian. Whilst it is easy to construct examples of different point processes with identical second-moment structure (see, for example, Diggle, 1983, chapter 4; Baddeley and Silverman, 1984), we remain convinced that the *K*-function is the single most useful tool in the exploratory analysis of spatial point process data. Nevertheless, it can and should be used in conjunction with other tools, including likelihood-based methods where available, or nearest neighbour methods, when there is a need to validate a parametric model.

A much more open methodological issue, is the whole question of how to detect individual unforeseen clusters, as opposed to general clustering or clusters around pre-specified points. As noted by Marshall (1991) in his useful review, the literature on this problem is vast, and growing. Most approaches to the problem seem to be cast ovetly in terms of significance testing, whilst recognising the need for a less formal interpretation than the classical accept/reject paradigm. For a recent example with stronger statistical credentials than most, see Besag and Newell (1991).

Our view, although the ideas are not yet properly developed, is that the cluster detection problem would be better cast as one of *estimating* anomalies in the local incidence, $\lambda(\mathbf{x})$ say, of the disease in question. This suggests estimating an incidence ratio $\rho(\mathbf{x}) = \lambda(\mathbf{x})/\lambda_0(\mathbf{x})$, where $\lambda_0(\mathbf{x})$ takes into account all *known* factors. A major practical difficulty with this idea is that anomalies are, almost by definition, likely to be both highly localised and involve small numbers of cases. So the precision of any nonparametric estimate of $\lambda(\mathbf{x})$ is likely to be low, and the precision of the corresponding estimate of the ratio $\rho(\mathbf{x})$ even lower. Nor is it true in a nonparametric setting that if $\hat{\lambda}(\mathbf{x})$ and $\hat{\lambda}_0(\mathbf{x})$ are good estimators of $\lambda(\mathbf{x})$ and $\lambda_0(\mathbf{x})$, then their ratio $\hat{\lambda}(\mathbf{x})/\hat{\lambda}_0(\mathbf{x})$ will be a good estimate of $\rho(\mathbf{x})$. Viewing clusters as neighbourhoods of high local intensity is explicit in some papers, for example Bithell (1990), and implicit in many others, including Openshaw *et al.* (1988) and Turnbull *et al.* (1990).

With this view of the cluster detection problem we can distinguish between the identification of local maxima in the historic incidence ratio $\rho(\mathbf{x})$, and the detection of unexplained *changes* in the space–time incidence ratio, $\rho(\mathbf{x}, t)$ say. The space–time problem is then somewhat akin to on-line quality-control monitoring, and spatial analogues of techniques such as CUSUMS (Page, 1954) or Kalman filtering (Smith and West, 1983) which have proven value in that context may also prove useful for cluster detection. This would require routine updating of spatial databases on disease incidence, ideally at the level of individual cases; an ambitious scenario, but surely an achievable one—if not now, then in the foreseeable future.

Visions of automatically updated spatial databases bring us to the question of data quality. All of the methods described in this chapter are interpretationally robust to random non-ascertainment of cases. They also react conservatively to random perturbation of case locations, in the sense that the evidence for spatial clustering is generally diminished by such perturbation. Another issue concerning data quality is the balance between spatial resolution and accuracy of information. Whilst we believe that in principle, point process methods using individual case locations are fundamentally superior to methods which use only counts in discrete regions, this advantage may be lost in practice if data on individual locations are less reliable than coresponding data on counts in regions, or if covariate information is only available at the regional level.

Another vital practical issue is the choice of controls for methods based on case-control comparisons. The practical difficulties raised are the same as in a classical case-control study, with the added problem that the need to select a spatially representative set of controls may make it harder to achieve the required matching of cases and controls on the relevant characteristics.

Finally, on a somewhat mundane note, the subject of environmental epidemiology still seems to be at the stage where most studies are retrospective. Somebody—perhaps a concerned group of local residents, a public health worker, or a pressure group—notices an unusual pattern and the statistician is then asked to decide whether or not this pattern is 'significant'. Whilst this may be inevitable, it does prevent a purely spatial analysis of historic data from doing any more than establishing a case for further investigation. There are exceptions. Gardner (1991), in the context of the Sellafield childhood leukaemia controversy, gives an excellent account of how spatial analysis of historic data can play a useful, albeit limited, role in the progression from anecdotal evidence, via exploratory analysis of observational data, to confirmatory studies using designed experiments.

5

Estimation of Quantiles in Airborne Pollution; Measuring Pollution around Gas Furnaces

Georg Lindgren, Lena Zetterqvist and Göran Holmstedt,
Lund Institute of Technology, Sweden

5.1 INTRODUCTION

In pace with the development of oil and natural-gas fields in the North Sea, local natural-gas furnaces have become increasingly common for the heating of homes in northern Europe. This has introduced a new type of local pollution source in many countries, which have traditionally relied on either electricity or district heating systems, with a few regional or district production units, or on local production units fired by what are regarded as traditional fuels, such as oil, coal, or wood. This has caused concern about a number of issues regarding emission of pollutants such as nitrogen oxides, NO and NO_2: denoted collectively NO_x. These issues include questions like:

- What is the total amount of substance emitted to the atmosphere?
- What is the increase in average concentration of the substance caused by the sum of many sources?
- What is the maximum short-time increase in concentration near any of the local sources?

Statistics for the Environment. Edited by Vic Barnett and K. Feridun Turkman

In order to obtain useful estimates of the increase in pollutant concentration caused by a local source one needs long-time coordinated upwind and downwind measurements near the source. We shall describe a five-month campaign for measuring the concentration of NO_x near homes with natural-gas furnaces in a residential area in southern Sweden. The aim was to estimate both the maximum concentration that could be experienced near the furnace, and the increase in concentration caused by a single furnace. The experiment is described by Vannerberg and Holmstedt (1989).

The interest is in moderately extreme quantiles in the concentration distributions. In long series of measurements one can use observed quantiles or parametrically estimated quantiles. When the measurements are corrupted by measurement errors which may perhaps be normal, but not negligible, statements about extreme quantiles can only be made after proper correction. This requires rather precise knowledge of the pollutant distribution, as well as of the measurement error distribution. In any case, the use of observed quantiles can be quite misleading. We shall discuss how to find the relation between the empirical quantiles in the observed measured distribution, and the quantiles in the true distribution without measurement error when the true distribution is assumed to be log-normal and the measurement errors are normal.

Log-normal distributions have been used extensively to describe pollutant concentrations; a survey paper on the use of these distributions in atmospheric sciences is given by Crow (1988).

5.2 NORMAL AND EXTREME NO_x-EMISSION

5.2.1 Background

The study is based on data collected over November 1987 to April 1988 in the village of Åkarp in southern Sweden in a residential area with small houses. The village is situated in the suburban region between the cities of Malmö and Lund and is surrounded by several highways with heavy traffic and by railways as well as vast areas of farmland.

Each home in the area is heated by its own natural gas furnace, fired with natural gas from the North Sea. Exhaust from this type of home furnace contains *c.* 70 ppm of nitrogen oxides, of which *c.* 20% is NO_2. As NO_2 this amounts to 130 mg NO_2/m^3. The normal concentration in the region during the winter months is *c.* 20 ppb, i.e. 40 μg NO_2/m^3. The requirement of the Swedish National Environmental Protection Board is that the maximum 1 hour average concentration of NO_x should not exceed 100 ppb.

5.2.2 The measuring system

One house was selected, located at the edge of the residential area, 500 metres west of highway E6. The house was equipped with 25 samplers which took continuous samples of the air for 35 minutes. Each air sample was then analysed for its content of NO_x in a chemiluminiscent analyser (AAL model 443). The analyses rotated between the 25 samplers with a cycle interval of *c.* 35 minutes, which means that each sample point gives approximately the 30-minute average NO_x concentration measured with an interval of 35 minutes. (The explanation for the 35-minute sample interval for the 30-minute average was that the air flow in the samplers was not constant but decreased at the end of the period.) In total the material consists of 111 565 measurements of the NO_x-content extended over a period of five months, with only brief interruptions.

Of the 25 samplers, 21 were placed on the house near the exhaust pipe, while 4 were placed 10–20 m from the house in north-west or south-east direction, near local streets; for a plan see Figure 5.1. The reason for the large number of samplers on the house was to get a detailed picture of the exhaust plume near the exhaust pipe. We shall study points number 9 and 25, located 15 metres south-east and north-west of the house, respectively.

5.2.3 Correction for drift

The NO_x analyser had a considerable and variable drift in its zero point. To compensate for this, measurements were taken of pure nitrogen gas immediately after each NO_x measurement, so a full series of measurements consists of values

$$NO_x^{(1)}, N_2^{(1)}, NO_x^{(2)}, N_2^{(2)}, \ldots, NO_x^{(25)}, N_2^{(25)}.$$

We assume that there is a linear trend in these data, i.e.

$$N_2^{(i)} = a + bt_i + e'_i = \mathrm{LR}^{(i)} + e'_i,$$
$$NO_x^{(i)} = \text{true}\, NO_x^{(i)} + a + bt_i + e''_i = \text{true}\, NO_x^{(i)} + \mathrm{LR}^{(i)} + e''_i,$$

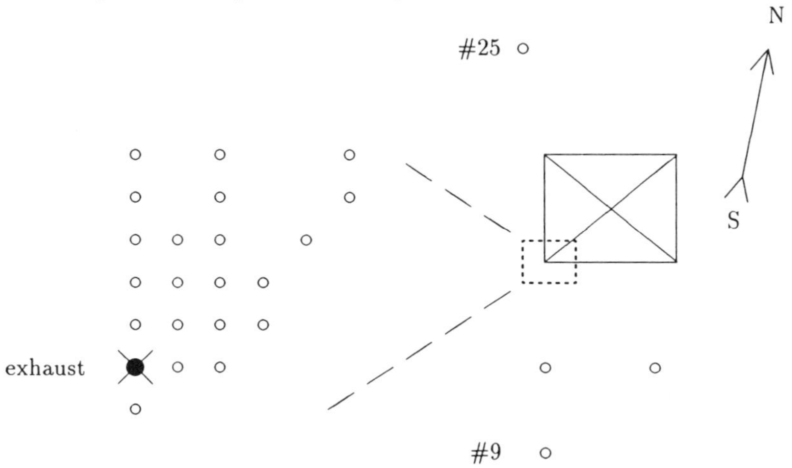

Figure 5.1 Plan of measuring system location

where e_i' and e_i'' are normal measurement errors with rather large standard deviation $D(e) = \sigma_0$ and $\mathrm{LR}^{(i)} = a + bt_i$.

Between each full cycle the instrument was flushed with nitrogen, which introduced further shifts in the zero level. To reduce the disturbances as far as possible, we therefore fitted separate linear regressions to each cycle of 25 measurements on N_2 obtained during a full measuring cycle, giving

$$\widehat{\mathrm{LR}}^{(i)} = \hat{a} + \hat{b}t_i.$$

The observation of the NO_x concentration was then given as the difference between the instrument reading on sample gas and the fitted N_2 line for that measurement cycle, i.e.

$$\widehat{\mathrm{NO}}_x^{(i)} = \mathrm{NO}_x^{(i)} - \widehat{\mathrm{LR}}^{(i)} = \text{true}\,\mathrm{NO}_x^{(i)} + \mathrm{LR}^{(i)} - \widehat{\mathrm{LR}}^{(i)} + e_i''. \tag{5.1}$$

The estimated standard deviation about the regression lines was used as a measure of the uncertainty in the NO_x measurements. In the analysis we found $\sigma_0 \approx 0.02$ ppm.

As an alternative to the linear trend analysis, one could use the pointwise difference between the instrument reading on sample gas and on pure nitrogen to represent the observed NO_x concentration. This would approximately double the variance of the measurement error. We shall in Section 5.3.2 see how this affects the estimates of the NO_x distribution.

5.2.4 Data

We shall analyse the overall distribution of NO_x from points number 9 and 25, located 15 metres south-east and north-west of the house, respectively. At these points, the influence of the surrounding sources dominate, except at winds from the south, when local and area sources increase the concentration. We shall confine ourselves to the 80-day period, 30 November, 1987 to 17 February, 1988, during which the system was rather stable. The series was checked for possible outliers due to errors in the chemiluminiscent analyser. After pruning, the series consisted of 2815 observations of NO_x obtained by comparison with a local linear trend, as described in Section 5.2.3. For comparison, we also considered the series of pointwise differences between the NO_x reading and the N_2 reading. In both series, *c.* 2% of the observations were above the critical level 100 ppb (= 0.1 ppm). One aim of the present analysis is to see how this proportion could be corrected by taking the measurement errors into account.

Figure 5.2 shows the regression-corrected NO_x measurements for the studied period. (The occurrence of the many negative values, *c.* 15%, is due to the presence of measurement errors.)

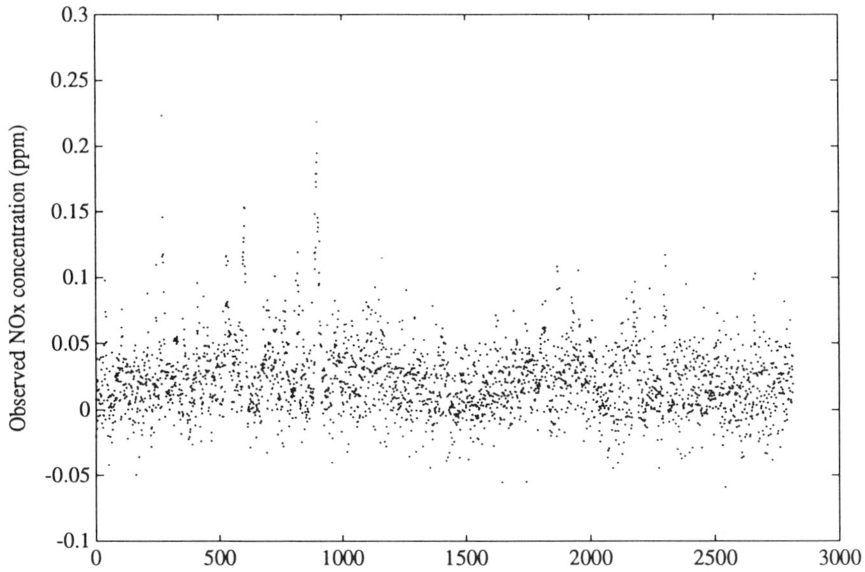

Figure 5.2 Observed NO_x at point 25 from 30 November 1987 to 17 February 1988

5.2.5 Influence of wind direction

One of the goals of the study was to estimate the increase in NO_x concentration that was caused by the local source. The test area was neighbouring an urban area with heavy traffic to the south, west and north, while farmland dominated between the north and north-east. To separate the influence of the local source from the urban traffic one, the wind direction and speed were also recorded. Figure 5.3 shows the observed NO_x data at point 9 in separate histograms for the four wind quadrants. As seen the data obtained with wind from the north-east contain mostly measurement errors with very few large values.

Statistical methods for analysing simultaneous measurements of wind directions and concentrations are discussed by Eilers (1991).

5.3 A STATISTICAL MODEL

5.3.1 The model

From now on we shall use the lognormal distribution to describe the true NO_x concentration, i.e.,

$$\text{true}\, NO_x^{(i)} = \exp(m_0 + \beta\widetilde{X}), \tag{5.2}$$

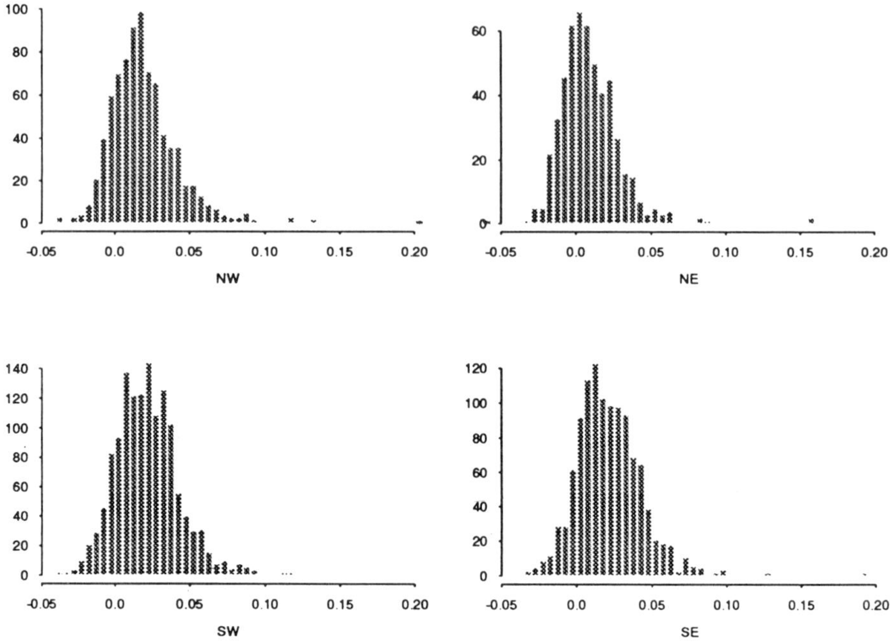

Figure 5.3 NO_x at point 9 separated for the four wind quadrants

$$\widehat{NO}_x^{(i)} = \exp(m_0 + \beta\tilde{X}) + \sigma\tilde{U}, \tag{5.3}$$

where $\tilde{X}$ and $\tilde{U}$ have standard normal distributions. The standard deviation σ depends on the measurement variance σ_0^2 and the type of correction that was made. For linear regression correction, we have $\sigma\tilde{U} = a - \hat{a} + (b - \hat{b})t_i + e_i''$, while for pointwise correction, $\sigma\tilde{U} = e_i'' - e_i'$. Since the sampling interval is constant, t_i is proportional to i, and hence, for point number i of the 25 consecutive points,

$$\sigma = \begin{cases} \sigma_{\mathrm{PC}} = \sigma_0\sqrt{2} & \text{(pointwise comparison)} \\ \sigma_{\mathrm{LR}} = \sigma_0\sqrt{1 + \dfrac{1}{25} + \dfrac{(i-13)^2}{S_{xx}}} & \text{(linear regression),} \end{cases}$$

where $S_{xx} = \sum_{i=1}^{25}(i-13)^2$.

5.3.2 Analysis

To analyse the data at point 25 described in Section 5.2.4 we used model (5.3) with σ_0 estimated from the linear regression (LR). From this particular series we obtained $\sigma_0 = 0.02$ ppm. The parameters m_0 and β in (5.2) were estimated by means of a maximum likelihood method applied to the grouped sample. A summary of the obtained estimates can be found in the Table 5.1. For the pointwise correction (PC) we incresed the estimate of σ_0 by 10% in order to obtain a better fit, and used $\sigma_0 = 0.022$.

Figure 5.4 shows the empirical distribution of NO_x, obtained by comparing with the linear regression, and the fitted distribution of type (5.3), with σ_0 estimated in the linear regression. For comparison, Figure 5.5 shows the NO_x distribution obtained by pointwise comparison between NO_x and N_2 values.

Table 5.1

Type of correction	σ	m_0	β	$P(NO_x > 0.1)$	Obs. frequency > 0.1
LR	$\sigma_{LR} = 0.021$	-3.99	0.75	0.010	0.018
PC	$\sigma_{PC} = 0.031$	-3.70	0.68	0.020	0.022

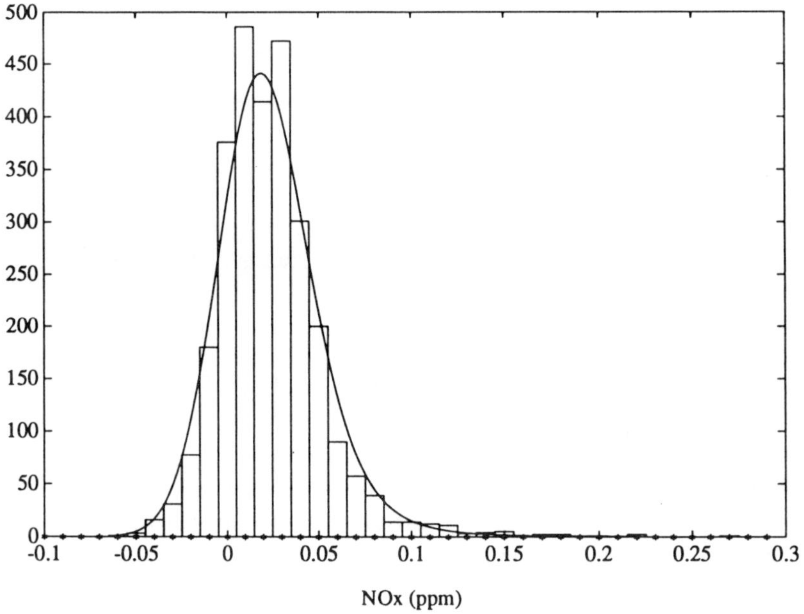

Figure 5.4 Observed NO_x distribution with fitted normal-plus-log-normal density; comparison with linear regression

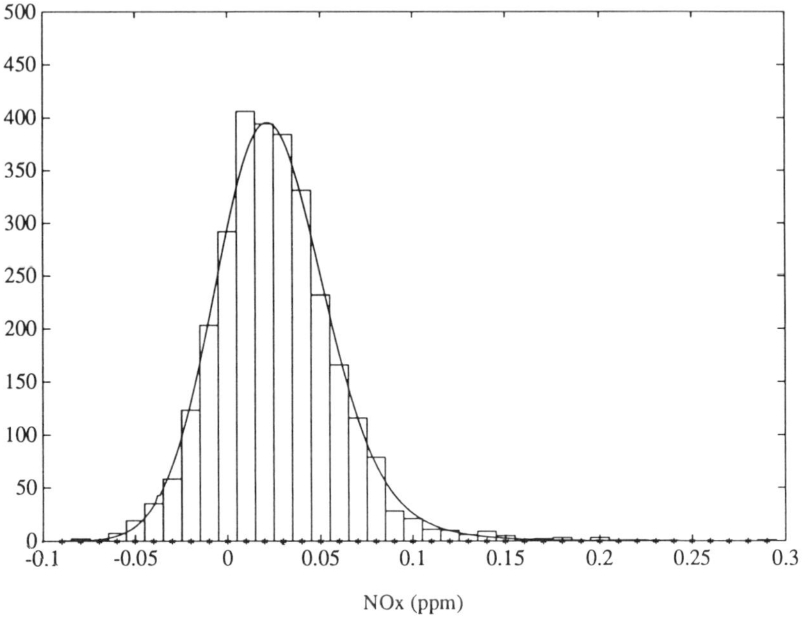

Figure 5.5 Observed NO_x distribution with fitted normal-plus-log-normal density: pointwise comparison with standard

According to the Environmental Protection Board, the critical level, which should not be exceeded, is 0.1 ppm. This level is exceeded in *c.* 2% of the cases in both the series. We can compare this with the calculated probability based on the parametric log-normal model. Using estimated parameters we get the probability of NO_x exceeding 0.1 ppm to be 0.010 in the linear regression model and 0.020 in the pointwise model. Thus, the refined analysis with linear regression and parametric fitting of an underlying distribution reduced the estimated probability by a factor of 2.

5.4 QUANTILE RELATIONS IN DISTURBED DISTRIBUTIONS

We shall now investigate the effect of measurement error on the empirical extreme quantiles in the emission distribution in order to relate the observed quantiles to the true quantiles. We use the basic model (5.3) with normal measurement errors and underlying log-normal distribution. We employ this parametric model only in a sensitivity analysis to illustrate how measurement errors affect the empirical quantiles. If one really wants to use these specific distributions, the analysis in Section 5.3.2 gives a more precise answer.

5.4.1 Log-normal case

Suppose the distribution in equation (5.3) to be log-normal, i.e.

$$\begin{aligned}\widehat{\mathrm{NO}}_x^{(i)} &= \exp(m_0 + \beta\tilde{X}) + \sigma\tilde{U}\\ &= \sigma\{\exp(m + \beta\tilde{X}) + \tilde{U}\},\end{aligned}$$

where $m = m_0 - \log\sigma$ and $\tilde{X}$ and $\tilde{U}$ are independent standard normal variables.

Let q_α be the α-quantiles in the log-normal distribution, so that

$$\alpha = P(\exp(m + \beta\tilde{X}) > q_\alpha),$$

and similarly let u_α be the quantiles in the disturbed distribution of $\exp(m + \beta\tilde{X}) + \tilde{U}$, i.e.

$$\alpha = P(\exp(m + \beta\tilde{X}) + \tilde{U} > u_\alpha).$$

The asymptotic relationship between the q_α and the u_α for small α and large q_α can then be expressed in terms of the quantiles λ_α of the standard normal distribution as in Theorem 5.1 below. (As usual, Φ denotes the standard normal distribution function with $\alpha = 1 - \Phi(\lambda_\alpha)$.)

Theorem 5.1 *For small α, let x_α be defined as the solution to the equation*

$$\lambda_\alpha^2 = x_\alpha^2 + \beta^{-2}x_\alpha^2\exp(-2(m + \beta x_\alpha)). \tag{5.4}$$

The corresponding values of q_α and u_α are then given by

$$q_\alpha = \exp(m + \beta\lambda_\alpha) \tag{5.5}$$

$$u_\alpha \approx \exp(m + \beta x_\alpha) + \beta^{-1}x_\alpha\exp(-m - \beta x_\alpha). \tag{5.6}$$

Proof: Write

$$g(x) = u_\alpha - \exp(m + \beta x).$$

The approximation (5.6) is obtained by linearising the boundary $\exp(m + \beta x) + u = u_\alpha$ near the point $(x_\alpha, g(x_\alpha))$ where the function $u = g(x)$ is the tangent to the circle with radius r_α, i.e. at the point where $x^2 + g(x)^2$ is minimum; see Figure 5.6. As is easily seen this point satisfies

$$x_\alpha = g(x_\alpha)\beta\exp(m + \beta x_\alpha), \tag{5.7}$$

$$g(x_\alpha)^2 = r_\alpha^2 - x_\alpha^2. \tag{5.8}$$

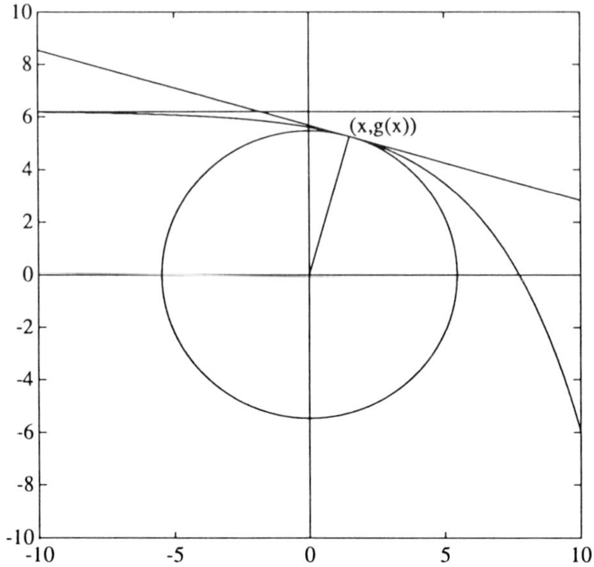

Figure 5.6 Linearisation of the curve $u = u_\alpha - \exp(m + \beta x)$

Thus we have the approximation

$$P(\exp(m + \beta\tilde{X}) + \tilde{U} > u_\alpha) \approx P\left(\frac{x_\alpha\tilde{X} + g(x_\alpha)\tilde{U}}{\sqrt{x_\alpha^2 + g(x_\alpha)^2}} > r_\alpha\right). \tag{5.9}$$

Since

$$\frac{x_\alpha\tilde{X} + g(x_\alpha)\tilde{U}}{\sqrt{x_\alpha^2 + gx_\alpha)^2}}$$

is standard normal, the probability in the right-hand side of (5.9) is equal to $1 - \Phi(r_\alpha)$. To make it equal to α we must have $r_\alpha = \lambda_\alpha$. This gives the relation (5.4) between λ_α and x_α.

The relations (5.5) and (5.6) then follow from the log-normal distribution, and the equations (5.7) and (5.8) for x_α.

Figure 5.7 shows the relation between the q-quantiles and the u-quantiles in the log-normal case for $m = -0.15$ and $\beta = 0.2(0.1)1.0$.

5.4.2 Application to NO_x data

The NO_x data at point 25 studied in Section 5.2.4, corrected by a linear regression, could be described by a log-normal-plus-normal distribution of type

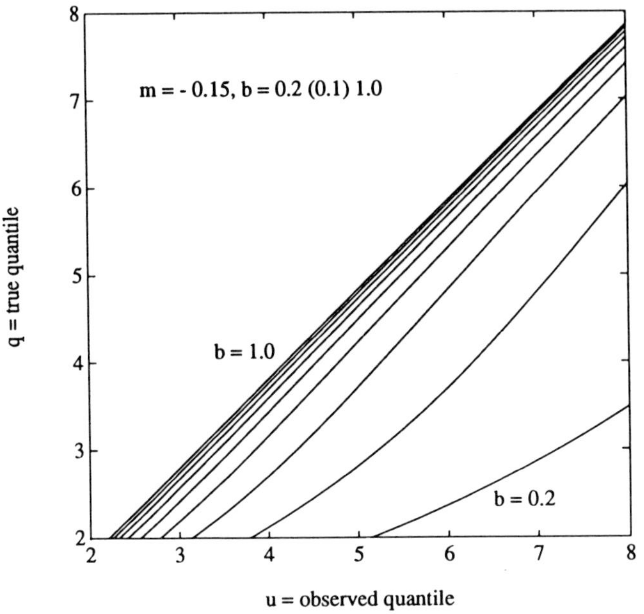

Figure 5.7 Observed and true quantiles of the log-normal distribution observed with normal errors (the parameter β is denoted by b in the figure)

(5.3) with parameters $\sigma_0 = 0.02$ ppm, $m_0 = -3.99$, $\beta = 0.75$. After normalisation, this gives $m = m_0 - \log \sigma_{LR} = -0.15$. The critical level, 100 ppb = 0.1 ppm, corresponds with $q_\alpha = 0.1/\sigma_{LR} = 4.66$, which is a rather extreme quantile. In a log-normal distribution with $m = -0.15$, $\beta = 0.75$ it is exceeded with probability 0.01. As seen from Figure 5.6 the observed quantile u_α is *c.* 10% larger.

5.5 CONCLUSIONS

When normal measurement errors disturb a low pollutant concentration the extreme quantiles of interest are affected in a way characteristic of the pollutant distribution, and the empirically observed quantiles can be misleading. There are two alternatives to remedy this. One is to use parametric models and estimate the parameters by standard statistical techniques. The other is to make a sensitivity analysis of the influence of measurement error on the empirical quantiles, under a few different alternative distributions.

When the true pollutant level is log-normal we have given a precise asymptotic relationship between the true and the observed quantiles. When the noise variance is estimated in a parametric model, independently of the true observations, maximum likelihood techniques work well. For very extreme quantiles, the effect of the error is shown to be small with either alternative. For the NO_x chemiluminiscent analysing technique, the true quantiles are overestimated by less than 10%.

6

A Stochastic Model for Times of Exposures to Air Pollution from a Point Source

A. C. Davison and N. I. Ramesh,
Oxford University, UK

6.1 INTRODUCTION

Statistical models for exposures to air pollution must deal with two related aspects, namely the sizes of exposures and the times at which they occur. For substances normally present in the atmosphere there may be no possibility of zero levels, and conventional time-series models may be suitable. This might be the case for substances such as ozone and carbon monoxide, though data analysis may be considerably complicated by the generally high proportion of missing values, the presence of covariate information, strong diurnal and seasonal components, and the level of aggregation of the data.The situation is quite different, however, when a pollutant not normally present in the atmosphere is released from a point source. At a receptor far from the source, there may be long periods with no contamination, followed by clusters of exposures arising from particular meteorological conditions. Models based on point processes are more natural for such data than those based on time series, and are the topic of this chapter.

The pattern of exposures at receptors far from the source of a release to the atmosphere depends on many factors, which can interact in a complicated way.

Statistics for the Environment. Edited by Vic Barnett and K. Feridun Turkman

Among the factors are the quantity of pollutant released, and the pattern of the release, the period over which the release occurs, and weather conditions over the trajectory that the material traverses. Exposure levels also depend on characteristics of the substances involved, such as their deposition velocities—the rate at which they are deposited on the ground as they pass overhead—and the effectiveness of rainfall and other wet deposition mechanisms, as well as the considerable complications introduced by chemical reactions.

Our work is motivated by data on simulated exposures due to atmospheric transport of radionuclides released from nuclear installations in Western Europe, in a study of the possible effects of such transport at ranges of roughly 100–1000 km from a potential source. These data have been used to construct a statistical model, largely based on regression ideas, as an aid in assessing the dose that might be received by an individual due to a release of a radionuclide at a distant source (ApSimon and Davison, 1986). This statistical model is static rather than dynamic: it does not take the temporal pattern of exposures into account. One difficulty with this is that temporal clustering is relevant to estimation of exposure levels for long releases. This prompts us to find a simple model for the clustering, so that its effect on return levels can be assessed.

Section 6.2 gives more details about the computer model that generated the data and about the data themselves. Although the model simulates data on exposures of various sorts, we concentrate on exposures to wet deposition, which can be an extremely effective removal mechanism for air borne pollutants. In the worst case a high proportion of the particles in a pollutant cloud may be scavenged from the atmosphere in the space of a few minutes, leading to dangerous exposures at ground level. We study particularly exposure times rather than the joint behaviour of times and sizes, because other work with these data (Davison, 1984) suggests that exposures at a particular point may be treated as independent and identically distributed. Another analysis from a different viewpoint is given by Davison and Smith (1990), who discuss statistical aspects of models for extreme exposure sizes.

One broad distinction in modelling is between empirical models, which can be rather general without necessarily much physical basis, and substantive models, in which some attempt is made to model the physical processes that underlie the data. The models used here, of the first type, are known as Markov-modulated Poisson processes. Section 6.3 describes some theoretical aspects of these models, including likelihood analyses in addition to the usual paraphernalia of second-order inference for series of events. In Section 6.4 we fit these models to the data. Section 6.5 contains a brief discussion.

6.2 THE MESOS MODEL AND DATA

The data that motivate our work were generated by members of the Environmental Safety Group at Imperial College, London, from a computer model of the meso-scale dispersion of pollutant from a point source (ApSimon, *et al.*, 1985). The model, MESOS, uses a simplified but reasonably realistic set of equations for the behaviour of the atmospheric boundary layer in combination with 'present weather' data averaged over squares of a grid covering Western Europe, to determine the paths taken by an imaginary series of consecutive three-hourly unit releases of pollutant at a point source. Each 'puff' of material follows a trajectory along which it experiences dispersion, dilution and depletion, determined by local conditions in the boundary layer at the time of passage. Puffs are followed until they leave the grid area, or for four days, after which time their material is generally highly diluted.

The effect of dilution due to synoptic spread of the trajectories is captured by interpolation between the paths taken by successive puffs of material. Depletion due to dry deposition of the radionuclide and wash-out in rainfall, as well as the effect of radioactive half-life, can be modelled for a variety of radionuclides released from the source, and a pattern of exposures is built up on a grid of exposures centred on the source. Although MESOS deals with series of imaginary releases, it has been validated on the Windscale 1957 and Chernobyl 1986 releases, for which it gave generally adequate results.

Wet deposition is modelled by means of a 'wash-out coefficient' which varies with the intensity of rainfall. No distinction is made between in-cloud processes such as rain-out, scavenging below cloud level in the form of wash-out, and occult deposition in, for example, land-lying cloud. Average rainfall rates are deduced from 'present weather' reports, and then applied uniformly to 100 km grid squares. This smearing of rainfall, usually a patchy phenomenon, is thought to increase wet deposition near the source, and to make it too uniform over the grid area.

The end result of a run of MESOS is a large number of time series of exposures at different receptors due to releases from the source. Figure 6.1 shows log exposures to wet deposition at sixteen receptors due to 2880 successive notional three-hour releases of the particulate form of iodine-131, $^{131}\mathrm{I}(p)$ through the year 1976. The series are laid out roughly according to their geographical position relative to the source, deemed to be at Ispra, in northern Italy. The sixteen exposures recorded for a particular three-hour period are those due to the puff of material released during that time, which may subsequently expose one or more of the receptors. Both exposure levels and the frequency of exposure drop rapidly with distance from the source. Comparison of the processes 300 km and 800 km north of Ispra shows that the effect of distance is not just to thin the process of exposure times: it may be dry when a puff passes over one receptor, but subsequently rain over a more distant one. Clustering of exposure times both at single receptors and across receptors is also evident. Ideally a suitable model will accommodate both these features of the data.

Plots of the cumulative numbers of exposures show no evidence of non-stationarity of the point processes, although if the series covered a number of years some sort of pattern might be expected due to seasonal changes in weather

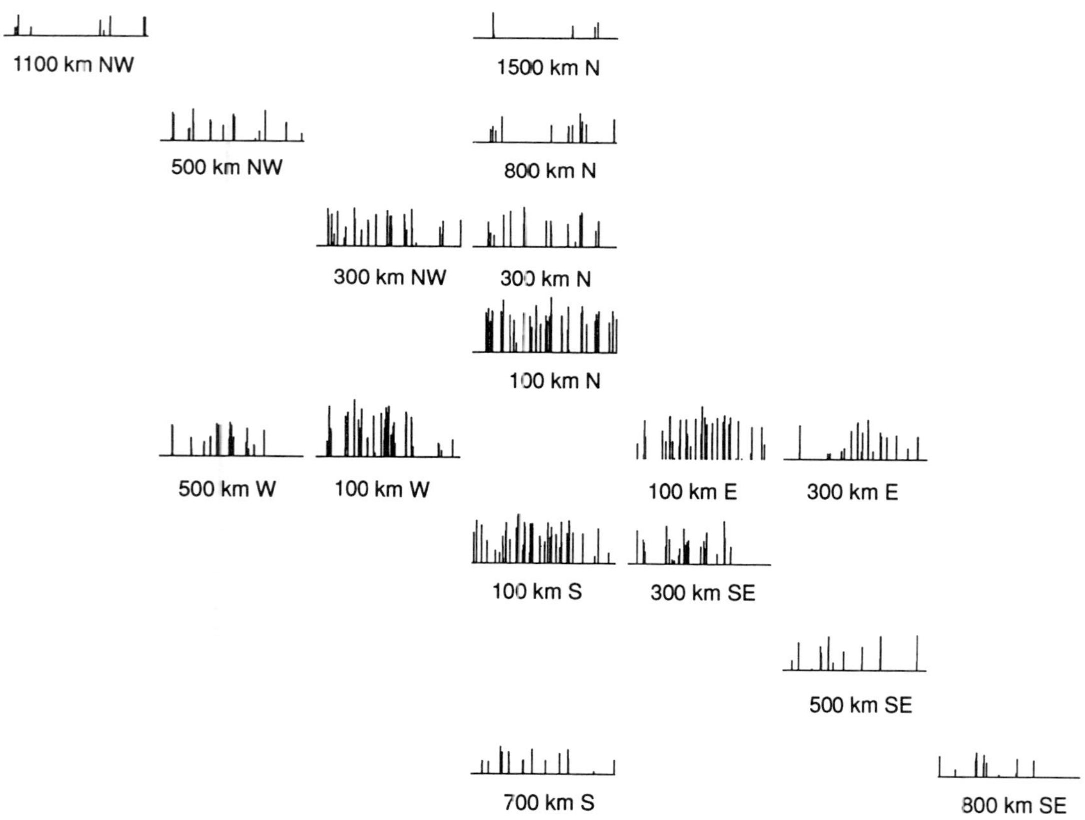

Figure 6.1 Log exposures to wet deposition of $^{131}I(p)$ due to releases from Ispra during 1976, at sixteen receptors. *Note the marked clustering of exposures both within and across receptors*

systems that influence Western Europe. In Section 6.4 we treat the series as stationary.

Table 6.1 shows the locations of the sixteen receptors and the numbers of exposures observed at each, together with various fitted numbers of exposures from models discussed later in the chapter. In Figure 6.2 the mean exposure rates are plotted as a function of distance from the source. Broadly speaking, rates drop with distance according to a power law, but there can be unexpected local effects, such as at the receptor 1500 km north of the source. There appear to be directional effects, and probably also interactions between distance and direction, but with the rather limited range of directions and distances for these data the scope for modelling interactions parsimoniously is limited.

6.3 MARKOV-MODULATED POISSON PROCESSES

6.3.1 Model

We model exposure times by a Markov-modulated Poisson process (Smith and Karr, 1983; Smith, 1984).

Let X_t denote the state of a stationary irreducible continuous-time Markov chain whose state space is $\{1, 2, \ldots, k\}$. The process has infinitesimal generator

Table 6.1 Numbers of exposures to wet deposition of $^{131}\mathrm{I}(p)$ released from Ispra during 1976

Receptor	Direction	Distance (km × 10^2)	Observed	Fitted M_1	M_2	M_3
1	N	1	125	111	119	103
2	N	3	59	61	65	57
3	N	8	32	36	33	34
4	N	15	34	25	24	24
5	E	1	94	90	96	85
6	E	3	52	50	53	47
7	SE	3	55	49	52	48
8	SE	5	33	37	34	36
9	SE	8	27	29	26	28
10	S	1	98	95	101	97
11	S	7	34	33	30	34
12	W	1	111	117	125	122
13	W	5	59	49	45	52
14	NW	3	63	66	70	65
15	NW	5	44	50	46	50
16	NW	11	29	32	30	33

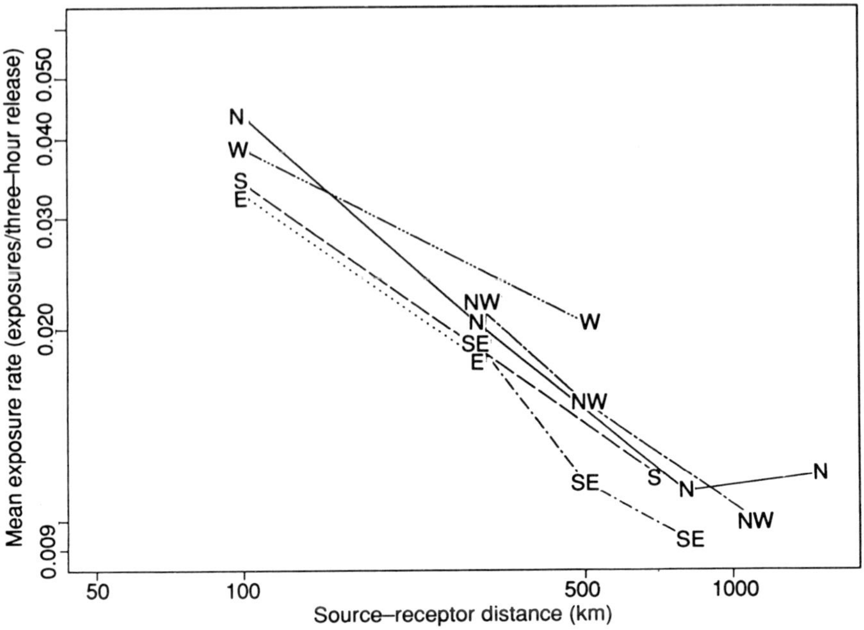

Figure 6.2 The effect of distance and direction on exposure rates at the sixteen receptors

G, a $k \times k$ matrix. The process X_t is taken to be unobserved, and may be thought of as representing a 'state of nature'. For example, state 1 might represent weather conditions not likely to lead to exposures at a particular receptor, whereas states $2, \ldots, k$ might represent conditions increasingly likely to give exposures. It might be impossible to determine from the exposure data alone whether such weather conditions are present, and hence X_t is treated as unobserved. What is observed, however, is a process of exposure times N_r at each of R receptors, $r = 1, \ldots, R$. We write $N_r(t, s)$ to denote the number of exposures at receptor r between time t and a later time s. We suppose that when $X_t = j$, the process of exposures at receptor r has non-negative rate $\lambda_{j,r}$, and we write $\mathbf{L}_r$ for the $k \times k$ matrix with diagonal elements $\lambda_{1,r}, \ldots, \lambda_{k,r}$ and zeros elsewhere.

Suppose that in the period $(0, t_0)$, exposures at receptor r are observed at times $0 < t_{1,r} < \cdots < t_{n_r,r} < t_0$, $r = 1, \ldots, R$. The times of events without regard to type in the pooled process obtained by superimposing the processes N_r are $0 < t_1 < \cdots < t_n < t_0$, where $n = \sum_r n_r$; simultaneous events cannot occur because the process is orderly. In a slight abuse of notation we suppose that $\mathbf{L}_{t_j}$ equals $\mathbf{L}_r$ if the event at time t_j is in the process of exposures at the rth receptor, N_r. A straightforward argument (Smith, 1984) establishes that the transition probability

$$\mathrm{pr}\{X_t = j,\ N_r(0,t) = 0,\ r = 1,\ldots,R \mid X_0 = i\}$$

is the (i, j) element of the $k \times k$ matrix $\boldsymbol{\Psi}(t) = \exp(t\mathbf{Q})$, interpreted as $\sum_{j=0}^{\infty}(t\mathbf{Q})^j/j!$, where $\mathbf{Q} = \mathbf{G} - \mathbf{L}_1 - \cdots - \mathbf{L}_R$. Then the joint likelihood of the processes N_r under our model, assuming that the underlying process X_t is initially in equilibrium, is

$$\mathbf{v}^{\mathrm{T}} \exp(t_1\mathbf{Q})\mathbf{L}_{t_1}\left[\prod_{j=2}^{n} \exp\{(t_j - t_{j-1})\mathbf{Q}\}\mathbf{L}_{t_j}\right]\exp\{(t_o - t_n)\mathbf{Q}\}\mathbf{1}, \qquad (6.1)$$

where the $k \times 1$ vector $\mathbf{v}$ is the equilibrium distribution of X_t and $\mathbf{1}$ is a $k \times 1$ vector of ones.

The likelihood (6.1) can be maximised numerically and estimates of $\mathbf{G}$ and the $\mathbf{L}_r$ obtained, together with their standard errors. For numerical purposes it is most convenient to use the spectral representation of $\mathbf{Q}$, namely $\mathbf{A}\mathbf{D}\mathbf{A}^{-1}$, where $\mathbf{A}$ is the matrix whose columns are the right eigenvectors of $\mathbf{Q}$, and $\mathbf{D}$ is the $k \times k$ diagonal matrix of eigenvalues, d_j, of $\mathbf{Q}$. Then

$$\exp(\mathbf{Q}t) = \mathbf{A}\,\mathrm{diag}\{e^{d_1 t},\ldots,e^{d_k t}\}\mathbf{A}^{-1},$$

which is easily calculated. More details are given by Ramesh (1992).

A special case of these models for a single process arises when there are two underlying states, $k = 2$, and events can occur only in one of them, i.e. $\lambda_1 = 0$. The inter-event intervals are then independent, and the points form a renewal process, called a renewal Cox model with Markovian intensity (Smith and Karr, 1983). This model has three parameters, g_{12}, g_{21} and λ_2, and is easily fitted by maximum likelihood.

6.3.2 Second-order properties

The standard statistical properties of series of events are described in Cox and Lewis (1966, Chapter 4), and are derived by Ramesh (1992) for Markov-modulated processes. There are two aspects: counting properties, which concern the numbers of events in a particular interval or intervals; and interval properties, which concern properties of the intervals between successive events, regarded as a time series.

The most basic counting property of N_r is its mean intensity, namely $\mathbf{v}^{\mathrm{T}}\mathbf{L}_r\mathbf{1}$, the average rate of arrival of events. The conditional intensity of N_r is

$$h_r(t) = \lim_{\delta_2,\delta_2 \to 0} \delta_1^{-1}\mathrm{pr}\{N_r(t, t+\delta_1) = 1 \mid N_r(-\delta_2, 0) = 1\} = \frac{\mathbf{v}^{\mathrm{T}}\mathbf{L}_r\mathbf{P}(t)\mathbf{L}_r\mathbf{1}}{\mathbf{v}^{\mathrm{T}}\mathbf{L}_r\mathbf{1}}, \qquad (6.2)$$

where $\mathbf{P}(t) = \exp(t\mathbf{G})$. This is the intensity of arrivals in N_r at time t conditional on the occurrence of an event at time zero; it approaches the mean intensity of N_r as $t \to \infty$. It follows from standard relations (Cox and Lewis, 1966, Section 4.5) that the spectrum of counts for N_r is

$$g_{+r}(\omega) = \frac{1}{\pi}\mathbf{v}^{\mathrm{T}}\{\mathbf{L}_r + \mathbf{L}_r\mathbf{P}^*(\mathrm{i}\omega)\mathbf{L}_r + \mathbf{L}_r\mathbf{P}^*(-\mathrm{i}\omega)\mathbf{L}_r\}\mathbf{1}, \qquad \omega \geqslant 0, \qquad (6.3)$$

where $\mathbf{P}^*(s)$ is the Laplace transform of $\mathbf{P}(t)$. For a homogeneous Poisson process, (6.2) equals $\mathbf{v}^{\mathrm{T}}\mathbf{L}_r\mathbf{1}$, and (6.3) equals $\mathbf{v}^{\mathrm{T}}\mathbf{L}_r\mathbf{1}/\pi$.

A useful tool when discussing the interval properties of N_r is its arbitrary-event initial distribution, $\mathbf{v}_r^* = \mathbf{v}^{\mathrm{T}}\mathbf{L}_r/(\mathbf{v}^{\mathrm{T}}\mathbf{L}_r\mathbf{1})$. This is the stationary distribution that corresponds to the stochastic matrix $\mathbf{P}_r = (\mathbf{L}_r - \mathbf{G})^{-1}\mathbf{L}_r$, the transition probability matrix of the embedded Markov chain of states X_t at the times of successive events in N_r. If we denote by Z_j the intervals between successive events of N_r, starting at an arbitrary event, we find that Z_0 and Z_h have covariance

$$c_h = \mathbf{v}^{*\mathrm{T}}(\mathbf{L}_r - \mathbf{G})^{-1}(\mathbf{P}_r^h - \mathbf{v}_r^*\mathbf{1}^{\mathrm{T}})(\mathbf{L}_r - \mathbf{G})^{-1}\mathbf{1}, \qquad h = 0, 1, \ldots$$

When $\mathbf{P}_r$ is diagonalisable we can use its spectral representation, $\mathbf{P}_r = e_1\mathbf{E}_1 + \cdots + e_k\mathbf{E}_k$, to write

$$\mathbf{P}_r^h = e_1^h\mathbf{E}_1 + \cdots + e_k^h\mathbf{E}_k, \qquad h = 0, 1, \ldots$$

Here $e_1, \ldots, e_k$, the eigenvalues of $\mathbf{P}_r$, are such that $e_1 = 1$, $|e_j| < 1$, $j = 2, \ldots, k$, and the $\mathbf{E}_j$ are formed from the eigenvectors of $\mathbf{P}_r$, with $\mathbf{E}_1 = \mathbf{1}\mathbf{v}_r^{*\mathrm{T}}$. Then the spectrum of intervals for the time series $\ldots, Z_{-1}, Z_0, \ldots$ derived from N_r can be written as

$$f(\omega) = \frac{1}{2\pi}\left(1 + \frac{1}{c_2}\left[\mathbf{v}^{*\mathrm{T}}(\mathbf{L}_r - \mathbf{G})^{-1}\left\{\sum_{j=2}^{k}\left(\frac{2e_j\cos\omega - 2e_j^2}{1 - 2e_j\cos\omega + e_j^2}\right)\mathbf{E}_j\right\}(\mathbf{L}_r - \mathbf{G})^{-1}\mathbf{1}\right]\right), \qquad (6.4)$$

where c_2 is the marginal variance of the series. This is just the usual spectral density of time-series analysis, which in this context is flat when N_r is a renewal process.

The interval properties for two related point processes are not generally useful because of the difficulties of relating intervals in one series to those in another. The most directly interpretable of the counting properties is the cross-intensity function for the processes N_r and N_s (Cox and Isham, 1980, Section 5.2)

$$h_{r|s}(t) = \lim_{\delta_1, \delta_2 \to 0} \delta_1^{-1}\mathrm{pr}\{N_r(t, t+\delta_1) = 1 \mid N_s(-\delta_2, 0) = 1\} = \frac{\mathbf{v}^{\mathrm{T}}\mathbf{L}_s\mathbf{P}(t)\mathbf{L}_r\mathbf{1}}{\mathbf{v}^{\mathrm{T}}\mathbf{L}_s\mathbf{1}}, \qquad r \neq s \qquad (6.5)$$

which converges to $\mathbf{v}^T\mathbf{L}_r\mathbf{1}$ as $t \to \infty$. Expression (6.5) represents the intensity of arrival of points in N_r, a period t after the occurrence of an event in N_s. Other counting properties of the bivariate process may be derived in terms of the components of $\mathbf{G}$ and the $\mathbf{L}_r$.

6.4 DATA ANALYSIS

6.4.1 Separate analyses

We first analyse the point processes at individual receptors separately. There is no pattern to the autocorrelations of the inter-event intervals, which strongly suggests that we may regard them as renewal processes. This impression is strengthened when separate two-state Markov-modulated Poisson processes are fitted, each with the four parameters g_{12}, g_{21}, λ_1, λ_2. The likelihood ratio statistic to test whether λ_1 equals zero is significant at the 5% level only for receptors 300 km north and 800 km south-east, and we conclude that the best-fitting two-state Markov-modulated Poisson process is the renewal Cox model. Figure 6.3 shows the conditional intensity functions (6.2) and spectral densities for successive inter-event intervals (6.4) for two typical series, confirming the impression of a good fit.

In order to understand how to proceed with a more comprehensive model, we plot in Figure 6.4 the parameter estimates for the renewal processes, together with the fitted stationary probability for state 2, $\hat{v}_2$. Regression of the estimates confirms the information in the figure, that each parameter depends on distance from the source and on direction, represented by $\sin \alpha_r$ and $\cos \alpha_r$, where α_r is the direction from the source to the receptor, measured as in an Argand diagram. The dependence of λ_2 on the covariates may be spuriously induced by the large rate at the most northerly receptor. The figure suggests that distance does not reduce the arrival rate in state 2, but instead that it reduces the probability that the process is in state 2. Thus clusters are equally intense while points are arriving, but of shorter duration further from the source.

6.4.2 Joint analysis

If we were interested purely in the dynamics of clustering at single receptors, we could fit regression models to the estimates shown in Figure 6.4, with due allowance for the differing numbers of events on which they are based, and stop. Such a model would have 12 parameters or so, with spuriously small standard errors because different receptors are regarded as independent when they are not, but this would probably be immaterial compared to the nonrandom errors from the crudeness of the representation of rainfall in the computer model. However, this statistical model would not account for the clustering across

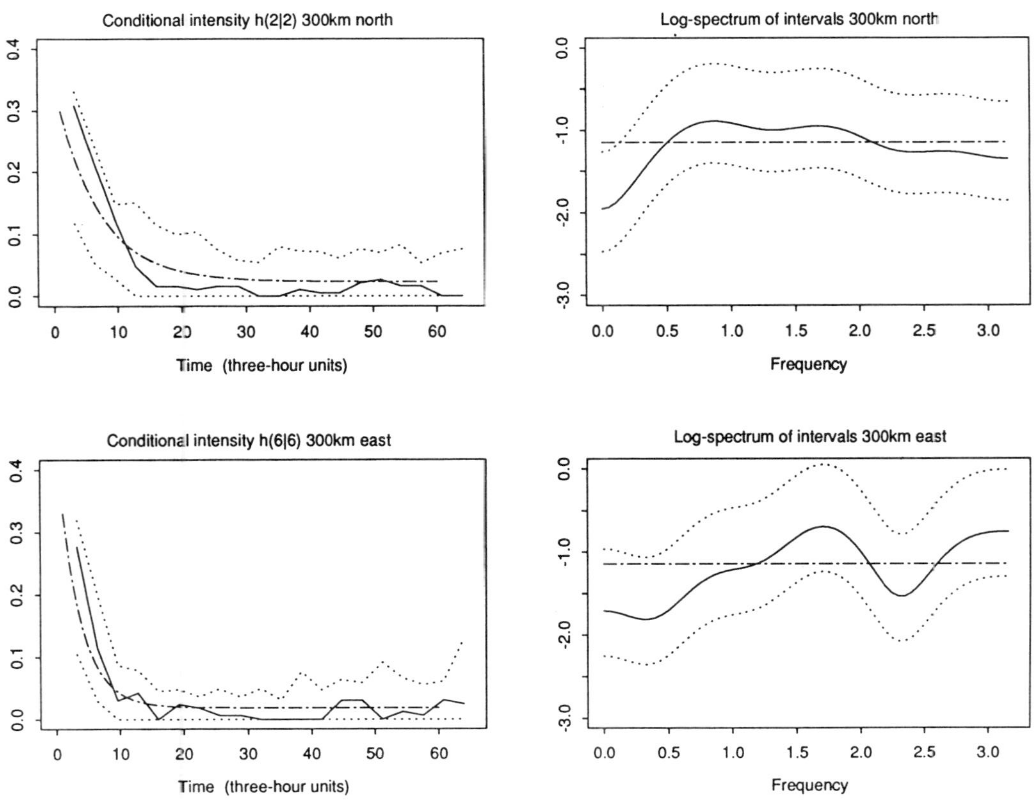

Figure 6.3 Conditional intensities $h_{r|s}(t)$ and estimated log-spectra of intervals at receptor 2, 300 km north, and at receptor 6,300 km east, for the renewal Cox models fitted separately. *In each case the dot-dashed line is the fitted curve, the solid line is the empirical curve, and the dotted lines are approximate pointwise 90% confidence bands for the true curve*

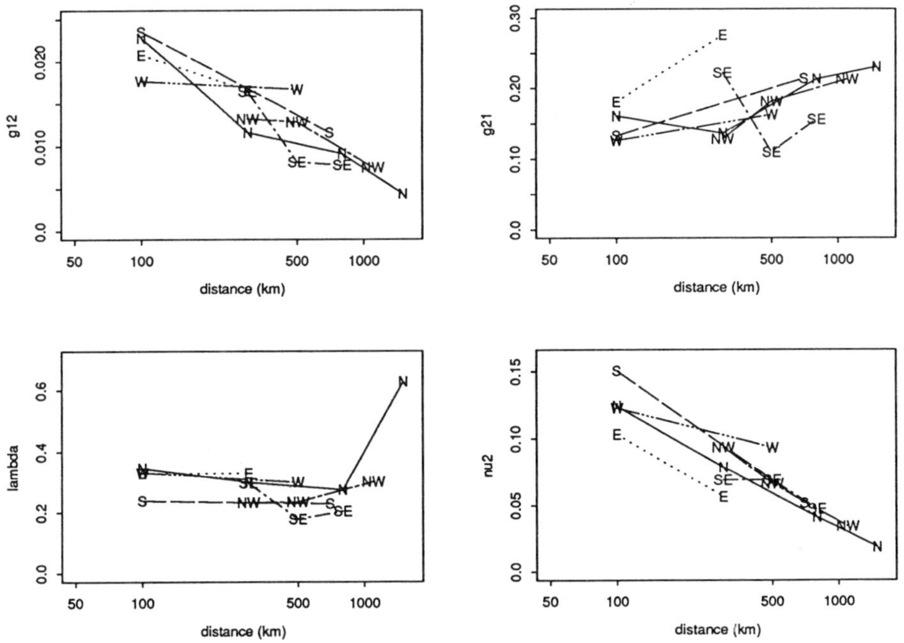

Figure 6.4 Parametes for separate renewal Cox processes at each receptor. *The upper panels are for the transition probabilities, q_{12} and g_{21}, the lower panels are for the rate, λ,* and the stationary probability for the state in which exposures can occur, ν_2

receptors—and here we hit a snag. The multivariate models described in Section 6.3 have the property that sojourn times in the states of the underlying process are the same throughout, but that events occur with different intensities in different processes N_r. This would correspond to dependence of λ_2 but not g_{12} and g_{21} on the covariates in Figure 6.4, and suggests that the model in Section 6.3 is unlikely to fit well. On the other hand Markov-modulated Poisson processes are fairly flexible, so we proceed, in hope rather than anticipation.

Our initial model has $k = 2$ states. Exposures to wet deposition cannot arise in state 1, so $\lambda_{1,r} \equiv 0$, whereas in state 2 they can arise at a receptor at distance d_r from the source and angle α_r with rate $\lambda_{2,r} = \exp(\beta_0 + \beta_1 \log d_r + \cos\alpha_r + \beta_3 \sin\alpha_r + \beta_4 \mathrm{d}_r)$. Under this model, exposure times at individual receptors form renewal processes. Power-law spread of a plume with distance is a common assumption in models for atmospheric dispersion over relatively short distance; we initially include a term in d_r in addition to the term in $\log d_r$ to account for possible attenuation of dispersion at longer distances. Crude directional effects are incorporated through the trigonometric terms. If data were available at more receptors, one might try to fit models of generalised additive type (Hastie and Tibshirani, 1990) to explain the distance and directional effects, but this would not be not worthwhile with the sixteen sites available here.

The likelihood ratio statistic to test whether $\beta_4 = 0$ takes value 0.02: clearly the term in d is redundant. When this term is dropped we have a model labelled M_1, whose maximum likelihood parameter estimates are given in Table 6.2. These correspond to mean sojourn times of 41 and 16 hours in states 1 and 2 respectively. The power-law decay of the rates with distance from the source, $d^{-0.55}$, is close to that found by Davison (1984) by an entirely different approach. The values of β_2 and β_3 correspond to a peak of exposures roughly equidistant between the major and minor peaks of the wet-weather wind-rose at Ispra during 1976.

The fitted numbers of exposures under M_1 are given in Table 6.1. On the whole the fit is very good, though the model underpredicts at receptor 1. However, comparison of the observed conditional intensities with those fitted under M_1 suggests that this model is not clustered enough and that we should explore further. We can allow for the evidence of Figure 6.4 that there are different values of the stationary probability for state 2, v_2, at different distances, by setting up models with the same rate parameter for each state, but with exposures occurring with lower probability further from the source. Our model allows no exposures in state 1, no exposures further than 400 km from the source in state 2, and exposures everywhere in state 3. Transitions are only possible between neighbouring states, with parameters: $g_{12}, g_{21}, g_{23}, g_{32}$; $\lambda_{1,r} \equiv 0$ and $\lambda_{2,r} = \lambda_{3,r} = \exp(\beta_0 + \beta_1 \log d_r + \beta_2 \cos \alpha_r + \beta_3 \sin \alpha_r)$. We call this model M_2.

The null hypothesis for comparison of M_1 and M_2 puts g_{23} and g_{32} on the boundary of the parameter space, and the appropriate asymptotic null distribution for the corresponding likelihood ratio statistic is $\frac{1}{4} + \frac{1}{2}\chi_1^2 + \frac{1}{4}\chi_2^2$ (Self

Table 6.2 Summaries of Markov-modulated Poisson process models fitted to exposure-time data

Parameter	M_1 Estimate (s.e.)	M_2 Estimate (s.e.)	M_3 Estimate (s.e.)
g_{12}	0.073 (0.026)	0.108 (0.018)	0.050 (0.007)
g_{21}	0.190 (0.082)	1.91 (0.039)	0.118 (0.014)
g_{23}	—	2.99 (0.870)	0.023 (0.012)
g_{32}	—	0.464 (0.091)	0.125 (0.048)
β_0	−2.05 (0.070)	−2.06 (0.063)	−2.72 (0.072)
β_1	−0.546 (0.042)	−0.479 (0.042)	−0.527 (0.054)
β_2	−0.130 (0.057)	−0.078 (0.056)	−0.372 (0.077)
β_3	0.080 (0.055)	0.105 (0.053)	0.039 (0.072)
γ_0	—	—	−1.18 (0.108)
γ_1	—	—	−0.551 (0.072)
γ_2	—	—	0.042 (0.010)
γ_3	—	—	0.027 (0.099)
2 × log-likelihood	−7784.6	−773.7	−7536.6

and Liang, 1987), where χ_k^2 denotes the chi-squared distribution on k degrees of freedom. This likelihood ratio statistic equals 10.9 and is convincing evidence of the need for a third state. Inclusion of the third state does not change the estimates of the β_j substantially, as Table 6.2 shows. Fitted stationary probabilities for the states are 0.70, .04 and 0.26 respectively, with corresponding mean sojourn times 28, 0.6 and 6.5 hours. Thus state 2 is rarely visited, and only for a short time, on average. The direction of highest exposures implied by the trigonometric terms is close to that for M_1.

Table 6.1 shows that M_2 gives better fitted counts than does M_1, except for receptors west of the source. But like M_1, it does not model the conditional intensities well. We must look further for a good model of the dynamic aspects of the data.

The improvement from M_1 to M_2 and the sharp decrease with distance in the fourth panel of Figure 6.4 suggests that a fourth state in which exposures can only occur 100 km from the source might be sensible, but a problem crops up. Likelihood estimation procedures for this model converge to parameter values corresponding to very high values of λ and a mean sojourn time of the order of a few minutes for one of the states. This implies that single points are treated as a cluster, giving a spike in the likelihood, and implies that models have to be chosen with care. Similar problems can arise in the more prosaic setting of likelihood fitting of mixture models to random samples. Our models are mixtures of Poisson processes, so with hindsight we see that we might have anticipated this problem.

A variety of other three- and four-state models were fitted, most of which could not model accurately the observed conditional intensities. Eventually we settled on the three-state model M_3, in which events cannot occur in state 1, occur at the rth receptor with rate $\lambda_{2,r} = \exp(\beta_0 + \beta_1 \log d_r + \beta_2 \cos\alpha_r + \beta_3 \sin\alpha_r)$ in state 2, and occur at the rth receptor with rate $\lambda_{3,r} = \exp(\gamma_0 + \gamma_1 \log d_r + \gamma_2 \cos\alpha_r + \gamma_3 \sin\alpha_r)$ in state 3. Likelihood ratio tests indicate that all the parameters in this model are needed; it has a likelihood roughly 120 higher than M_2. The parameter estimates and their standard errors are given in Table 6.2. The stationary probabilities of the states are 0.67, 0.28 and 0.05, with corresponding mean sojourn times 60, 21 and 24 hours. Here state 3 is rarely visited, but the average time spent in it is about one day. The rates of occurrence are highest in state 3, and in both states they decline with distance from the source at about the same rate as in M_1. The state-2 directional effect is in roughly the same direction as the major peak in the wet-weather wind-rose at the source. Curiously, the state-3 directional effect is strongest in the direction opposite to the minor peak of this wind-rose. Neither of these directional effects is very well-determined by the data, however.

The fitted counts for M_3 given in Table 6.1 show that the marginal aspects are somewhat less well fitted by this model. In particular, mean exposure rates at three of the receptors closest to the source are considerably worse than for M_1 or M_2. The fit elsewhere is fairly comparable. None of the models accounts for the increase in exposure rate 1500 km north of the source.

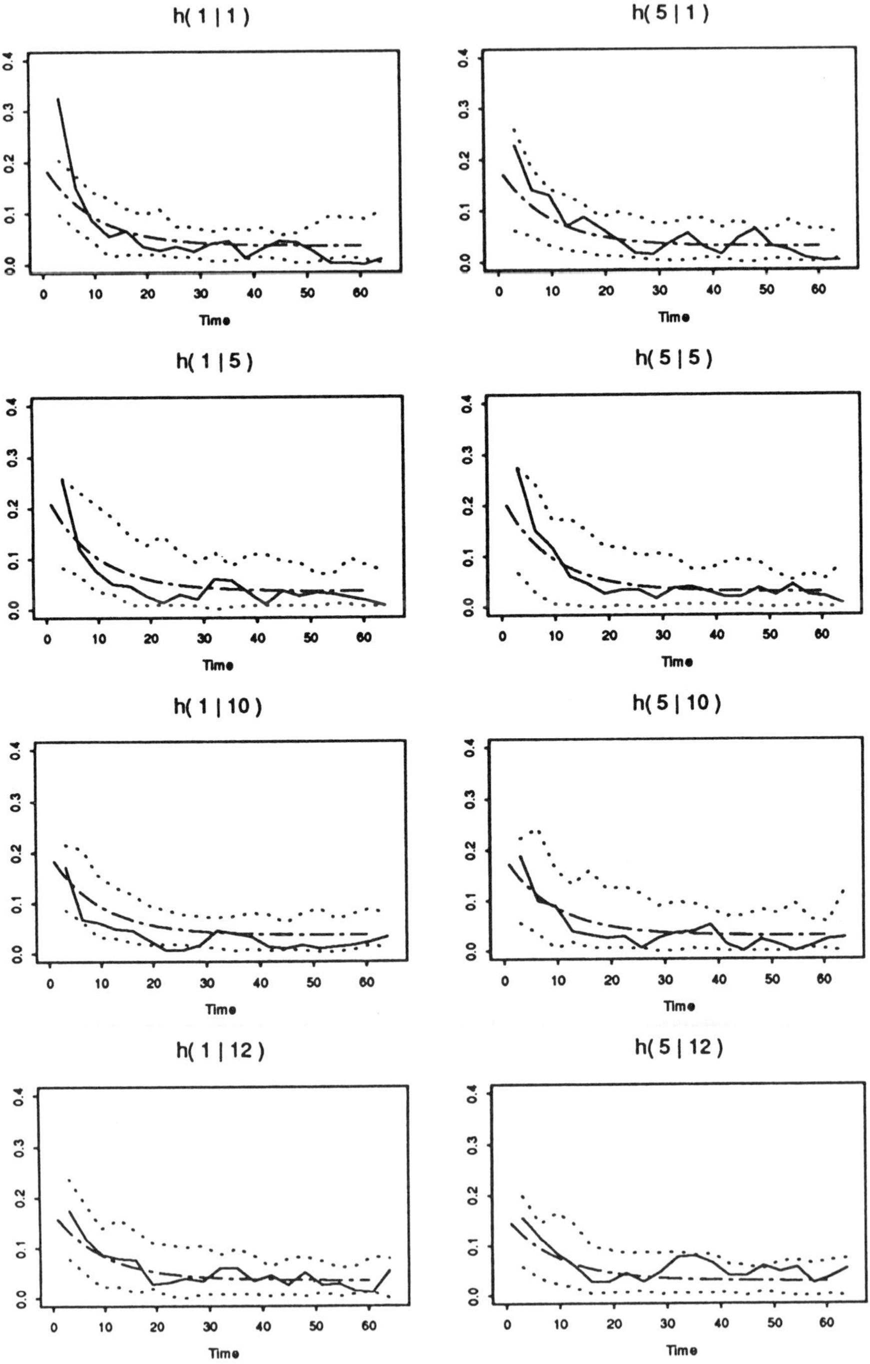

Figure 6.5 Matrix plot of conditional intensities $h_{r|s}(t)$ at the four nearest receptors for the best three-state model, M_3. *In each case the dot-dashed line is the fitted curve, the*

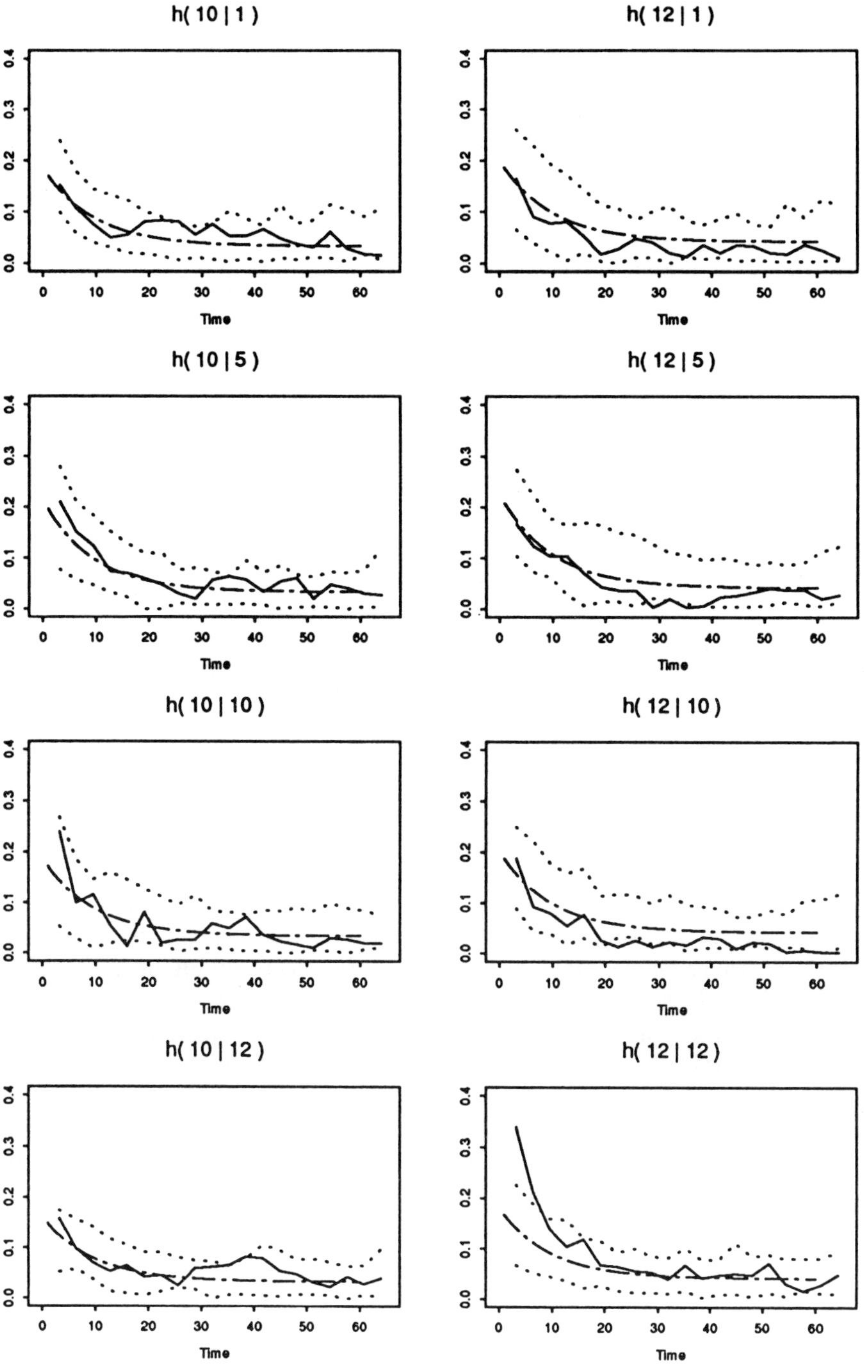

solid line is the empirical curve, and the dotted lines are approximate pointwise 90% confidence bands for the true curve

Figure 6.5 shows the observed and fitted conditional intensity functions $h_{r/s}(t)$ for the four receptors closest to the source. The confidence bands are simulation envelopes generated as the pointwise maximum and minimum conditional intensities for 39 simulations from the fitted processes, and are narrower than overall bands for the true conditional intensities would be. The fitted intensities, particularly $h_{1|1}(t)$ and $h_{12|12}(t)$, lie systematically below the observed ones for small t, but the broad pattern is adequate otherwise, and this is confirmed by plots for other subsets of the receptors.

The twelve-parameter model M_3 seems on the whole to fit the data adequately. It improves on the model described in Section 6.4.1 by modelling the joint structure of the exposure times without increasing the number of parameters. Once a distribution has been fitted to the exposure sizes, return levels at receptors for which data are not available can be estimated by Monte Carlo simulation from the fitted model, thus taking into account the effect of clustering on exposure levels.

6.5 DISCUSSION

Environmental data are often aggregated. In our application, for example, the occurrence or not of exposures at each receptor was in fact recorded over three-hour intervals, leading to multiple occurrences at the same epoch. We dealt with this empirically by adding small random displacements to the exposure times in order to obtain an orderly point process. A better but more complicated approach is to model exposure times as a binary time series. Ramesh (1992) describes the properties of such series based on aggregated Markov-modulated Poisson processes, for which likelihood and second-order calculations may be performed in a manner similar to that described above. Pilot analysis suggests that these models fit the data somewhat better than do unaggregated processes.

Full, partial, or covariate information may be available related to the underlying process X_t. In principle time-varying covariate effects might be included by allowing the generator $\mathbf{G}$ of X_t to vary between intervals—allowing seasonal effects by having a different $\mathbf{G}$ for each month, for example. If there is a well-established physically based stochastic model for the X_t, and if it is possible to simulate data from that model, it seems more fruitful to use likelihood estimation methods based on Markov-chain Monte Carlo (Hastings, 1970) to estimate the associated parameters. A good deal more must be known about such methods before they can be used reliably in problems such as that outlined above.

ACKNOWLEDGMENTS

This work was supported by the Science and Engineering Research Council Complex Stochastic Systems Initiative.

PART III

Climatological and Meteorological Issues

7

Long-range Dependence and Global Warming

Richard L. Smith,
University of North Carolina, USA

7.1 INTRODUCTION

One of the scientific issues connected with 'global warming' is the detection of trends in observed series of temperature data. This is important for providing evidence that global warming has actually occurred during the 20th century, but also and perhaps more crucially as a means of validating general circulation models (GCMs) which are the main means of predicting future temperature rises that will occur if the emission of greenhouse gases is allowed to continue unhindered. However, the interpretation of trends in observed temperature data is complicated by numerous factors, including the fact that measured trends are different for different parts of the Earth's surface, and that they are different for different time periods. For example, global temperatures showed a steady rise from about 1910 to 1940, but there followed a period when temperatures were static or even decreasing until about 1975, since when a further rise has taken place. These kinds of effects raise the possibility that the observed temperature fluctuations are merely part of long-term natural variability and cannot be taken as indicative of any man-made activity.

In this chapter, we address these issues from the point of view of time series with long-range dependence, i.e. a spectral density which is proportional to ω^{1-2H} as $\omega \to 0$ where ω is frequency and the parameter H (the Hurst coefficient) is between 0.5 and 1. A method of estimating H directly from the periodogram is proposed and applied to the temperature data. The effect of such dependence on trend estimates is then analysed.

Statistics for the Environment. Edited by Vic Barnett and K. Feridun Turkman

The methods are applied to a number of series, (a) the Central England temperature data of Manley (1974), (b) US data from the Historical Climatological Network (HCN), and (c) global averages from the report of the Intergovernmental Panel on Climate Change (IPCC). The proposed method is compared with a more straightforward technique based on high-order autoregressive series. In all cases the standard error appears to be higher when calculated under the long-range dependence model than under an autoregressive model. However, the effect is different for different series, and greatest for the IPCC global data. The results for the Central England and US data do not show significant global warming over the 20th century. For the global data, the results are consistent with other statistical studies that have claimed that the global rise is significant, but they do confirm the importance of taking long-range dependence into account.

7.2 GLOBAL WARMING AND THE DETECTION OF TRENDS IN TEMPERATURE SERIES

We begin by reviewing some basic facts about global warming. These are taken from numerous recent publications of which a comprehensive summary is in the report of the Intergovernmental Panel on Climate Change (1990), henceforth IPCC. There is a natural greenhouse effect which keeps the Earth warmer than it otherwise would be. Human activities are substantially increasing the proportions in the atmosphere of a group of gases collectively known as greenhouse gases. These include carbon dioxide, methane, chlorofluorocarbons (CFCs) and nitrous oxide. Some gases are more effective than others at changing climate, but carbon dioxide is generally considered the most important single factor. For this reason, the level of carbon dioxide is widely use as a measure of the concentration of the greenhouse gases in the atmosphere.

A major scientific issue connected with global warming is to determine the magnitudes of the temperature effects resulting from an actual or hypothesised increase in greenhouse gases. To facilitate this, atmospheric scientists have created a number of mathematical models collectively known as general circulation models (GCMs). These try to take account of the many interactions between the physical and dynamical processes in the global climate. However, many of the descriptions of these processes are quite crude and consequently there is substantial uncertainty attached to the predictions of the GCMs.

The results of GCMs are often compared in terms of their predictions of the rise in temperature associated with a doubling of atmospheric carbon dioxide (the climate sensitivity). This is an artificial criterion—for example, it assumes equilibrium conditions, which certainly will not be attained for a very long time. Nevertheless, it serves for making comparisons. The range of results from current models for this scenario is from 1.9 to 5.2 °C. Most results are close to 4.0 °C. After taking into account all the model results and comparisons with observational evidence over the last century (which has tended to support values in the lower portion of this range), the IPCC adopted the value of 2.5 °C as its best estimate of the climate sensitivity. When translated into actual predictions of global warming on the assumption that greenhouse gases continue to rise at the currently projected rate (the 'business-as-usual scenario'), this leads to a prediction of an overall global warming of 1.8 °C by 2030 and 4 °C by 2100.

Observed temperature series thus play an important role in calibrating the GCMs. There is evidence of an actual rise in temperatures in the range 0.3–0.6 °C over the last century. However, the rise has not been steady. Much of the warming since 1900 has been concentrated in two periods, the first between 1910 and 1940 and the second since 1975. During the period 1940-75 the Northern Hemisphere was actually cooling whereas the Southern Hemisphere remained at a constant temperature; some regions have continued to cool until very recently. This substantial temporal and spatial variability shows that future temperature changes are likely to differ substantially from a global average. 'Global mean temperature alone is an inadequate indicator of greenhouse-gas-induced climate change. Identifying the causes of any global mean temperature change requires examination of other aspects of the changing climate, particularly its spatial and temporal characteristics.... Any changes to date could be masked by natural variability and other (possibly man-made) factors, and we do not have a clear picture of these' (IPCC 1990, p. xxix).

From these remarks we can see that a major *statistical* issue connected with global warming is the separation of effects due to natural variability, such as may for example be represented by a stationary time-series model, and those that are due to a trend in the data. In this chapter we consider particularly how these issues will work out in the case of *long-range dependence*, which may be represented by a power-law spectral density proportional to ω^{1-2H} as $\omega \to 0$ where ω is frequency and the parameter H (known as the Hurst coefficient after the pioneering study of the River Nile by the English physicist H. E. Hurst in the 1950s) is between 0.5 and 1. The correlations from such a process decrease with lag k at a rate k^{2H-2}, i.e. a polynomial decrease in k compared with the exponential decrease that occurs with more familiar processes such as the ARMA family.

In the present chapter we consider methods for estimating H based on the spectral density (developed in more theoretical detail in Samarov *et al.* (1992)), and consider the effects of long-range dependence on the estimation of a linear trend.

7.3 LONG-RANGE DEPENDENCE AND THE ESTIMATION OF A LINEAR TREND

In this section we review some basic relations among the autocovariances, spectral density and variances of sample means of a stationary time series, and derive an extension to the asymptotic variance of the least squares estimator of the slope of a linear regression.

Let $\{Y_n, n = 1, \ldots, N\}$ denote a sample of N observations from a stationary time series with mean 0. Let $\gamma_k = E\{Y_n Y_{n+k}\}$ denote the kth autocovariance and $f(\omega)$, $-\pi < \omega < \pi$, the spectral density. The relations between the autocovariances and the spectral density are

$$f(\omega) = \frac{1}{2\pi} \sum_{n=-\infty}^{\infty} \gamma_n e^{-in\omega} = \frac{\gamma_0}{2\pi} + \frac{1}{\pi} \sum_{n=1}^{\infty} \gamma_n \cos(n\omega), \tag{7.1}$$

$$\gamma_n = \int_{-\pi}^{\pi} e^{in\omega} f(\omega) d\omega = \frac{1}{2} \int_0^{\pi} \cos(n\omega) f(\omega) d\omega. \tag{7.2}$$

The spectral density at a particular frequency ω may be estimated by means of the periodogram

$$I_N(\omega) = \frac{2}{N} \left| \sum_{n=1}^{N} Y_n e^{in\omega} \right|^2 = \frac{2}{N} \left[\sum_{n=1}^{N} Y_n^2 + 2 \sum_{k=1}^{N-1} \sum_{m=1}^{N-k} Y_m Y_{m+k} \cos(k\omega) \right] \tag{7.3}$$

whose mean

$$2\gamma_0 + 4 \sum_{k=1}^{N-1} \left(1 - \frac{k}{N}\right) \gamma_k \cos(k\omega)$$

converges as $N \to \infty$ to

$$2\gamma_0 + 4 \sum_{k=1}^{\infty} \gamma_k \cos(k\omega) = 4\pi f(\omega). \tag{7.4}$$

The formulae (7.1)–(7.4) have been spelt out because there seems to be no unanimity about the definition of either the spectral density or the periodogram, published formulae often differing from the above by factors of 2π or 4π. They are, however, consistent with those in Olshen (1967) and Cox (1984), and we take these as our starting-point.

Suppose now we have long-range dependence, which we may define by either of the relations

$$\gamma_k \sim ak^{2H-2}, \quad k \to \infty, \tag{7.5}$$

$$f(\omega) \sim b\omega^{1-2H}, \quad \omega \downarrow 0 \tag{7.6}$$

where $\frac{1}{2} < H < 1$. The relations (7.5) and (7.6) are also meaningful for other values of H outside this range, but it is for these value of the parameter H that a process is stationary with long-range dependence and it is therefore on these that we concentrate.

The relation between the positive constants a and b is

$$b = \frac{a}{\pi}\Gamma(2H-1)\sin(\pi - \pi H) \tag{7.7}$$

which may be derived from the asymptotic relationships

$$\sum_{n=1}^{\infty} n^{-\beta}\cos(nx) \sim x^{\beta-1}\Gamma(1-\beta)\sin(\pi\beta/2),$$

$$\sum_{n=1}^{\infty} n^{-\beta}\sin(nx) \sim x^{\beta-1}\Gamma(1-\beta)\cos(\pi\beta/2),$$

both valid as $x \downarrow 0$ for $0 < \beta < 1$ (Zygmund, 1959, Section V.2). See also Beran (1989, pp. 261–2), after correcting for a factor 2π.

One problem that has been studied in this area is the effect of long-range dependence on the estimation of a sample mean $\bar{Y}_N = (1/N)\sum_{n=1}^{N} Y_n$. We have

$$\operatorname{var} \bar{Y}_N = \frac{\gamma_0}{N} + \frac{2}{N}\sum_{k=1}^{N-1}\left(1 - \frac{k}{N}\right)\gamma_k.$$

Under (7.5) we have, for large N,

$$\begin{aligned}\sum_{k=1}^{N}\left(1 - \frac{k}{N}\right)\gamma_k &\sim \frac{a}{N}\int_0^N (N-x)x^{2H-2}\mathrm{d}x \\ &= \frac{aN^{2H-1}}{2H(2H-1)}\end{aligned}$$

from which

$$\operatorname{var} \bar{Y}_N \sim \frac{a}{H(2H-1)}N^{2H-2} = \frac{b\pi}{H\Gamma(2H)\sin(\pi - \pi H)}N^{2H-2}. \tag{7.8}$$

This equation is the same as the formulae on pages 192 and 194 of Samarov and Taqqu (1988), which are given in a different form in Beran (1989).

In the present chapter our focus is on an extension of this idea, to the case when $\{Y_n\}$ is regressed on a covariate $\{x_n\}$. the ordinary least squares estimator of the slope β is

$$\hat{\beta} = \frac{\sum Y_n(x_n - \bar{x}_N)}{\sum (x_n - \bar{x}_N)^2}.$$

For simplicity we shall assume $\bar{x}_N = 0$. The variance of $\hat{\beta}$ is then

$$\left(\sum_{n=1}^{N} x_n^2\right)^{-2}\left\{\gamma_0 \sum_{n=1}^{N} x_n^2 + 2\sum_{k=1}^{N-1}\gamma_k \sum_{m=1}^{N-K} x_m x_{m+k}\right\}.$$

Let us specialise to the case of a linear trend, $x_n = n - (N+1)/2$, $1 \leqslant n \leqslant N$. We then approximate

$$\sum x_n^2 \sim \int_0^N \left(x - \frac{N}{2}\right)^2 \mathrm{d}x = \frac{N^3}{12}$$

and

$$\sum_{k=1}^{N-1}\gamma_k \sum_{m=1}^{N-k} x_m x_{m+k} \sim a\int_0^N x^{2H-2}\int_0^{N-x}\left(y - \frac{N}{2}\right)\left(x + y - \frac{N}{2}\right)\mathrm{d}y\,\mathrm{d}x.$$

After some manipulation the right-hand integral can be shown to equal

$$\frac{a(1-H)}{8H(1+H)(2H-1)} N^{2H+2}$$

and it follows that

$$\operatorname{var}\hat{\beta} \sim \frac{36a(1-H)}{H(1+H)(2H-1)} N^{2H-4} = \frac{36b\pi(1-H)}{H(1+H)\Gamma(2H)\sin(\pi - \pi H)} N^{2H-4}. \quad (7.9)$$

The formulae (7.8) and (7.9) may also be written n the form

$$\operatorname{var}\bar{Y}_N \sim \theta_1 b N^{2H-2}, \quad \operatorname{var}\hat{\beta} \sim \theta_2 b N^{2H-4} \quad (7.10)$$

where the constants θ_1 and θ_2 are given (as a function of H) in Table 7.1. Note that, although the value of θ_1 becomes infinite as $H \to 1$, the value θ_2 remains well-defined for $H \geqslant 1$. It is not clear whether the result is meaningful in this case—the derivation just given is not then valid, since the processs is nonstationary and neither the autocovariances nor the spectral density are well-defined.

The preceding analysis is a special case of a more general result due to Yajima (1988) in the case of polynomial regression and Yajima (1991)for general regressors. In particular it can be shown, with some work, that the formula (7.9) is a special case of Theorem 2.2 of Yajima (1988). The derivation given here has the advantage of being a direct calculation. Yajima (1988) also considered the efficiency of the least squares estimate (LSE) as compared with the best linear unbiased estimate (BLUE). We have recomputed his Table 7.1 to consider the asymptotic efficiency (expressed as a ratio of variances of the

Table 7.1 Variation of θ_1 and θ_2 with H

H	θ_1	θ_2
0.5	6.28	75.40
0.55	6.08	63.54
0.6	6.00	53.97
0.65	6.04	46.16
0.7	6.25	39.72
0.75	6.68	34.38
0.8	7.48	29.91
0.85	8.96	26.15
0.9	12.13	22.98
0.95	21.98	20.29
1.0		18.00
1.1		14.38
1.2		11.74
1.3		9.81
1.4		8.44
1.5		7.54

BLUE to LSE) of $\hat{\beta}$. Efficiencies were obtained of 1, 0.991, 0.969, 0.942, 0.915 and 0.889 for $H = 0.5$ to 1 in increments of 0.1 (corresponding to $d = 0, 0.1, \ldots, 0.5$ in Yajima's table). These seem to to confirm that the efficiency is near 1 so that there is no need to do the extra work required to compute the BLUE. On the other hand, taking the opposite viewpoint, these figures also quantify the potential benefit of a fully model-based approach against the approach adopted in this chapter.

7.4 CENTRAL ENGLAND TEMPERATURE DATA

In this section we consider a monthly series of mean temperatures in Central England for 1659–1973, collected by Manley (1974), and updated to 1989. It should be pointed out that there are numerous sources of possible nonstationarity due to changes in measurement methods, and in particular, Manley drew attention to the fact that the data values before 1723 were much less reliable than later values. However, in this analysis we consider the original data set as reported by Manley.

Figure 7.1 shows a plot of the whole data set. Data are smoothed by calculating a 49-month centred moving average; it is this moving average that is plotted. The most obvious feature is the remarkable dip in temperature around 1700. In the light of the reservations just mentioned, there must be some doubt about the accuracy of the precise values, though presumably not about the fact that there was a spell of very cold weather at that time. At the opposite end, there is some suggestion of a temperature rise over the period 1920–1960, but the evidence for that is not clear-cut.

Figure 7.2 shows *monthly* values for 1907–73. In this case the seasonal effect was eliminated by subtracting, from the data for each month of the year, the overall average for that month. There is no visual evidence of any trend.

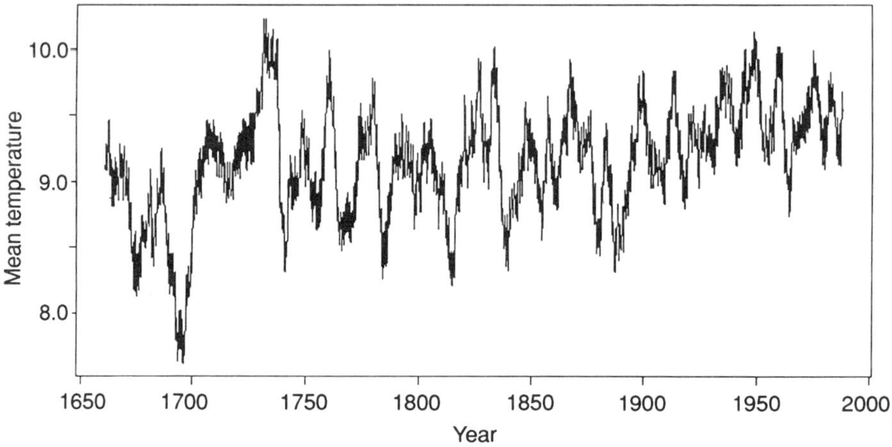

Figure 7.1 Smoothed plot of annual mean temperatures for Central England data series (°C), 1659–1989

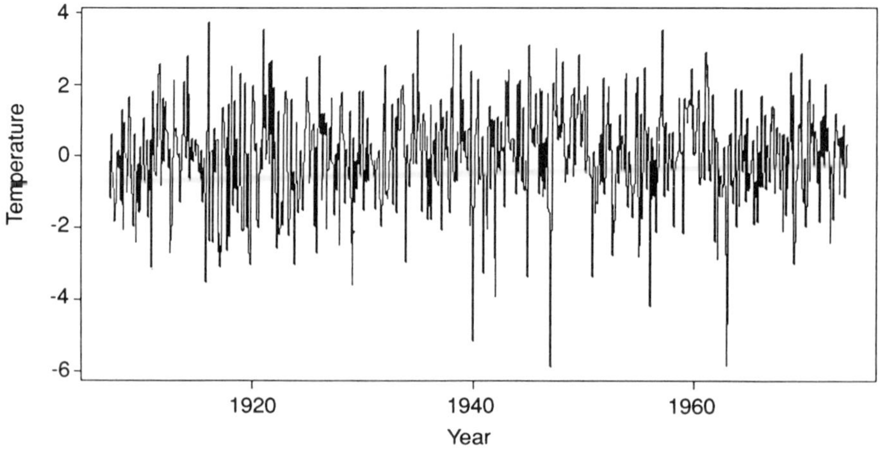

Figure 7.2 Plot of monthly mean temperatures for Central England data series expressed as deviations from overall monthly means (°C), 1907–1973

In the spirit of Section 7.2 we can consider a model of the form

$$Y_n = \alpha + \beta x_n + \eta_n \tag{7.11}$$

where Y_n is the observed value in month n and $x_n = (n - 402.5)/1200$. The value 402.5 was subtracted from n so that the $N = 804$ values of x_n are centred about 0, and the result was divided by 1200 to express it as a change in temperature per century (the most widely used unit). The residual η_n may be taken as a stationary time series. A natural short-range dependent model for $\{\eta_n\}$ is the AR(p) model:

$$\eta_n = \sum_{j=1}^{p} a_j \eta_{n-j} + \varepsilon_n \tag{7.12}$$

where $\{\varepsilon_n\}$ is white-noise sequence which we take to be independent $N(0, \sigma^2)$ for unknown σ^2. The simplest way to fit the model defined by (7.11) and (7.12) is by conditional maximum likelihood: for some $p_0 \geqslant p$ use equations (7.11) and (7.12) to express $\{\varepsilon_n, n > p_0\}$ as a function of the data and parameter values, and then minimise $D = \sum_{j=p_0+1}^{N} (\varepsilon_j/\sigma)^2 + (N - p_0)\log(\sigma^2)$ where N is total sample size and D is twice the negative log-likelihood, or deviance. In the analysis to follow we consider values of p up to 10 and $p_0 = 10$, the value of p_0 being fixed so as to make direct comparison between fits for different values of p. This method is called conditional maximum likelihood because the likelihood being calculated is the conditional distribution of the last $N - p_0$ data values given the first p_0.

A simple estimate assuming independence (i.e. $p = 0$) produced a value $\bar{\beta} =$ 0.275 (°C per century) with standard error 0.17. This is reasonably consistent with the global-warming of 0.3–0.6, but the standard error casts considerable doubt on the significance of this. This is confirmed when higher value of p are fitted: for $p = 1, 2, 3$ the estimates of β are 0.274, 0.274, 0.273 to three decimal places, with standard errors 0.22, 0.22, 0.23. There is a sharp drop in deviance when passing from AR(0) to AR(1), but only a small drop thereafter, and the Akaike information criterion selects $p = 1$ as the most suitable value.

7.5 LONG-RANGE DEPENDENCE IN ENGLISH DATA

We now consider the effect of possible long-range dependence on these data. For the purpose of the present Section, it was considered desirable to analyse as long a series as possible. Therefore the full data set (up to 1973) was used without any allowance for possible nonstationarity. In further analysis, it would be preferable to consider the robustness of the following analysis against various forms of departure from stationarity.

The first step was to form the periodogram (7.3). This was evaluated at the

Fourier frequencies $\{\omega_j = 2\pi j/N, \ 1 \leqslant j \leqslant N/2\}$. In Figure 7.3, the first 500 log periodogram ordinates are plotted against frequency, and in Figure 7.4 against log frequency. There is a prominent spike at frequency 1/12 (the annual seasonal effect) but no other indication of deterministic seasonality. In addition, the behaviour of the log periodogram near frequency 0 does suggest long-range dependence.

According to well-known theory, $I_n(\omega_j)$ for $j = 1, 2, \ldots$ are approximately independent exponential random variables with means $4\pi f(\omega_j)$. When combined with the assumption(7.6) this suggests a regression of $\log I_n(\omega_j)$ on $\log \omega_j$ for $1 \leqslant j \leqslant n$, where $n \ll N$. Such a procedure was proposed by Geweke and Porter-Hudak (1983). Subsequent work (Samarov *et al.*, 1992) has suggested the following refinements of the Geweke–Porter-Hudak procedure:

(a) Geweke and Porter-Hudak suggested fitting the regression by least squares. However, the approximate exponentiality of periodogram ordinates suggests that a superior method would be a maximum likelihood fit based on the Gumbel error distribution function $1 - \exp(-e^x)$, which arises from the logarithm of a standard exponential random variable.
(b) The value of n needs to be much smaller than N and carefully chosen to balance bias and variance.
(c) In addition, it may be desirable to restrict the regression to Fourier frequencies $m \leqslant j \leqslant n$ with $m > 1$. This was suggested by Figure 7.5 and the accompanying discussion in Haslett and Raftery (1989), in which the linearity of the plot of log periodogram against log frequency broke down in the very smallest ordinates.

As an indication of the sensitivity of the results to m and n, we show both the least squares and maximum likelihood estimates for the data under discussion, for a variety of values of m and n.

First, let us arbitrarily fix $n = 100$ and try several values of m (see Table 7.2). The tabulated parameter is the slope $1 - 2H$. It can be seen that the values tend to settle down from $m = 4$ onwards, suggesting that the first three Fourier frequencies are high-leverage points which have a distorting effect on the results. Similar behaviour has been observed in a number of other series to which this technique has been applied. The difference between the maximum likelihood and least squares estimates reflects, largely, the effect of a few very small periodogram ordinates which are downweighted by a Gumbel fit.

Henceforth the value $m = 5$ was fixed, and a number of values of n tried (see Table 7.3). In this case there seems a steady drift in slope as n increases, with no indication of where to draw the line. However, it is also possible to fit an extension of the model (7.6) to include a next-order term:

$$\log f(\omega) = \log b + (1 - 2H) \log \omega + c\omega^2. \tag{7.13}$$

With intelligent application, this can provide further indication of the correct

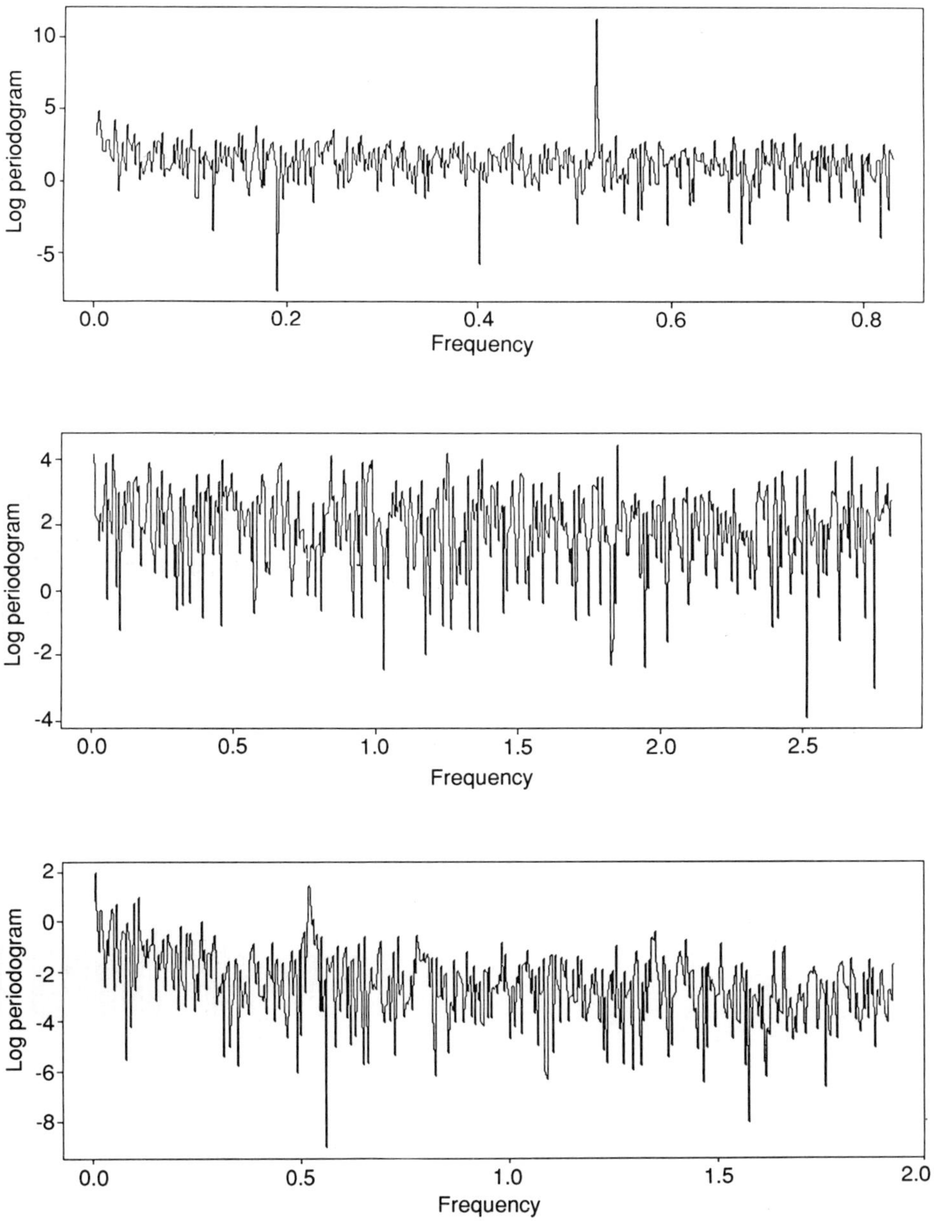

Figure 7.3 Log periodogram v. frequency for Central England series (top), USA (middle) global (bottom)

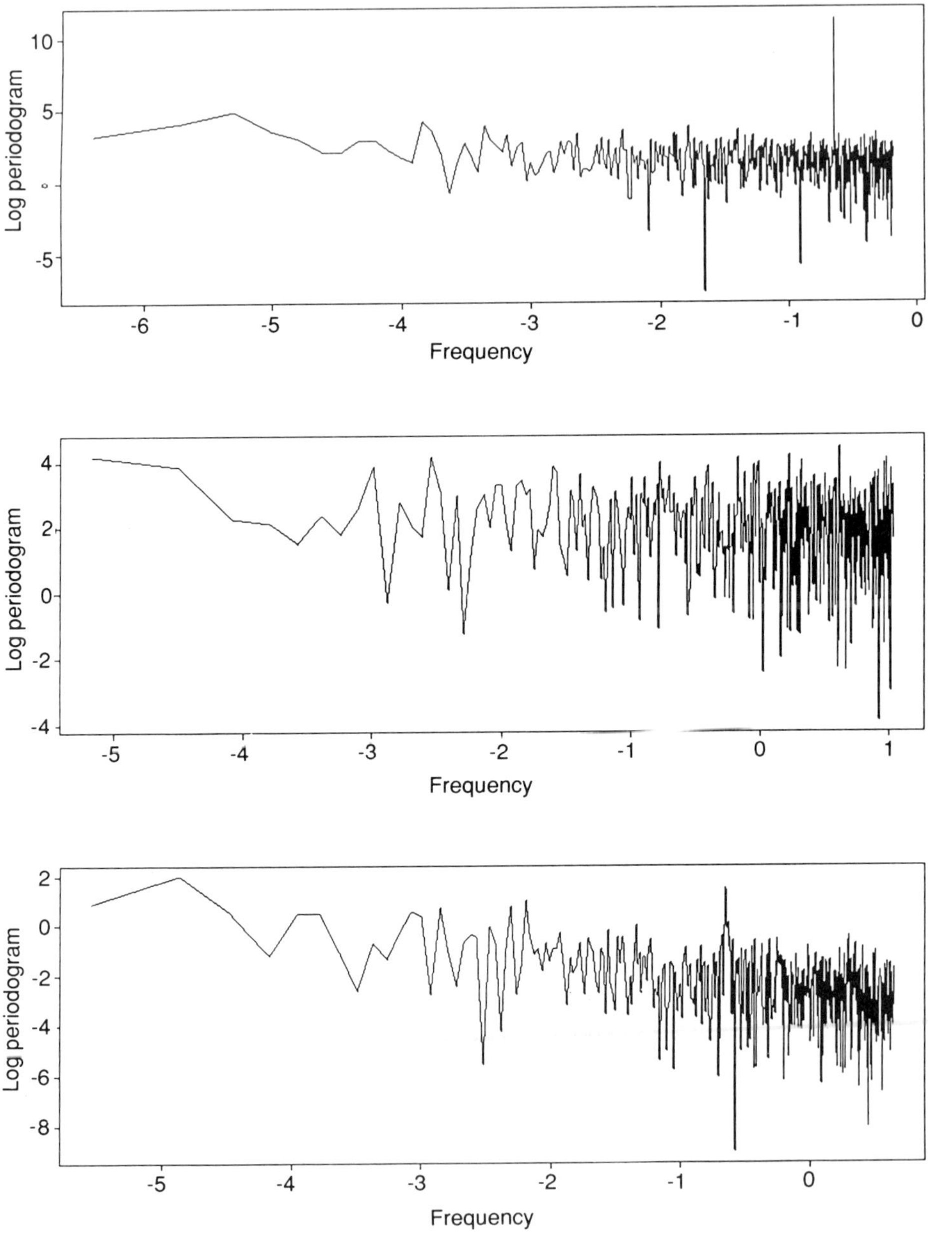

Figure 7.4 Log periodogram v. log frequency for Central England series (top), USA (middle), global (bottom)

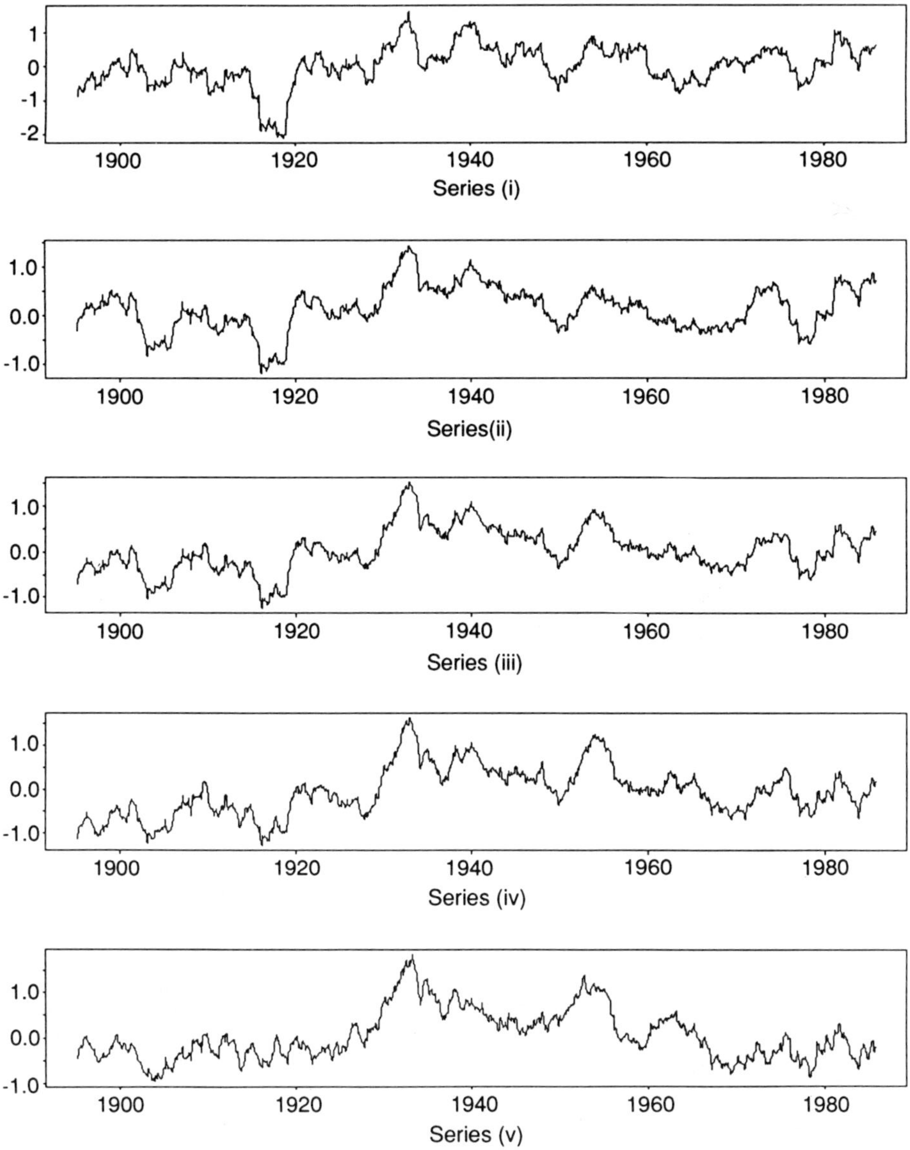

Figure 7.5 USA average temperature series expressed as anomalies from monthly means 1950–1979. The five plotted series are all based on 49-month moving averages. Top to bottom: monthly minima of daily minima, monthly averages of daily minima, monthly averages of daily averages, monthly averages of daily maxima, monthly maxima of daily maxima

Table 7.2 Slope estimates and standard errors for fixed n

m	n	LS est.	Std. err.	ML est.	Std. err.
1	100	−0.58	0.13	−0.52	0.10
2	100	−0.62	0.14	−0.55	0.11
3	100	−0.59	0.15	−0.54	0.12
4	100	−0.51	0.16	−0.43	0.13
5	100	−0.48	0.17	−0.41	0.14
6	100	−0.47	0.18	−0.41	0.15
7	100	−0.49	0.19	−0.44	0.16
8	100	−0.51	0.20	−0.46	0.16
9	100	−0.49	0.20	−0.47	0.17
10	100	−0.46	0.21	−0.47	0.17

Table 7.3 Slope estimates and standard errors for fixed m

m	n	LS est.	Std. err.	ML est.	Std. err.
5	50	−0.55	0.25	−0.54	0.26
5	60	−0.49	0.21	−0.50	0.21
5	70	−0.52	0.20	−0.44	0.19
5	80	−0.59	0.20	−0.48	0.18
5	90	−0.49	0.18	−0.46	0.15
5	100	−0.48	0.17	−0.41	0.15
5	110	−0.42	0.16	−0.32	0.13
5	120	−0.53	0.18	−0.34	0.14
5	130	−0.50	0.17	−0.36	0.13
5	140	−0.47	0.16	−0.33	0.12
5	150	−0.34	0.15	−0.23	0.11
5	160	−0.36	0.14	−0.26	0.10
5	170	−0.33	0.14	−0.25	0.09
5	180	−0.33	0.13	−0.27	0.09
5	190	−0.28	0.12	−0.25	0.09
5	200	−0.29	0.12	−0.27	0.09
5	250	−0.31	0.10	−0.31	0.08

choice of n. In this case, fitting (7.13) by maximum likelihood with $n = 200$, 250 and 300 respectively produced estimates of $1 - 2H$ of -0.34, -0.32, -0.30 with standard errors 0.17, 0.15, 0.07. Given that many of the maximum likelihood estimates in the Table 4.2 are around these values, there seems to be strong evidence that the true value of $1 - 2H$ is around -0.3, i.e. H is about 0.65. Fixing $n = 250$ and again trying different m produced Table 7.4.

Table 7.4 Slope estimates and standard errors for $n = 250$

m	n	LS est.	Std. err.	ML est.	Std. err.
1	250	−0.39	0.09	−0.39	0.06
2	250	−0.39	0.09	−0.40	0.06
3	250	−0.37	0.10	−0.39	0.07
4	250	−0.33	0.10	−0.32	0.07
5	250	−0.31	0.10	−0.31	0.08
6	250	−0.30	0.11	−0.31	0.08
7	250	−0.30	0.11	−0.32	0.08

Again there is stability for $m \geqslant 4$. The final estimates used $m = 5$, $n = 250$, leaving $1 - 2\hat{H} = -0.31$, and hence $\hat{H} = 0.655$, and also $\hat{b} = 0.035$, after correcting for a factor 4π to allow for the fact that the periodogram estimates $4\pi f(\omega)$ rather than $f(\omega)$ itself.

To conclude this section, we discuss the effect of long-range dependence on the trend estimates reported in Section 7.3. Note that this assumes that the model just fitted for the whole data series is also applicable to the 1907–73 data series analysed in Section 7.3. For the reasons outlined in the Introduction, this may not be a valid assumption. For $H = 0.655$ we have $\theta_2 = 45.5$, and hence for $N = 800$ (the actual number of data points used in Section 7.3) the approximate variance in (7.10) becomes 2.47×10^{-8}, or standard deviation 1.57×10^{-4}. When multiplied by 1200 to be consistent with the correct units, the estimated standard error of the parameter β becomes 0.19, comparable with the earlier calculation of around 0.22.

By analogy with the method of Beran (1989), it should be possible to 'studentise' the estimate of β by taking into account the distribution of $(\hat{H}, \hat{b})$. This will presumably increase the effective standard error of $\hat{\beta}$ still further.

7.6 US DATA

The US data are part of the Historical Climatological Network (HCN) and comprise daily maximum and minimum temperatures over a network of 95 stations over the continental USA. Most of the series started from between 1890 and 1930 and contain some missing values. For the purpose of this analysis, the following were computed for each station:

(i) monthly average daily average (note that the daily average is *defined* to mean the average of the daily maximum and daily minimum values);
(ii) monthly average daily maxima;
(iii) monthly average daily minima;

(iv) monthly maximum daily maxima (i.e. the absolute maximum temperature for each month);
(v) monthly minimum daily minima.

All five series were expressed as *anomalies* from the monthly averages of 1950–79; in other words, for each of the 12 months of the year, the 1950–79 average for that month was subtracted from the observations. This is standard practice in analysing climate series and has the advantage of removing overall differences due to different mean temperatures at different locations, and also removing the dominant seasonal effect, though the reader should be cautioned that neither spatial nor seasonal effects are eliminated by the crude device of substracting off the means! However, in the following analysis we shall proceed as if they are, and shall consider in particular the US average series formed by averages each of the series corresponding to (i)–(v) over the 95 stations.

Two recent papers in particular have drawn attention to fresh issues of global warming based on analysis of (different portions of) the same data set. Karl *et al.* (1991a) fitted linear trends with autoregressive errors to the data, and then used simulation to assess the standard errors of the estimated trend coefficients. They concluded that trend estimates based on average temperatures, though consistent with those produced by IPCC, were not significant as judged by the standard errors. They pointed out that this could have serious implications for public policy: if the predictions of the GCMs are correct, then according to their calculations it will be another 30 years or so before the observed temperature changes are sufficient to be judged significant using statistical techniques (by which time, by implication, it may be too late to do anything about it). However, they also suggested that the conclusion may be different when applied to maximum or minimum temperatures. In a further development, Karl *et al.* (1991b) suggested specifically that daily minima may be rising, while daily maxima remain roughly constant. This is consistent with a rough notion of the greenhouse effect acting to prevent night-time cooling while not having any particular influence on day-time temperatures. This approach makes clear the need to analyse the entire distribution of temperatures, as opposed to just mean values, which has been the main focus of attention in the global-warming literature to data.

In Figure 7.5, 49-month moving averages are plotted for each of the five series. All of these show the same pattern of a sharp rise between 1920 and 1940, followed by a steady decline up to 1975 and a recent rise. Visual inspection of the plots suggests that the top two (corresponding to the first and the third series in Figure 7.5) do indeed show the greatest rate of warming since 1975 and the smallest rate of cooling during 1940–75, consistent with Karl *et al.* (1991b).

The autoregressive model of Section 7.3 was fitted to each of the five series for the period 1895–1987. In each case, inspection of log-likelihoods suggests that either an AR(1) or an AR(2) model is a perfectly good fit to the data—for consistency, we shall take AR(2) in each case. For these data sets, the estimated linear trends (with standard errors in parentheses) are 0.63 (0.22), 0.81 (0.25), 0.42 (0.21), 0.21 (0.27) and 0.87 (0.30) respectively for the series (i)–(v). The overall rise, represented by the first figure, does appear to be significant, but the results also suggest that the sharpest rise is in the minimum daily minima, consistent with the hypothesis of Karl *et al.* (1991b).

Let us now consider the effect of long-range dependence in this case. For brevity we only consider the overall mean series (i). The periodogram is plotted against frequency in Figure 7.3 and against log frequency in Figure 7.4. From the latter it is hard to see any evidence of long-range dependence (we are looking for a negative slope on the left-hand side of the plot) but if this does happen at all, it seems to be on the log frequency range up to about 0.3, which is at about the 120th Fourier frequency. In the notation and terminology of Section 7.4, the estimate was the ML estimate with $m = 1$, $n = 120$. Note that the detailed exploration of Section 7.4, to select the best values of m and n, was not repeated here, though a brief study of alternatives showed no great sensitivity to the precise choice of these parameters. On the basis of this we deduce a slope in the log-log periodogram of -0.1452 (standard error 0.08), which suggests that there is indeed some evidence for long-range dependence, though not so strong as to be definitely indentified by a significance test. This leads to an estimate of $H = 0.5726$, and also of $b = 0.5704$. After applying (7.10) with $\theta_2 = 59$, $n = 1116$, we deduce a standard error of the slope of the regression to be about 0.31. The standard error in this case is somewhat larger than that obtained from the autoregressive fit (0.22), and serves to reinforce doubts about the significance of the estimated trend.

As things stand, it appears that there are too many uncertainties in this analysis to make any definitive conclusions. The most obvious drawback of this analysis is that the trend is clearly not linear (see Figure 7.5). Indeed, repeating the same analysis for the years 1920–87 produces evidence of an overall cooling in all five series! However, in the spirit of Karl *et al.* (1991a), the analysis does permit discussion of what rises would need to be observed to be judged statistically significant. For this, it is relevant that all the standard errors quoted are of the same order of magnitude as the claimed overall greenhouse effect of the IPCC estimates.

7.7 GLOBAL DATA

Having failed to detect a truly significant trend in either the English or the US data, we now consider some of the IPCC data, divided into (i) mean overall (land and sea) temperatures for the Northern Hemisphere 1854–1989, (ii) mean overall (land and sea) temperatures for the Southern Hemisphere 1854–1989, (iii) mean land temperatures for the Nothern Hemisphere 1851–1989, (iv) mean

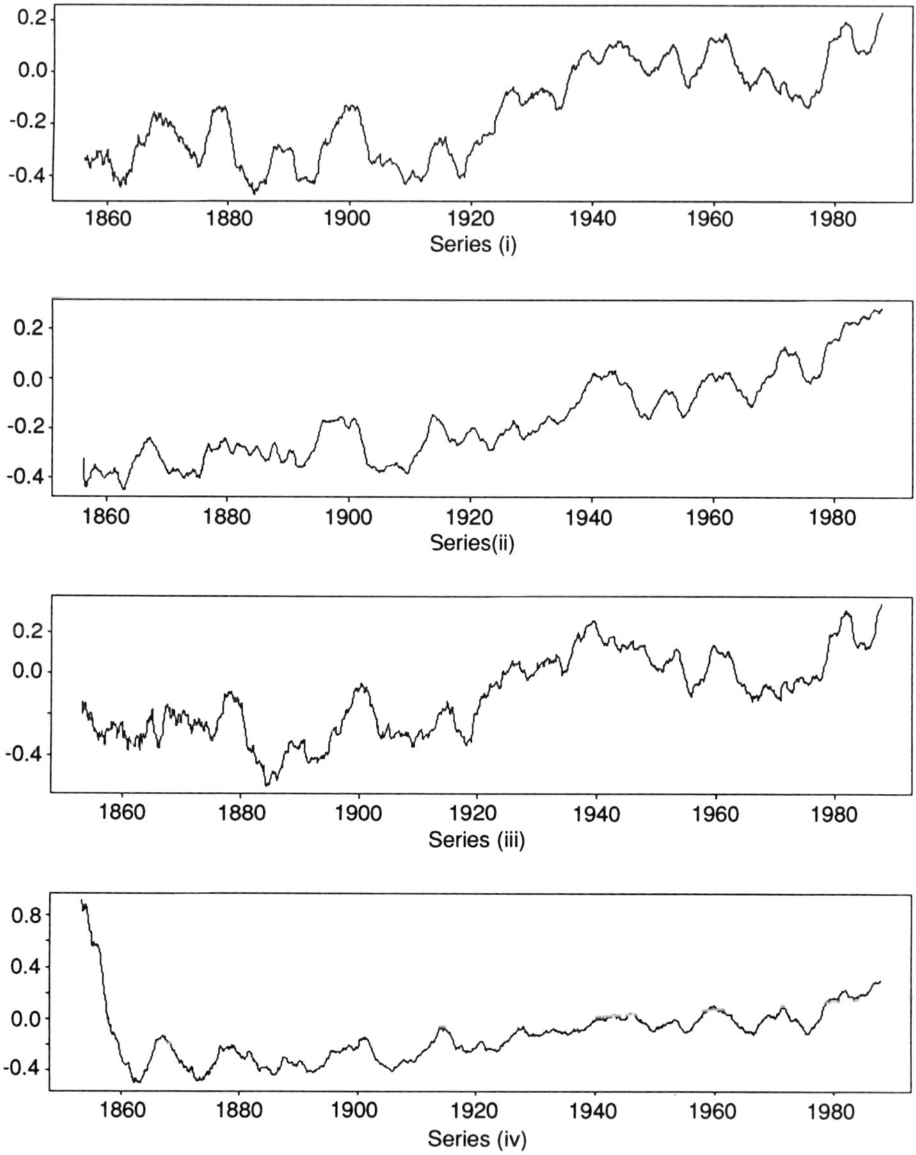

Figure 7.6 Global temperatures relative to 1950–1979 monthly averages. Plotted are 49-month moving averages. Top to bottom: Land and sea averages, Northern hemisphere, 1854–1989; Land and sea averages, Southern hemisphere, 1854–1989; Land only averages, Northern hemisphere, 1851–1989; Land only averages, Southern hemisphere, 1851–1989

land temperatures for the Southern Hemisphere 1851–1989. The four data sets, again smoothed by a 49-month moving average, are shown in Figure 7.6. These confirm the general pattern described at the beginning of the chapter, i.e. a sharp rise in temperature between 1920 and 1940, followed by a levelling off or even a cooling up to 1975, and then a further rise. The pattern is not quite the same in all four series. In the following analysis we concentrate on series (i).

In this case the attempt to fit an autoregressive model is not so successful. Models of order up to 20 were fitted with no clear identification of a model. However, all the fitted models showed an estimated trend of about 0.40 with a standard error of 0.02–0.03, which ought to put the significance of the estimated trend beyond doubt. Consider, however, the periodogram in Figures 7.3 and 7.4. In this case the evidence for long-range dependence seems clear, and in Figure 7.4 this appears to persist up to a log frequency of about -0.8, which is again near the 120th Fourier frequency. Based on $m = 1$, $n = 120$ and an ML fit, the estimated slope of the log-log periodogram plot is -0.806 (standard error 0.11), corresponding to $H = 0.903$, and also an estimate of b of 0.0033. In this case we have $\theta_2 = 22.8$ (and $n = 1632$), which leads via (7.10) to an estimated standard error of 0.10 for the rate of global warming in °C per 100 years. In this case the choice of m and n seems very highly subjective, but other values lead to similar standard errors for rate of global warming even if rather different values of H. For example, $m = 4$, $n = 100$ yields a slope of -0.751 (standard error 0.15), which corresponds to $H = 0.875$ and $\theta_2 = 24.5$. In this case we also have the estimate of $b = 0.0038$ which leads to a standard error of 0.09 for rate of global warming. Analysis with $m = 5$, $n = 50$ leads to a final standard error of 0.07 via the estimate $H = 0.79$.

Although these results are somewhat sensitive to the choice of m and n, the following points do seem clear: (i) there is significant evidence of long-range dependence (i.e. that $H > 0.5$), and (ii) the standard error is well in excess of its value computed under an autoregressive model. The overall conclusion, that there is significant evidence of global warming, is unchanged by this analysis, though the increase in standard error does show that this is more in doubt than previously.

Bloomfield (1992) considered the use of autoregressive and long-range-dependent models for the IPCC temperature series studied here, and also for the Hansen and Lebedeff (1987, 1988) data set. In this case autoregressive models were fitted to *annual* data, the best order being claimed as $p = 4$ for the IPCC data and $p = 8$ for the Hansen and Lebedeff data. Bloomfield also fitted the fractional ARIMA model of Hosking (1981), which is a parametric model satisfying (7.6). Using this model, also, there was very clear evidence of long-range dependence. For the IPCC data, the estimated rates of global warming were around 0.4 °C per century, with a standard error of about 0.1, which is consistent with our results. The somewhat shorter Hansen and Lebedeff data produced a warming estimate of 0.6 °C per century with a standard error again around 0.1—the difference between the two series being partly explainable in terms of the differences in period covered. Haslett and Raftery (1989) contained a very detailed discussion of fractional ARIMA models. A more recent paper by Bloomfield and Nychka (1992) has compared a number of different time-series approaches with particular regard for the spectrum of the series and its effect on the uncertainties of estimated trend. Yet another approach was taken in Kuo *et al.* (1990), which used frequency-domain nonparametric methods for the estimation of trend. For the Hansen–Lebedeff data, they again obtained estimates of trend and its standard error which were comparable with those of Bloomfield.

7.8 DISCUSSION

The conclusions of this chapter are necessarily highly tenative. There remain doubts about both the reasonableness of fitting linear trends to series with such erratic behaviour, and about the details of the proposed estimation. Our main purpose has been to present the method as an alternative to the ones currently available for assessing the uncertainties about global warming as estimated from observational data. It does seem clear, however, that long-range dependence is a legitimate cause for concern, and that the standard errors obtained under an assumption of long-range dependence can be considerably larger than those obtained under an autoregressive model. For this reason, we do not believe that the autoregressive model provides an acceptable method for assessing these uncertainties.

The comparison with the fractional differencing method, mentioned briefly at the end of Section 7.7 but really well established as a method since the paper of Haslett and Raftery (1989), is more problematical. The fractional differencing model is a finite-parameter model whose estimation procedures via maximum likelihood are by now well understood, whereas the method proposed here still leaves open many theoretical questions. On the other hand, there are advantages to a nonparametric method. One can argue that since the principal formula (7.9), for the asymptotic variance of the least squares estimator, depends only on the lower tail of the spectral density, then for reasons of parsimony the method of estimation should also depend only on the lower tail, a property not satisfied by the fractional ARIMA approach. In other words, the method given here may be more robust. This is to be judged against the efficiency of the LSE as compared with the BLUE (the latter representing what is possible when the assumed ARIMA model is correct) that was mentioned at the end of Section 7.3.

The issues raised by Karl, *et al.* (1991b) have drawn attention to the need to consider trends in the whole distribution rather than just the mean values. This is intended to be a focus of future work. Particular emphasis will be given to extremal properties of the distributions.

The analysis based on spatial averages also has the disadvantages of not allowing any modelling of spatial features, such as spatial correlations or the undoubtedly true fact that the rate of global warming is different in different locations. This raises the possibility of a comprehensive spatial model. A model combining spatial features with long-range dependence, within the fractional differencing framework, was proposed for wind-speed data by Haslett and Raftery (1989), and this is also expected to be a focus of future work.

ACKNOWLEDGMENTS

For the data sources we thank: Ms T. Bishop, formerly of the University of Surrey (Mainly data), Dr T. R. Karl of the National Climatic Data Center, Asheville, N.C. (US data) and Dr P. D. Jones of the Climatic Research Unit, University of East Anglia (global data). Ms A. Grady helped with data processing. We also thank Peter J. Robinson and Peter Bloomfield for their remarks. The work was partly supported by the NSF under contract DMS-9115750 as part of the Mathematics–Geosciences initiative (jointly with P. J. Robinson).

8

Environmental Factors Affecting Reservoir Safety

C. W. Anderson and S. Nadarajah,
University of Sheffield, UK

8.1 INTRODUCTION

If a reservoir receives a sudden and large enough inflow of water—perhaps during a very violent storm—then it is conceivable that the usual provision for outflow at the reservoir's dam could prove inadequate, that the resulting unplanned overflow could cause serious damage to the dam structure, and that it could possibly lead to catastrophic failure, with serious economic and environmental consequences, and perhaps even loss of life. Though such a scenario is unlikely, it is nevertheless something to be taken seriously in design standards for dams.

In Great Britain there are in fact strict regulations concerning the safety of dams. The Reservoirs Act 1975 (updating the Reservoirs (Safety Provisions) Act of 1930) requires that each reservoir of capacity greater than $25\,000\,\mathrm{m}^3$ should be inspected at least once every 10 years by a qualified and experienced engineer drawn from a panel appointed for the purpose, and also that there should be ongoing supervision by Supervising Engineers. The precise requirements for safety of a dam are not all laid down rigidly by statute, but include matters for the professional judgement of the inspecting engineer, in the light of the accumulated collective experience of the profession. As a summary of this experience in relation to floods, and as a guide both to panel engineers and to designers of new dams, the Institution of Civil Engineers published in 1978 a booklet entitled *Floods and Reservoir Safety: an Engineering Guide* (2nd edition with minor amendments: ICE 1989), recommending standards and assessment procedures.

Statistics for the Environment. Edited by Vic Barnett and K. Feridun Turkman

The rules put forward in the *Guide* suggest ways in which data on such factors as inflow to the reservoir, windspeed and initial water level can be combined to arrive at a maximum water level at the dam wall rare enough to be acceptable. On the whole the guidelines appear to have been broadly satisfactory, in the sense that only limited damage due to overtopping has occurred. However, there have been isolated incidents of unexpectedly high levels associated with wind-induced waves, which suggest that *interactions* between determining factors might nevertheless be having a significant effect. To investigate this possibility a research project has been set up by the UK Department of the Environment, and it is this project, entitled 'Joint Probability Studies for Reservoir Flood Safety', which is described here. The aim is to assess current design recommendations, and, if necessary, to suggest modifications. The study exploits recently developed methodology for the modelling of multivariate extremes.

In this chapter we outline the strategy of the study and some of its methodology. As well as being of some direct environmental interest, we hope the description will illustrate the increasing maturity of statistical extreme-value theory and its growing ability to contribute to the understanding of complex environmental problems. A further discussion of some of the hydrological aspects of the study is contained in Reed and Anderson (1992).

The plan of the rest of the chapter is as follows. In Section 8.2 we describe the problem in more detail, and sketch the current approach to reservoir safety assessment. In Sections 8.3 and 8.4 we outline respectively the broad strategy and the statistical methodology of the study, and in Section 8.5 report on some preliminary findings.

8.2 RESERVOIR SAFETY AND THE CURRENT METHOD OF ASSESSMENT

The key variable for flood safety is the height L of water at the dam wall. If L exceeds the *top water level* (the level of the dam's weir or spillway, which will generally be below the top of the dam wall) then there will be an overflow. Most dams have some provision for overflow in the form of spillways, but if L is sufficiently high then spillway capacity may be exceeded or the dam wall may be overtopped, and the rush of flood water may wash away foundations or otherwise weaken the downstream slope of the dam and ultimately lead to catastrophic failure. Water level L will vary through time in response to reservoir usage and weather factors. We are interested in high levels of L, brought about by episodes of wild weather which we will call *storm events*. (A more precise definition of storm is given in Section 8.4.2 below.)

At these storm events, L is thought to be a function of five main factors:

- **R**, rain falling during the storm over the area draining into the reservoir (the reservoir's *catchment*). Specification of **R** needs to include not only the total depth of rainfall but also the duration of the storm and its temporal profile.

- *C*, wetness of the ground at the onset of the storm—a function of earlier rainfall, and possibly also of wind, temperature and season.
- *S*, snowmelt—principally a function of prior precipitation and temperature, and of concurrent temperature and wind speed.
- **W**, wind speed and direction at the time when the reservoir water level is responding to flood inflow.
- *F*, reservoir fullness—a function of prior rainfall, reservoir usage and/or season.

The peak flood entering the reservoir during a storm will be governed by factors **R**, *C* and *S*, in a way which will depend on the size of the reservoir's catchment, the steepness of the ground, the type of soil, etc. The current recommended standard is framed in terms of a specified size for this flood inflow—called the *design flood*—together with a *design wind speed* and a *design initial reservoir level.* Different settings of these design variables are suggested for different categories of dam. An earthen embankment dam, for example, positioned where a breach would endanger lives in a community, should be capable of passing the *probable maximum flood* inflow occurring at the same time as a 10-year wind speed (that is, a wind speed equal to the hourly value expected only once in 10 years), and when the reservoir is already full and spilling its long-term average daily inflow. (The probable maximum flood is a value calculated according to prescribed rules (see the Flood Studies Report (NERC, 1975) and Sutcliffe (1978)), and is intended to represent a physical upper bound to the size of any inflow.) For dams where a breach would pose less of a threat to life, or where the structure is designed to withstand some overtopping without breaching, alternative design standards framed in terms of floods of 10 000-, 1000- and 150-year return periods are suggested, together with less extreme wind speeds and initial reservoir conditions.

The design flood for a particular reservoir usually has to be estimated indirectly from rainfall data, because direct records of flood inflows are typically available only over relatively short periods. The recommended approach for estimation of both probable maximum flood and floods of high return periods is the rainfall–runoff method of the Flood Studies Report (NERC, 1975). This uses a simple numerical model of flows to convert a notional fall of rain over a catchment into a time series of inflows to the reservoir, taking into account detailed catchment characteristics such as size, slopes, vegetation, and prior soil wetness. The 10 000-year design flood, for example, is obtained by use of the numerical model with rain input equal to the estimated 10 000-year rainfall for the locality in question (usually estimated from regional rainfall data). The design flood is then the resulting time series of inflow. The rain input to the numerical flow model is itself taken as a temporal profile over some duration (typically between 1 and 14 hours), which the engineer determines, either from characteristics of the system or by experimentation with the model, as that producing the largest flood inflow.

The design flood, in combination with the design initial conditions, allows calculation of a still-water level at the dam wall. To this is added an allowance for wave run-up derived by an empirical formula from the design wind. The result is the extreme water level which the dam must be capable of withstanding.

In the light of this existing framework for safety assessment, the Joint Probability Studies aim to:

(i) study the dependence between hydrometeorological variables relevant to reservoir flood safety,
(ii) assess the present rules for judging safety,
(iii) advise on any modifications to these rules which might seem desirable.

8.3 STRATEGY OF THE JOINT PROBABILITY STUDIES

In this section we outline the broad strategy adopted in pursuit of the aims given above.

The peak water level L at the dam wall during a storm is a function of the value of rainfall $\mathbf{R}$, catchment wetness C, snowmelt S, wind $\mathbf{W}$ and reservoir fullness F during the storm: say,

$$L = \lambda(\mathbf{R}, C, S, \mathbf{W}, F),$$

where λ will be referred to as the *structure function* (cf Coles and Tawn, 1992, World Meteorological Organization, 1988). It will be a complicated function, depending in detail on the specific characteristics of the reservoir under study. In practice we will take it to be defined implicitly by numerical models for catchment and reservoir responses, of the kind described in Section 8.2. These models, though imperfect, are well understood and are based on the physical processes known to be important in flood formation.

Through λ, joint probabilities attached to $(\mathbf{R}, C, S, \mathbf{W}, F)$ will induce a probability distribution on L, and this distribution will be obtainable explicitly by simulation from the joint model. Accordingly, the broad approach of the study is to estimate the joint probability distribution of the variables $\mathbf{R}, C, S, \mathbf{W}$ and F, and, with the help of simulation and appropriate numerical models, check whether the design floods, wind speeds and initial reservoir conditions specified by the existing rules lead to values of L of acceptable rarity. (If the inputs specified by the rules are put through the models, will they produce a 10 000-year value of L, for example?) Also from the induced probability distribution of L it will be possible—though with some difficulty—to study how the rules might be modified to yield values of L of specified rarity.

It will be important to obtain results for a representative variety of reservoir locations and conditions. On the other hand, data of the extent and quality needed for estimation of the relevant joint distributions are not very widely available, and are expensive to obtain. The Study's proposal for overcoming this difficulty is as follows. The joint probability distribution of the determining variables **R**, C, S, **W** and F can be specified in terms of the marginal distributions of the variables, and the dependence structure of the joint distribution after margins have been transformed to some convenient standard form. The proposal is to estimate this *dependence structure for standardised margins* from high-quality hourly data from one principal meterological observatory, the Meteorological Office's station at Eskdalemuir in the Southern Uplands of Scotland, Values of **R**, C, S, **W** and F for hypothetical or real reservoirs at any other location will then be obtained by simulation from the distribution which combines the Eskalemuir standardized dependence structure with the univariate marginal distributions of extreme rainfall, wind speed, etc. appropriate for the particular catchment and reservoir being considered. (See *Appendix 8.1* for a more formal development of this distributional argument.) Thus if the reservoir were at a low altitude in the east of England, for example, the appropriate distributions of extreme wind speeds and rainfall would be concentrated on rather lower values than would be used, say, for an upland reservoir in the north-west.

A crucial assumption in the foregoing is that the structure of dependence between standardized transformations of determining variables such as extreme rainfall and wind speed is transferable from site to site. Since the same physical processes are at work in different places, there are some qualitative grounds for expecting this to be true. Nevertheless, checks on the assumption will be made using more limited data from other sites, and sensitivity of results to the assumption will be investigated. The restriction of the data-investigation phase of the study to a limited number of sites, none of which is actually a reservoir, is due to the high cost of good-quality hourly meteorological data and the non-availability of such data from reservoirs (at which records of *daily* rain may be kept, but not usually detailed wind and temperature measurements).

8.4 STATISTICAL METHODOLOGY

The statistical methodology used in the programme outlined above is based on recent advances in the modelling of multivariate extremes. We describe this modelling now in the present context, indicating where extensions are called for by the reservoirs problem.

8.4.1 Primary and secondary variables

Some of the components of $\mathbf{X} = (\mathbf{R}, C, S, \mathbf{W}, F)$ are *primary* variables, in the sense that it is their extremes which are almost certainly important in creating large values of L. The remaining components of **X** will be called *secondary*. Thus snowmelt S and the first component of **R**, total rainfall during the storm, will certainly be primary variables. Wind speed will also be considered to be primary, though in fact some function of wind speed and direction, moderated by the size, shape and orientation of the reservoir to give an effective fetch at the dam wall, will govern wave run-up. Detailed local features, however, will be best incorporated through the numerical models specific to particular reservoirs. The temporal profile of rainstorms (crudely classified according to the position of the peak) and wind direction will be taken as secondary variables, central values of which, and not exclusively extremes, affect extremes of L. Prior reservoir fullness F is, in one sense, the prime example of a primary variable, since its extremes are directly related to extremes of L. On the other hand, *actual* values for F will be highly specific to the particular reservoir under study, and there will be no data to enable it to be treated directly in the same way as the other meteorological primary variables, since there is no reservoir at Eskdalemuir. We return to the handling of F in Section 8.4.4 below.

8.4.2 Dependence between extremes of primary variables

Suppose $X_1, \ldots, X_p$ denote the primary variables, and $X_{p+1}, \ldots, X_q$ the secondary variables. We concentrate first on extremes of the primary variables. Let $u_1, \ldots, u_p$ denote high threshold values, one for each primary variable. Then a storm event will be taken as an occurrence of a value of any one of $X_1, \ldots, X_p$ over its corresponding threshold. (Thus in some storm events, X_1 may exceed its threshold value, whilst the other variables $X_2, \ldots, X_p$ may remain below theirs. In this case the values of $X_2, \ldots, X_p$ for that storm will be the values of these variables recorded at the same time as X_1. In other storm events more than one primary variable might exceed its threshold, and again the values of the remaining variables will be determined by simultaneity.) We suppose that data are collected from a sequence of independent storms. The dependence between variables $X_1, \ldots, X_p$ in storm events will be modelled by a multivariate point process model for extremes (Coles and Tawn, 1991; Joe *et al.*, 1992). We now describe these models, concentrating for simplicity on the bivariate case.

The models rest on a convergence result for point processes. Suppose that $\{(Z_{1i}, Z_{2i}): i = 1, \ldots, n\}$ are independent identically distributed random vectors with marginal unit Fréchet distributions:

$$P(Z_{ji} \leqslant z) = \exp(-1/z), \qquad z > 0,$$

for $j = 1, 2$ and $i = 1, \ldots, n$. Then it is known (de Haan, 1985; Resnick, 1987, Chap. 5) that the set of n points

$$\{(Z_{11}/n, Z_{21}/n), \ldots, (Z_{1n}/n, Z_{2n}/n)\},$$

regarded as a point process in the plane, converges weakly as $n \to \infty$ to an inhomogeneous Poisson process on the positive quadrant $\mathbb{R}_+^2 \setminus \{\mathbf{0}\}$ excluding the origin, with some intensity measure Λ. Moreover Λ has the property that for any set B in $\mathbb{R}_+^2 \setminus \{\mathbf{0}\}$ and any positive t,

$$\Lambda(B/t) = t\Lambda(B), \tag{8.1}$$

reflecting the stability property of the unit Fréchet margins. We denote points in the limit space $\mathbb{R}_+^2 \setminus \{\mathbf{0}\}$ by (ζ_1, ζ_2). The intuitive content of the limiting result is simply that as $n \to \infty$ the scaling by $1/n$ drags down to the origin all points except those with unusually large values of either Z_1 or Z_2 or both. Points with unusually large Z_1, say, but not large Z_2, will move under the scaling to the ζ_1 axis, and those with large Z_2 but not large Z_1 will move to the ζ_2 axis: only points with both components unusually large will survive in $\mathbb{R}_+^2$ away from the axes.

If a transformation is made to pseudo-polar coordinates then the intensity Λ takes on a particularly simple form. Let

$$r = \zeta_1 + \zeta_2; \qquad \omega = \zeta_1/r.$$

Then it is found that

$$\Lambda(\mathrm{d}r \times \mathrm{d}\omega) = \frac{\mathrm{d}r}{r^2} H(\mathrm{d}\omega), \tag{8.2}$$

where H is a measure on the unit interval with the properties

$$\int_0^1 \omega H(\mathrm{d}\omega) = \int_0^1 (1-\omega) H(\mathrm{d}\omega) = 1.$$

Evidently the measure H embodies the dependence structure of extremes of (Z_1, Z_2).

Imagine now that the original primary variables X_{1i} and X_{2i} have been transformed by monotone increasing transformations t_1 and t_2 so that $Z_{1i} := t_1(X_{1i})$ and $Z_{2i} := t_2(X_{2i})$ have marginal unit Fréchet distributions. The probability results above suggest that, away from the origin, the pattern of points $\{(Z_{1i}, Z_{2i})/n\}$ could be modelled by an inhomogeneous Poisson process with an intensity density of the form in equation (8.2). It seems reasonable too to regard the region outside a rectangle $B = [0, t_1(u_1)/n] \times [0, t_2(u_2)/n]$ for large enough thresholds u_i (corresponding to thresholds u_1 and u_2 for the X_{1i} and X_{2i} respectively) as being 'away from the origin'.

If t_1 and t_2 are completely known then we may estimate the dependence structure of extremes of (X_1, X_2) by using the transformed points $(Z_{1i}, Z_{2i})/n$ to estimate the measure H in equation (8.2). A parametric model for H with density $h(\omega{:}\ \theta)$ for some vector of parameters θ, for example, may be estimated by maximising the likelihood based on the (Z_1, Z_2)-Poisson process:

$$L(\theta) = \exp[-\Lambda(B^c: \theta)]\frac{1}{n^2}\prod_{\{Z_i/n \in B^c\}} \frac{h(\omega_i: \theta)}{r_i^3}, \tag{8.3}$$

where $r_i = (z_{1i} + z_{2i})/n$ and $\omega_i = z_{1i}/nr_i$ are calculated from the observed values z_{ji}.

In practice t_1 and t_2 are not usually known. However, if the form of the marginal distribution function, G_j say, of the primary variable X_j were known up to some marginal parameters η_j, then it is easily seen that

$$t_j(X_j) = -1/\log(G_j(X_j: \eta_j))$$

would have a unit Fréchet distribution. Thus a *parametric form* for the transformation t_j would be known. Coles and Tawn (1991) note that the univariate theory of extremes suggests a generalised Pareto (GPD) model for the conditional distribution of excess of a primary variable X_j over a high enough threshold u_j (Davison and Smith, 1990). In this model

$$P(X_j > u_j + x | X_j > u_j) = \left(1 - \frac{k_j x}{\sigma_j}\right)_+^{1/k_j}$$

where $\sigma_j > 0$ and k_j are GPD parameters, and x_+ denotes $\max\{0, x\}$: hence

$$G_j(y) = 1 - p_j\left(1 - \frac{k_j(y - u_j)}{\sigma_j}\right)_+^{1/k_j}, \quad y > u_j,$$

where $p_j = P(X_j > u_j)$. Therefore, above threshold u_j, X_j may be transformed to Z_j by

$$t_j(x) = -1/\log\left\{1 - p_j\left(1 - \frac{k_j}{\sigma_j}(x - u_j)\right)_+^{1/k_j}\right\}, \quad x > u_j.$$

Below u_j, Coles and Tawn (1991) suggest that G_j may be estimated by an empirical distribution function $\tilde{G}_j$, for which $\tilde{G}_j(X_{ji}) = (R_{ji} - 0.5)/n$, where R_{ji} is the rank of X_{ji} in the sample, counting from the lowest. Thus

$$t_j(x) = \log\{n/(R_j(x) - 0.5)\}, \quad x \leqslant u_j, \tag{8.4}$$

$R_j(x)$ being, for a data value x, the rank of x in the sample.

When these transformations are used the likelihood based on data (X_{1i}, X_{2i}) becomes

$$L(\theta, \eta_1, \eta_2) = \exp[-\Lambda(B^c: \theta)]\frac{1}{n^2}\prod_{\{Z_i/n \in B^c\}} \frac{h(\omega_i:\theta)}{r_i^3}\frac{\partial t_1(x_{1i})}{\partial x_1}\frac{\partial t_2(x_{2i})}{\partial x_2},$$

and, from this, estimates of both the dependence parameters θ and the marginal transformation parameters η_1 and η_2 may be found. (Note that the terms $\partial t_j(x_{ji})/\partial x_j$ depend on the marginal parameters η_j only if $x_{ji} > u_j$, so can be omitted when $x_{ji} \leqslant u_j$.)

Explicit parametric models (logistic, bilogistic, negative bilogistic, asymmetric logistic, asymmetric negative logistic, Dirichlet) for h are given by Coles and Tawn (1991, 1992) and Joe *et al.* (1992): see *Appendix 8.2*.

Though the description above refers to the case of two primary variable, the approach generalizes readily to more than two dimensions (although in higher dimensions fewer explicit parametric models for h have been worked out).

8.4.3 Inclusion of secondary variables

The foregoing gives a way of generating joint values of primary variables for storm events. However, to be able to simulate extreme values of L in accordance with the programme sketched in Section 8.3 we need to be able to generate corresponding values of secondary variables too. The difference between the two kinds of variable is that in a storm no secondary variable need be extreme. (Indeed for some of the secondary variables (wind direction, storm profile) the notion of extremes may not be applicable at all.) Accordingly the secondary variable values will be treated as marks or concomitants attached to the points in the primary-variable point-process representation, and their conditional joint distributions—the Palm distributions—estimated conditionally on values of the primary variables. One simple approach to the estimation is to use smoothed empirical distributions of secondary values for a sequence of bands in the primary-variable space. An alternative is parametric modelling. If models are needed, it will be important to base them as securely possible on recognised physical mechanisms. We return to this point in Section 8.5 below.

8.4.4 Further aspects of modelling

The fitting described in the last two subsections will produce a model for the dependence structure of all the variables $(X_1, \ldots, X_p, X_{p+1}, \ldots, X_q) = (\mathbf{R}, C, S, \mathbf{W}, F)$ relevant to individual storm events (see *Appendix 8.1*). For estimation of the distribution of large values of the structure variable L we need, in addition, information about the rate of occurrence of storms through time. A Poisson process model with seasonally varying intensity will be used for this.

Seasonality is also likely to be apparent in the extremal characteristics of primary variables. It can be accommodated by allowing the parameters in the point process models of Section 8.4.2 to depend on a seasonal covariate. Possible climatic trends can be treated similarly by use of a covariate. (This is in fact more conveniently done via a modification due to R. L. Smith *et al.* of the likelihood in Section 8.4.2; see Smith *et al.*, 1992.)

Catchment wetness C, reservoir fullness F and snowmelt S are potentially important variables which must feature in the modelling outlined in Sections 8.4.2 and 8.4.3. However, their values will be highly specific to the particular catchment under study, and no data are available to enable them to be treated directly in the same way as other meteorological variables. A solution is to use numerical models, of the same kind intended for the simulation phase of the study, to generate time series of values for these variables from the basic meteorological hourly data on precipitation, wind and temperature, which are available. Parameters in the numerical models will be set to values which would be typical for a reservoir at the Eskdalemuir site, if one were there. From the generated series, *indices* of catchment wetness, reservoir fullness and snowmelt will be constructed as functions of the basic meteorological variables and of season, augmented in the case of F by crude information on the pattern of reservoir usage (for example, is the reservoir kept more-or-less full for water supply purposes, or is it frequently drawn down?). In the case of the primary variables F and S the resulting indices will then be incorporated into the dependence modelling of Section 8.4.2 in the standard way, and, similarly the index corresponding to C will be treated with the other secondary variables as in Section 8.4.3. In the simulation phase of the study, values for the indices will be generated with values for the other variables detemining L by combining marginal distributions appropriate to the particular hypothetical reservoir under study with the dependence structure fitted as above.

8.4.5 Declustering

An important assumption in the model-fitting described in Sections 8.4.2 and 8.4.3 is that data from different storms are independent. In peaks-over-threshold modelling of *univariate* extremes, techniques for the declustering of a series of observations to obtain independent values are well established (NERC, 1975; Davison and Smith, 1990), but the multivariate nature of the present problem introduces new complications. To explain the problems and the study's method of addressing them we first describe in more detail the effect of a storm on the water level L.

As mentioned in Section 8.2, extreme levels of L are likely to be generated for a particular reservoir by storms of a particular *duration*, d say, characteristic of that reservoir. Total rainfall during a storm is therefore the aggregate of hourly rainfall totals over d hours. There is some delay between the fall of rain over a reservoir's catchment and its peak effect on L, and this lag, l say, again is a characteristic of the reservoir and catchment under study. Extreme values of L are therefore generated by d-hour aggregated values of rain, and the values of other primary variables such as wind *at a time l hours later*. The declustering procedure must therefore extract values of primary variables linked by an appropriate time lag l (and in the case of total rainfall, aggregated over d hours) from episodes when one or more of these variables exceeded its threshold; and it must ensure that the assumption of independence of the resulting vectors is credible. (Coles and Tawn, 1992 face a similar, though simpler, problem in connection with waves and surges at sea.)

For a pilot problem in which only the rainfall and wind-speed primary variables are considered, the following procedure has been used to meet these requirements:

(i) Apply a standard univariate declustering method to the d-hour aggregated rainfall series (in which the d-hour aggregate is located say at the central time) and to the hourly wind-speed series, shifting the time for the wind series backwards by l hours. The declustering method produces peak values from clusters of exceedances of the relevant threshold with the property that the clusters are separated in time by a specified interval, s say. The separation intervals are chosen to be the same for the two series, and large enough to guarantee that clusters derive from physically independent storms.
(ii) For each rain peak choose the 'concurrent' wind value, and vice versa.
(iii) Pool the resulting series to produce a set of *potential* storms. Carry out a further filtering of the pooled series to ensure that there is adequate separation between consecutive values. This is done by amalgamating the two extreme components of any potential storms occurring within a time s of each other to produce a new observation. The amalgamation is carried out in such a way that it is time-reversible.

In procedures of this kind the choice of the separation interval s is important: if s is chosen too small then independence of the filtered series is compromised, whilst if s is chosen too large, independent observations are omitted and bias is introduced into the resulting data.

8.5 SOME PRELIMINARY RESULTS

Some interim results on the fitting of dependence models for primary variables and on some features of the inclusion of secondary variables are given in this section. They are based on preliminary data and oversimplified assumptions, and so are best regarded at this stage as illustrative of the analysis described earlier rather than in any way definitive. The basic data used here consist of concurrent hourly values of rainfall and mean wind speed at the Eskdalemuir Observatory for the calendar years 1970–86.

8.5.1 Dependence between rain and wind

Figure 8.1 shows, a plot of mean hourly wind speed and 11-hour aggregate rain, for independent storms obtained by the declustering procedure of

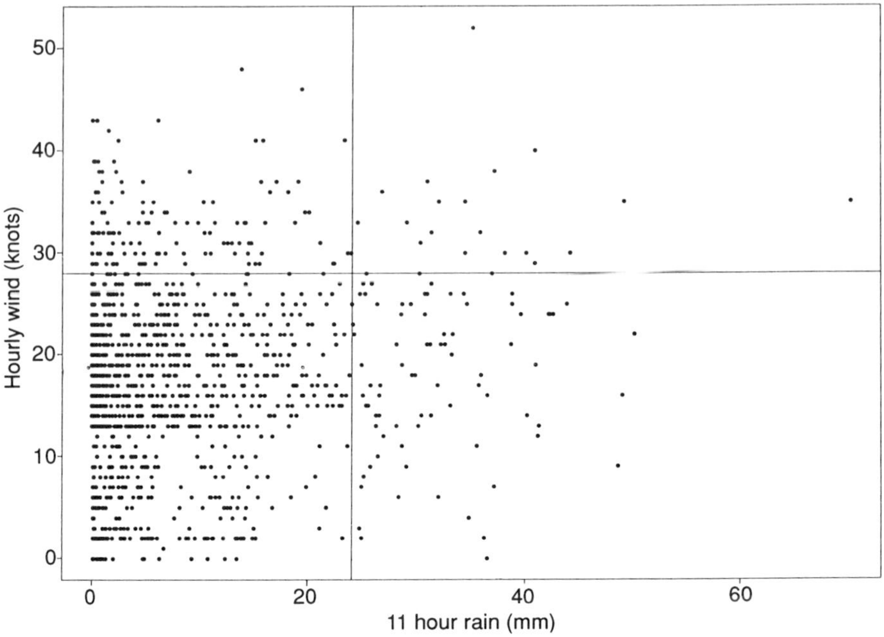

Figure 8.1 Hourly wind v. 11-hour rain

Section 8.4.5 with the lag l set equal to 5 and the minimum storm separation s equal to 24 hours. The plot gives an informal impression of rather weak dependence at high values. For illustration (although not directly needed for the model-fitting procedure of Section 8.4.2) Figure 8.2 shows the corresponding plot after standardization of the margins to unit Fréchet form. Choice of marginal thresholds was made here by reference to mean-excess plots (Davison and Smith, 1990), which are shown in Figures 8.3(a) and 8.3(b). Pointwise approximate 95% confidence limits are superimposed. Linearity of the plots above about 24 mm and 28 knots respectively suggested the tentative threshold values used in Figures 8.1 and 8.2. Figures 8.4(a) and 8.4(b) show estimated return levels for rain and wind obtained from marginal fitting above these levels to generalised Pareto distributions. Again pointwise approximate 95% confidence limits are superimposed.

Table 8.1 gives parameter estimates for a variety of point process models for the dependence measure density h, and Figure 8.5 shows probability-density-function contour plots of extreme wind and rain for one of these models, the logistic, for which

$$h(\omega:\theta) = \frac{(\theta - 1)}{\{\omega(1-\omega)\}^{1+\theta}}\{\omega^{-\theta} + (1-\omega)^{-\theta}\}^{1/\theta - 2}, \quad 0 \leqslant \omega \leqslant 1,$$

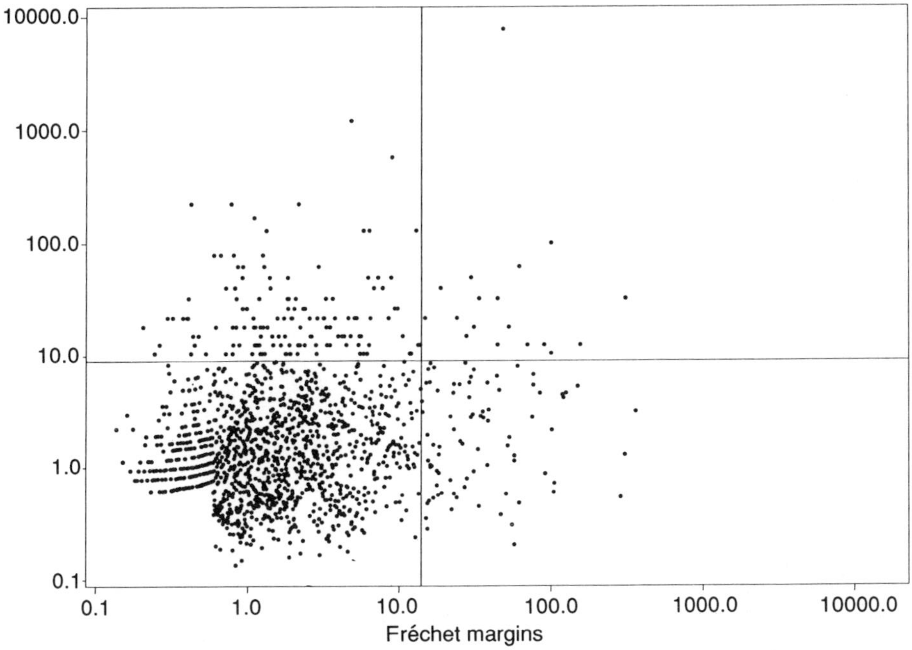

Figure 8.2 Hourly wind v. 11-hour rain

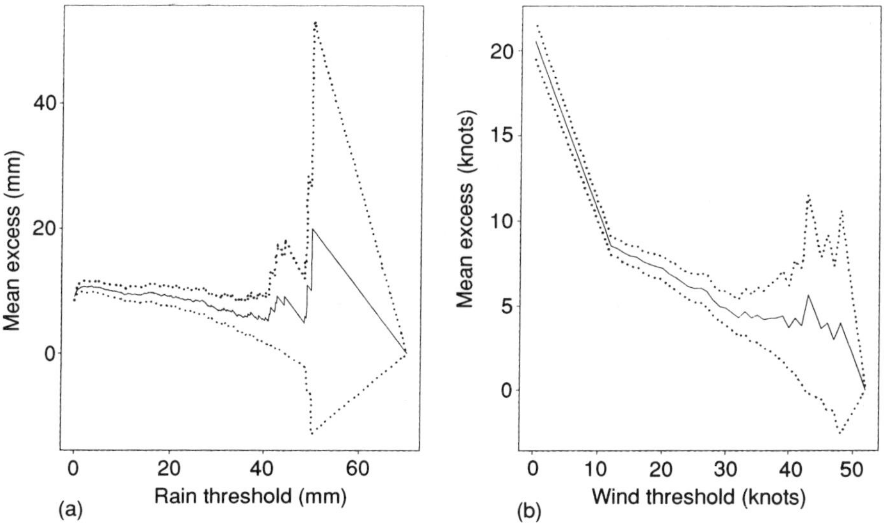

Figure 8.3 Marginal mean-excess plots

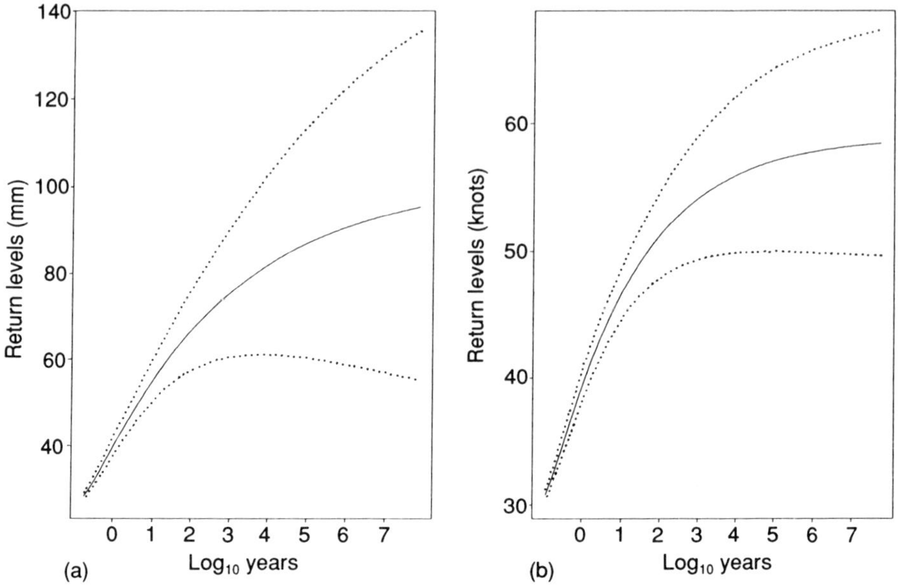

Figure 8.4 (a) Return levels for 11-hour rain. (b) Return levels for hourly wind

Table 8.1 Parameter estimates for point process models

Model	Margin Parameters and (s.e.) σ_1	σ_2	k_1	k_2	Dependence Parameters and (s.e.)	NLLH
Logistic	0.098 (0.012)	0.062 (0.006)	0.119 (0.067)	0.185 (0.057)	$\theta = 1.101$ (0.024)	424.9
Negative logistic	0.096 (0.011)	0.062 (0.006)	0.116 (0.064)	0.181 (0.057)	$\theta = -0.330$ (0.032)	424.4
Bilogistic	0.095 (0.012)	0.063 (0.006)	0.105 (0.072)	0.191 (0.056)	$\theta_1 = 1.179, \theta_2 = 1.064$ (0.122)(0.029)	424.4
Negative bilogistic	0.095 (0.011)	0.063 (0.006)	0.100 (0.067)	0.189 (0.056)	$\theta_1 = -0.193, \theta_2 = -0.574$ (0.112)(0.353)	423.9
Symmetric dirichlet	0.098 (0.011)	0.062 (0.006)	0.118 (0.068)	0.184 (0.057)	$\theta = 0.107$ (0.026)	424.9
Dirichlet	0.097 (0.012)	0.063 (0.006)	0.113 (0.070)	0.188 (0.056)	$\theta_1 = 0.081, \theta_2 = 0.159$ (0.027, 0.061)	423.9

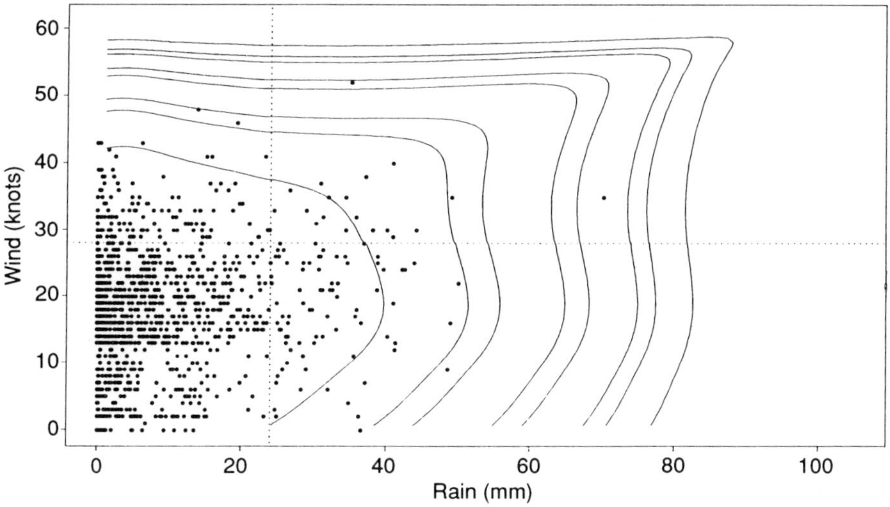

Figure 8.5 Estimated joint p.d.f. contours

$\theta \geqslant 1$ being here a parameter governing the degree of dependence between the two marginal variables, with $\theta = 1$ corresponding to independence. (For a similar plot, which suggested Figure 8.5, see Coles and Tawn, 1992.) Figure 8.5 is therefore a model analogue of the empirical plot in Figure 8.1. If the structure function λ depended solely on wind and rain, then estimates of probabilities

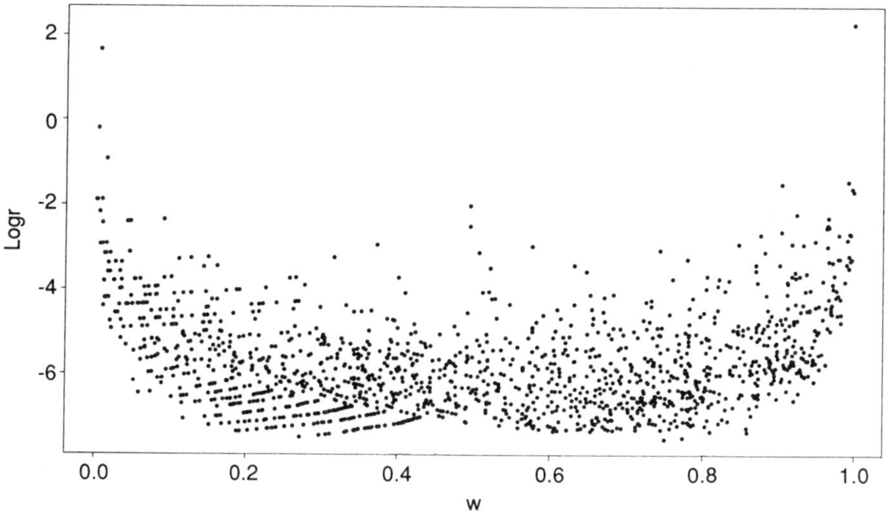

Figure 8.6 Point process in pseudo-polar coordinates

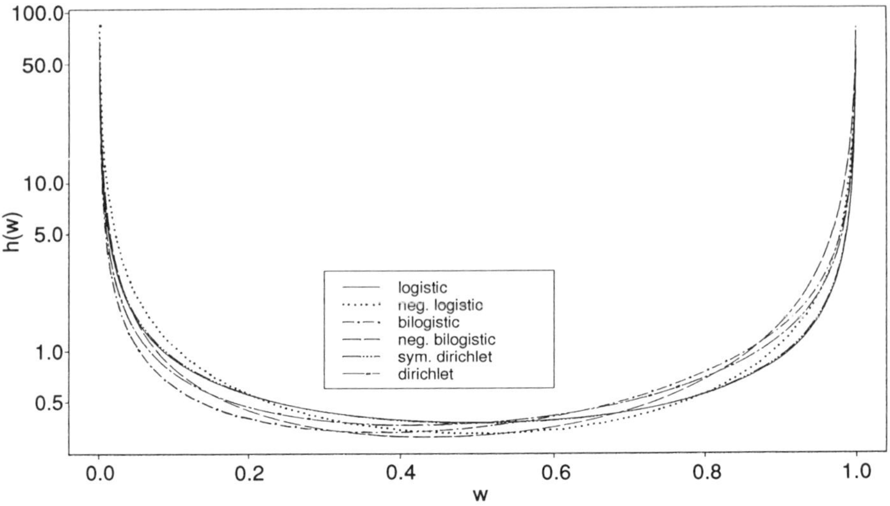

Figure 8.7 Fitted measure densities

such as $P(L > l_0)$ would correspond to integrals of the fitted density function over the region $L > l_0$, which would usually be the simple region outside the curve $L = l_0$.

Figure 8.6 represents the point process of Figure 8.2 in pseudo-polar co-ordinates, and Figure 8.7 shows corresponding estimated densities h for the distribution of the angular component ω of these points. The estimated densities are based on the parametric models listed in *Appendix 8.2*. An immediate and important feature is the apparent similarity of the different fitted models. Figure 8.6 also gives an indication of dependence. If extremes of the two variables were independent then h would put all its mass at the end-points of the interval $[0, 1]$, so the extent to which h is nonzero in the interior of the interval is an indicator of the degree of dependence. Bearing in mind the logarithmic vertical scale in Figure 8.7, we conclude that there is evidence of only weak dependence.

The parameter θ provides a numerical measure of dependence in a logistic model: θ must lie in the range $\theta \geqslant 1$, with $\theta = 1$ corresponding to independence, and large values corresponding to increasing degrees of dependence. Table 8.2 shows the degree of dependence, as indicated by the value of θ in the logistic model, for various values of storm duration d and lag l (relative to the middle of the rainstorm). Apart from a small increase in dependence for durations $d > 1$ compared to $d = 1$, there is little evidence of sensitivity to either d or l. We emphasise that this is a preliminary finding, which may be modified after a more complete analysis of the data. Also, further investigation of the sensitivity of extreme water level L to these levels of dependence (taking into account the appropriate marginal distributions of wind and d-hour rain in particular reservoirs) is needed, and will be undertaken in the simulation phase of the study.

Table 8.2 Estimates of dependence parameter in a logistic model

Duration d (hours)	Lag l (hours after the middle of the storm)					
	0	1	2	3	4	5
1	1.014 (0.010)	1.014 (0.010)				
3	1.117 (0.031)	1.099 (0.028)	1.091 (0.027)			
5	1.141 (0.033)	1.132 (0.032)	1.121 (0.030)	1.107 (0.029)		
7	1.122 (0.030)	1.131 (0.032)	1.132 (0.032)	1.120 (0.030)	1.120 (0.030)	
9	1.120 (0.030)	1.112 (0.029)	1.120 (0.030)	1.128 (0.031)	1.113 (0.029)	1.112 (0.029)

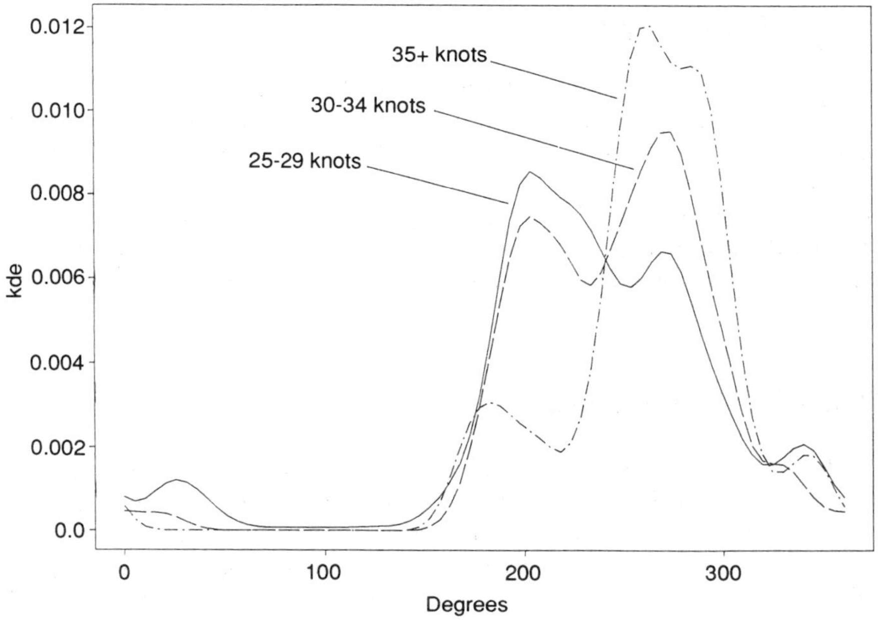

Figure 8.8 Directions of Eskdalemuir peak winds, 1970–89

8.5.2 Wind speed and direction

Figure 8.8 illustrates some of the issues raised in Section 8.4.3 about inclusion of secondary variables in the model. It shows kernel density estimates of wind direction for storms in which mean hourly wind speed was at least 24 knots (1 knot = 0.515 m/s). Evidently, strong winds mostly blew from westerly directions between 180° and 330°. The directional distributions show some bimodality, and, perhaps more importantly, a tendency for the peak winds in more violent storms to blow from points further round the compass than those in milder storms. This phenomenon has also been observed in similar data from Sheffield, but not in data collected at sea. A possible explanation is that, at higher speeds, wind blows in a direction closer to the high-altitude geostrophic wind, which in UK latitudes is at an angle of up to +40° to the wind over land, but at a much smaller angle to wind over the sea, because of smaller surface-frictional effects there. The possibly related phenomenon of the veering of wind-direction at increasing altitudes (at which wind speeds are higher too) is well known to meteorologists as the *Ekman spiral* (see e.g. Haltiner and Martin, 1957, Chap. 14). If this is a correct explanation of the observations, then it will be desirable to take account of the appropriate physical equations relating direction to speed in the estimation of the Palm distributions.

8.6 CONCLUSIONS

The studies described above are far from complete, and it is too early to try to draw definite conclusions about the interaction of environmental variables which affect reservoir safety, but we hope that the chapter serves to illustrate that new methods in extremal statistics, in combination with physical modelling, give a coherent and flexible approach to rather complex multi-factor environmental problems. A subsidiary theme of the chapter, as with almost any study of extremes, is that more data are always desirable.

ACKNOWLEDGMENTS

The Joint Probability Studies for Reservoir Flood Safety are supported by the UK Department of the Environment. The authors gratefully acknowledge the guidance and assistance of their colleagues on the project team: Ian Dwyer, Duncan Reed and Jonathan Tawn. They are grateful too to Colin Wright of DoE for helpful comments. Data for Eskdalemuir were supplied by the Meteorological Office Edinburgh Climate Office. Some of the work used equipment provided by SERC under the Complex Stochastic Systems Initiative.

APPENDIX 8.1 A DISTRIBUTIONAL OVERVIEW

Let $\mathbf{X} = (\mathbf{X}_p, \mathbf{X}_s)$ denote the vector of variables defined in Section 8.2, with $\mathbf{X}_p$ the primary variables and $\mathbf{X}_s$ the secondary variables. We will use a superscript to distinguish variables from different sites, so that, for example, $\mathbf{X}^E$ will correspond to Eskdalemuir variables, and $\mathbf{X}^H$ to variables appropriate for some hypothetical reservoir to be studied in the simulation phase of the project. This appendix presents the distribution theory which underlies the method of Section 8.5 for the generation of values of $\mathbf{X}^H$ for the simulation study.

Note first that if the marginal distribution functions F_{X_i} of each of the components X_i of $\mathbf{X}$ are known, and if

$$t_i(x) = -1/\log(F_{X_i}(x)), \quad i = 1, \ldots, q, \tag{8.5}$$

then the transformed vector

$$\mathbf{U} = \mathbf{t}(\mathbf{X}) = (t_1(X_1), \ldots, t_q(X_q))$$

has unit Fréchet margins.

The distribution functions $F_{\mathbf{X}}$ and $F_{\mathbf{U}}$ of $\mathbf{X}$ and $\mathbf{U}$ respectively are related by

$$F_{\mathbf{X}}(\mathbf{x}) = F_{\mathbf{U}}(\mathbf{t}(\mathbf{x})). \tag{8.6}$$

Thus if $F_{\mathbf{U}^H}$ and the marginal distributions $F_{X_i^H}$ for $i = 1, \ldots, q$ are known, then

$$F_{\mathbf{X}^H}(\mathbf{x}) = F_{\mathbf{U}^H}(\mathbf{t}(\mathbf{x})),$$

where $\mathbf{t}$ is the transformation whose components are given by (8.5) with $\mathbf{X}^H$ written for $\mathbf{X}$. A key assumption of the Joint Probability Studies is that $F_{\mathbf{U}^H} = F_{\mathbf{U}^E}$, and so

$$F_{\mathbf{X}^H}(\mathbf{x}) = F_{\mathbf{U}^E}(t(\mathbf{x})).$$

Once the joint distribution of the standardized Eskdalemuir variables $\mathbf{U}^E$ is known, therefore, $F_{\mathbf{X}^H}$ may be found from the marginal distributions of the components of $\mathbf{X}^H$, which will be obtainable from local meteorological and resevoir-usage data.

The distribution $F_{\mathbf{U}^E}$ is decomposed as

$$F_{\mathbf{U}^E} = F_{\mathbf{U}_p^E} F_{\mathbf{U}_q^E | \mathbf{U}_p^E}, \tag{8.7}$$

and

$F_{\mathbf{U}_p^E}$ is estimated via the point process modelling of Section 8.4.2,
$F_{\mathbf{U}_q^E | \mathbf{U}_p^E}$ estimated as the Palm distribution described in Section 8.4.3.

For the latter a transformation of the secondary variables $\mathbf{X}_s^E$ to standardized form will be required, and may be based on empirical distribution functions as in (8.4) of Section 8.4.3.

APPENDIX 8.2 DEPENDENCE MODELS

Logistic model

$$h(\omega:\theta) = (\theta - 1)(\omega(1-\omega))^{\theta-2}(\omega^\theta + (1-\omega)^\theta)^{\theta^{-1}-2}, \quad \theta > 1.$$

Negative logistic model

$$h(\omega:\theta) = (1 - \theta)(\omega(1-\omega))^{\theta-2}(\omega^\theta + (1-\omega)^\theta)^{\theta^{-1}-2}, \quad \theta < 0.$$

Bilogistic model

$$h(\omega:\theta_1,\theta_2) = \frac{\left(1-\dfrac{1}{\theta_1}\right)(1-u)u^{1-(1/\theta_1)}}{(1-\omega)\omega^2\left(\dfrac{1}{\theta_1}(1-u)+\dfrac{u}{\theta_2}\right)}, \quad \omega\in(0,1), \theta_1 \geqslant 1, \theta_2 \geqslant 1,$$

where $u = u(\omega;\theta_1,\theta_2)$ is the root of

$$\left(1-\frac{1}{\theta_1}\right)(1-\omega)(1-\omega)^{1/\theta_2} - \left(1-\frac{1}{\theta_2}\right)\omega u^{1/\theta_1} = 0. \tag{8.8}$$

Negative bilogistic model

$$h(\omega:\theta_1,\theta_2) = -\frac{\left(1-\dfrac{1}{\theta_1}\right)(1-u)u^{1-1/\theta_1}}{(1-\omega)\omega^2\left(\dfrac{1}{\theta_1}(1-u)+\dfrac{u}{\theta_2}\right)}, \quad \omega\in(0,1), \theta_1 < 0, \theta_2 < 0,$$

where u is the root of equation (8.8).

Dirichlet model

$$h(\omega:\theta_1,\theta_2) = \frac{\theta_1^{\theta_1}\theta_2^{\theta_2}\Gamma(\theta_1+\theta_2+1)}{\Gamma(\theta_1)\Gamma(\theta_2)}\frac{\omega^{\theta_1-1}(1-\omega)^{\theta_2-1}}{(\theta_1 w + \theta_2(1-\omega))^{\theta_1+\theta_2+1}};$$

$$\omega\in(0,1), \theta_1 > 0, \theta_2 > 0.$$

Setting $\theta_1 = \theta_2$ gives a symmetric version of this model.

9

Rainfall Depth-Duration-Frequency Curves; a Problem of Dependent Extremes

T. A. Buishand,
Royal Netherlands Meteorological Institute, The Netherlands

9.1 INTRODUCTION

Information about the probability distribution of precipitation amounts is often presented as rainfall depth–duration–frequency curves. The duration D refers to the length of the time-interval over which the precipitation amount is considered. Each rainfall depth–duration–frequency curve represents a quantile of the distribution of the precipitation amounts for a range of durations, e.g. the precipitation amounts that are exceeded once in 10 years for durations between 1 and 10 days (Figure 9.1).

There are different methods of constructing rainfall depth–duration–frequency curves. Quite often the annual maximum amounts for a limited number of durations are abstracted from a record of observations. For each selected duration a theoretical distribution is fitted to these annual maxima. Let $\theta_{1D}, \ldots, \theta_{pD}$ denote the parameters of this distribution. Rainfall depth–duration–frequency curves can then be obtained by describing the change of these parameters with duration by a functional relation.

A problem in this approach is that the parameter estimates are correlated as a consequence of the dependence between the annual maxima over different

Statistics for the Environment. Edited by Vic Barnett and K. Feridun Turkman

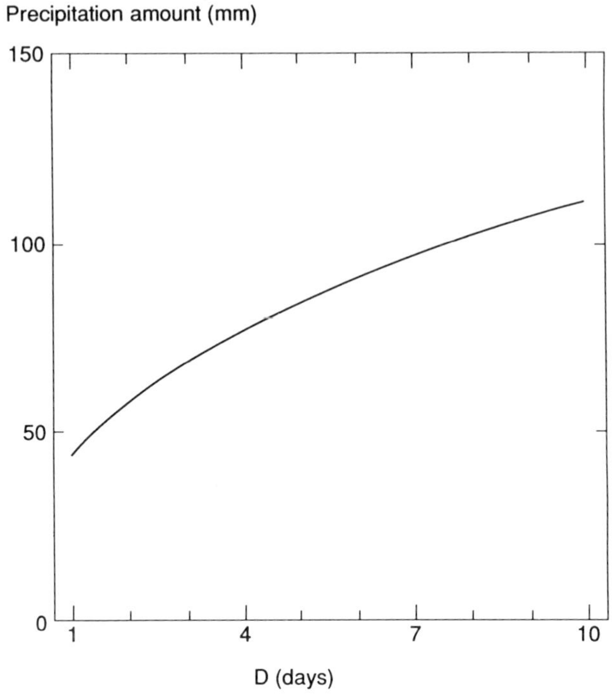

Figure 9.1 Rainfall depth-duration-frequency curve for De Bilt. The curve represents the amount of precipitation being exceeded once in 10 years as a function of duration D

durations. For instance, in most records there are many years in which the highest 1-day and 2-day precipitation amounts can be ascribed to the same rain shower. Up to now this dependence has been ignored in the determination of rainfall depth–duration–frequency curves.

In this chapter rainfall depth–duration–frequency curves for the rainfall record at De Bilt (52°06′N, 5° 11′E) are discussed. The two-parameter Gumbel distribution is fitted to the annual maxima. Various methods are presented to obtain the covariance matrix of the parameter estimates. With this covariance matrix the change of the parameters with duration can be estimated by the generalised least squares method. Realistic approximations to the standard errors of the resulting quantile estimates are also obtained.

9.2 ANNUAL MAXIMA AT DE BILT

From the 72-years daily record for the period 1906–77 the annual maxima were abstracted for $D = 1, 2, 4, 7$ and 10 days. Table 9.1 shows the correlation coefficients between these annual maxima. For two neighbouring durations the

Table 9.1 Correlation coefficients between the *D*-day annual maximum precipitation amounts at De Bilt (1906–77) for $D = 1, 2, 4, 7$ and 10

D	1	2	4	7	10
1	1.00				
2	0.86	1.00			
4	0.72	0.85	1.00		
7	0.52	0.69	0.87	1.00	
10	0.47	0.58	0.74	0.91	1.00

correlation coefficient amounts to 0.85 or more and even for the 1-day and 10-day annual maxima the correlation coefficient is still about 0.5.

For each duration the annual maximum distribution can reasonably be described by the Gumbel distribution:

$$F(x) = \Pr(X \leqslant x) = \exp[-\exp\{-(x-\mu)/\sigma\}]. \tag{9.1}$$

The index *D* for duration is left here for convenience. It should be noted that the existing bivariate and multivariate extensions of the Gumbel distribution cannot be used to describe the joint distribution of the maxima at various durations. A problem is that such an extension must have the property that the annual maximum at a particular duration never exceeds the largest precipitation amounts at the longer durations in the same year.

The method of maximum likelihood (ML) was used to obtain estimates of μ and σ. The resulting estimates are presented in Table 9.2. The logarithms of these estimates roughly vary linearly with $\ln D$.

For the construction of rainfall depth–duration–frequency curves it is useful

Table 9.2 Maximum likelihood estimates of Gumbel distribution parameters for *D*-day annual maximum precipitation amounts at De Bilt (1906–77)

D	$\hat{\mu}$ (mm)	$\hat{\sigma}$ (mm)	$\hat{\tau} = \ln \hat{\mu}$	$\hat{c} = \hat{\sigma}/\hat{\mu}$
1	29.3	7.04	3.38	0.241
2	38.8	8.87	3.66	0.228
4	49.3	11.35	3.90	0.230
7	62.5	14.00	4.14	0.224
10	75.6	16.40	4.33	0.217

to consider the following reparametrisation:

$$\tau = \ln \mu; \quad c = \sigma/\mu. \tag{9.2}$$

The parameter c is related to the coefficient of variation ($CV = \pi c/[(1 + \gamma c)\sqrt{6}]$, where $\gamma = 0.5772\ldots$ is Euler's constant). Table 9.2 shows that this parameter hardly varies with D.

The asymptotic second-order moments of the ML estimates $\hat{\tau}$ and $\hat{c}$ are given by:

$$\begin{aligned}
\text{var}(\hat{\tau}) &\approx A\frac{c^2}{n} \approx 1.109\frac{c^2}{n},\\
\text{var}(\hat{c}) &\approx [Ac^2 + 12(\gamma - 1)c/\pi^2 + 6/\pi^2]\frac{c^2}{n}\\
&\approx [1.109c^2 - 0.514c + 0.608]\frac{c^2}{n},\\
\text{cov}(\hat{\tau}, \hat{c}) &\approx -[Ac + 6(\gamma - 1)/\pi^2]\frac{c^2}{n} \approx -[1.109c - 0.257]\frac{c^2}{n}
\end{aligned} \tag{9.3}$$

with n the sample size and $A = 1 + 6(\gamma - 1)^2/\pi^2$. From the last equation it follows that $\hat{\tau}$ and $\hat{c}$ are asymptotically uncorrelated if $c = 0.232$. The ML estimates of c in Table 9.2 are close to this value and thus for the annual maxima at De Bilt $\hat{\tau}$ and $\hat{c}$ are nearly uncorrelated. A consequence of the slow variation of $\hat{c}$ with duration is that the standard errors of $\hat{\tau}$ and $\hat{c}$ are almost constant over the considered range of durations.

9.3 ESTIMATION OF THE CORRELATION BETWEEN PARAMETER ESTIMATES

Recently Smith (1992) proposed a method of approximating the covariance matrix of parameter estimates for spatially dependent data. Although independence between the series at different sites was assumed to construct the likelihood function, the method allows for the effect of spatial dependence on the variances and covariances of the parameter estimates. In this approach the estimates and their covariance matrix are obtained without specifying a joint distribution. Smith's method is also applicable to the problem considered here. The jack-knife and the bootstrap are alternative methods to obtain the covariance matrix of the parameter estimates.

9.3.1 Smith's method

Let $\boldsymbol{\theta} = (\tau_1, c_1, \ldots, \tau_k, c_k)'$ denote the vector of Gumbel parameters for k different durations. In case of independence between the annual maxima, the log-likelihood for all durations combined would read:

$$L(\boldsymbol{\theta}) = \sum_{i=1}^{n} \sum_{j=1}^{k} \ln f(x_{ij}; \tau_j, c_j), \tag{9.4}$$

where x_{ij} is the maximum precipitation amount in year i for the jth duration and $f(\cdot)$ is the Gumbel density. The ML estimates $\hat{\tau}_j, \hat{c}_j$ for the individual durations maximise $L(\boldsymbol{\theta})$ with respect to $\boldsymbol{\theta}$. Let $\hat{\boldsymbol{\theta}}$ denote the vector of these estimates. Then $\nabla L(\hat{\boldsymbol{\theta}}) = \mathbf{0}$, and a Taylor expansion of $\nabla L(\hat{\boldsymbol{\theta}})$ about $\boldsymbol{\theta}$ yields the standard approximation:

$$\hat{\boldsymbol{\theta}} \approx \boldsymbol{\theta} + \{-\nabla^2 L(\boldsymbol{\theta})\}^{-1} \nabla L(\boldsymbol{\theta}), \tag{9.5}$$

where ∇ and ∇^2 denote gradient and Hessian respectively. The Hessian matrix $\nabla^2 L(\boldsymbol{\theta})$ contains many zero values in this application. The nonzero components can be approximated by their expected values from equation (9.3). The covariance matrix of $\hat{\boldsymbol{\theta}}$ then becomes

$$\mathrm{cov}(\hat{\boldsymbol{\theta}}) \approx \mathbf{H}^{-1} \mathbf{V} \mathbf{H}^{-1}, \tag{9.6}$$

where $\mathbf{H} = -E\{\nabla^2 L(\boldsymbol{\theta})\}$ and $\mathbf{V} = \mathrm{cov}\{\nabla L(\boldsymbol{\theta})\}$. Because of the correlation between the annual maxima at different durations the equality $\mathbf{V} = \mathbf{H}$ no longer applies. An estimate of $\mathbf{V}$ can, however, be obtained as follows.

A reasonable assumption is that the contributions $h_1(\boldsymbol{\theta}), \ldots, h_n(\boldsymbol{\theta})$ from the separate years to $L(\boldsymbol{\theta})$ are independent and identically distributed (i.i.d.) random variables. Consequently,

$$\nabla L(\boldsymbol{\theta}) = \sum_{i=1}^{n} \nabla h_i(\boldsymbol{\theta}) \tag{9.7}$$

expresses $\nabla L(\boldsymbol{\theta})$ as a sum of n i.i.d. terms, and thus

$$\mathbf{V} = \mathrm{cov}\{\nabla L(\boldsymbol{\theta})\} = n \mathbf{V}_{\nabla h} \tag{9.8}$$

where $\mathbf{V}_{\nabla h} = \mathrm{cov}\{\nabla h_i(\boldsymbol{\theta})\}$. The estimate $\hat{\mathbf{V}}$ of $\mathbf{V}$ is obtained by replacing $\mathbf{V}_{\nabla h}$ by the empirical covariance marix $\hat{\mathbf{V}}_{\nabla h}$ of the observed $\nabla h_1(\hat{\boldsymbol{\theta}}), \ldots, \nabla h_n(\hat{\boldsymbol{\theta}})$. In this application each $\nabla h_i(\hat{\boldsymbol{\theta}})$ is a vector of the form

$$\nabla h_i(\hat{\boldsymbol{\theta}}) = \left[\frac{\partial}{\partial \tau_1} \ln f(x_{i1}), \frac{\partial}{\partial c_1} \ln f(x_{i1}), \ldots, \frac{\partial}{\partial \tau_k} \ln f(x_{ik}), \frac{\partial}{\partial c_k} \ln f(x_{ik}) \right]',$$

Table 9.3 Estimated correlation matrices of the Gumbel parameters τ and c for D-day annual maximum precipitation amounts at De Bilt (1906–77) according to Smith's method, the jack-knife and the bootstrap

	D	1	2	4	7	10	1	2	4	7	10
Smith	1	1.00				τ	1.00				c
	2	0.83	1.00				0.72	1.00			
	4	0.68	0.80	1.00			0.43	0.56	1.00		
	7	0.53	0.66	0.90	1.00		0.23	0.37	0.78	1.00	
	10	0.50	0.60	0.81	0.90	1.00	0.15	0.23	0.62	0.86	1.00
Jack-knife	1	1.00				τ	1.00				c
	2	0.83	1.00				0.72	1.00			
	4	0.68	0.80	1.00			0.41	0.54	1.00		
	7	0.53	0.66	0.90	1.00		0.22	0.36	0.79	1.00	
	10	0.50	0.59	0.81	0.90	1.00	0.14	0.22	0.62	0.86	1.00
Bootstrap	1	1.00				τ	1.00				c
	2	0.80	1.00				0.70	1.00			
	4	0.67	0.80	1.00			0.40	0.52	1.00		
	7	0.53	0.64	0.89	1.00		0.21	0.35	0.79	1.00	
	10	0.50	0.60	0.83	0.93	1.00	0.15	0.23	0.64	0.86	1.00

where the partial derivatives are evaluated at the ML estimates. The matrix $\hat{\mathbf{V}}_{\nabla h}$ is then given by

$$\hat{\mathbf{V}}_{\nabla h} = \frac{1}{n} \sum_{i=1}^{n} \{\nabla h_i(\hat{\boldsymbol{\theta}}) \cdot \nabla h_i(\hat{\boldsymbol{\theta}})'\}. \tag{9.9}$$

Table 9.3 shows the resulting correlation matrices of the ML estimates $\hat{\tau}$ and $\hat{c}$ for the annual maximum amounts at De Bilt. The correlation coefficients between the ML estimates of τ are roughly equal to those of the annual maximum amounts in Table 9.1. This similarity between the correlation coefficients could be expected because τ is closely related to the mean and the sample means for the various durations have the same correlation matrix as the individual annual maxima as a result of the independence between years. Table 9.3 also shows that the correlation coefficients for the parameter c are considerably lower than those for τ.

9.3.2 The jack-knife

In the jack-knife method the Gumbel parameters are also estimated for the n possible situations in which one year is deleted from the complete data set. Let

$\hat{\boldsymbol{\theta}}_{(i)}$ denote the vector of ML estimates for the case where the ith year is omitted ($i = 1, \ldots, n$). From these estimates the pseudo-values $\mathbf{P}_i$ are calculated as

$$\mathbf{P}_i = n\hat{\boldsymbol{\theta}} - (n-1)\hat{\boldsymbol{\theta}}_{(i)}, \quad i = 1, \ldots, n. \tag{9.10}$$

The standard jack-knife estimate $\hat{\boldsymbol{\theta}}_{\mathrm{J}}$ of $\boldsymbol{\theta}$ is given by $\hat{\boldsymbol{\theta}}_{\mathrm{J}} = \sum_{i=1}^{n} \mathbf{P}_i / n$. This estimate is generally less biased than $\hat{\boldsymbol{\theta}}$. For a sample size of 72 the bias in the ML estimates of the Gumbel parameters is very small and there is no advantage in using $\hat{\boldsymbol{\theta}}_{\mathrm{J}}$ instead of $\hat{\boldsymbol{\theta}}$. The jack-knife, however, also offers the possibility of estimating the covariance matrix of $\hat{\boldsymbol{\theta}}$. This covariance estimate is obtained as

$$\mathbf{S}_{\mathrm{J}} = [n(n-1)]^{-1} \sum_{i=1}^{n} (\mathbf{P}_i - \hat{\boldsymbol{\theta}}_{\mathrm{J}})(\mathbf{P}_i - \hat{\boldsymbol{\theta}}_{\mathrm{J}})'. \tag{9.11}$$

The resulting correlation matrices of the ML estimates $\hat{\tau}$ and $\hat{c}$ for the annual maximum amounts at De Bilt are presented in Table 9.3. There is a striking correspondence between the jack-knife estimates and those from Smith's method.

9.3.3 The linear jack-knife

In the jack-knife method the ML equations have to be solved $n + 1$ times for each duration. The aim of the linear jack-knife is to perform this iteration procedure only once. The idea of a linear jack-knife is well known in nonlinear regression (Simonoff and Tsai, 1986). Apart from a large reduction in computer time, it also provides an explanation for the similarity between the approximated correlation coefficients using Smith's method and the jack-knife estimates.

Let $L_{(i)}(\boldsymbol{\theta})$ be the combined log-likelihood for all durations for the case where the ith year is omitted. Obviously $\nabla L_{(i)}(\hat{\boldsymbol{\theta}}_{(i)}) = \mathbf{0}$. An approximation to $\hat{\boldsymbol{\theta}}_{(i)}$ can be obtained from a Taylor expansion of $\nabla L_{(i)}(\hat{\boldsymbol{\theta}}_{(i)})$ about $\hat{\boldsymbol{\theta}}$. This results in

$$\hat{\boldsymbol{\theta}}_{(i)}^{\mathrm{lin}} = \hat{\boldsymbol{\theta}} - \{\nabla^2 L_{(i)}(\hat{\boldsymbol{\theta}})\}^{-1} \nabla L_{(i)}(\hat{\boldsymbol{\theta}}) = \hat{\boldsymbol{\theta}} + \{\nabla^2 L_{(i)}(\hat{\boldsymbol{\theta}})\}^{-1} \nabla h_i(\hat{\boldsymbol{\theta}}). \tag{9.12}$$

Here use has been made of the fact that $L(\boldsymbol{\theta}) = L_{(i)}(\boldsymbol{\theta}) + h_i(\boldsymbol{\theta})$ and $\nabla L(\hat{\boldsymbol{\theta}}) = 0$. Equation (9.12) is just the result of the first iteration in the Newton–Raphson procedure with starting value $\hat{\boldsymbol{\theta}}$. The use of this linear approximation to $\hat{\boldsymbol{\theta}}_{(i)}$ in equation (9.10) yields for the annual maximum amounts at De Bilt almost the same covariance estimates as the full jack-knife. The Hessian matrix $\nabla^2\{L_{(i)}(\hat{\boldsymbol{\theta}})\}$ can be approximated by $-(n-1)\mathbf{H}/n$, giving $\mathbf{P}_i \approx \hat{\boldsymbol{\theta}} + n\mathbf{H}^{-1}\nabla h_i(\hat{\boldsymbol{\theta}})$, $\hat{\boldsymbol{\theta}}_J \approx \hat{\boldsymbol{\theta}}$, and thus:

$$\mathbf{S}_{\mathrm{J}} \approx \frac{n^2}{n-1} \mathbf{H}^{-1} \hat{\mathbf{V}}_{\nabla h} \mathbf{H}^{-1} = \frac{n}{n-1} \mathbf{H}^{-1} \hat{\mathbf{V}} \mathbf{H}^{-1}. \tag{9.13}$$

The jack-knife estimates of the correlation coefficients are therefore almost identical to those from Smith's method.

9.3.4 The bootstrap

In the bootstrap method, B new samples of n years are drawn from the original set of annual maxima by choosing years randomly with replacement, cf. Zwiers and Ross (1991). Every time when a particular year is selected the maximum precipitation amounts for the k durations are added to the bootstrap sample. Let $\hat{\boldsymbol{\theta}}_b$ be the vector of the ML estimates of the Gumbel parameters for the bth bootstrap sample ($b = 1, \ldots, B$). Then, the estimate $\mathbf{S}_B$ of the covariance matrix of $\hat{\boldsymbol{\theta}}$ is obtained from the $\hat{\boldsymbol{\theta}}_b$'s as

$$\mathbf{S}_B = \frac{1}{B-1} \sum_{b=1}^{B} (\hat{\boldsymbol{\theta}}_b - \hat{\boldsymbol{\theta}}_{(\cdot)})(\hat{\boldsymbol{\theta}}_b - \hat{\boldsymbol{\theta}}_{(\cdot)})', \tag{9.14}$$

where $\hat{\boldsymbol{\theta}}_{(\cdot)} = \sum_{b=1}^{B} \hat{\boldsymbol{\theta}}_b / B$.

For the annual maxima at De Bilt 2500 bootstrap samples were drawn. The resulting estimated correlation matrices of $\hat{\tau}$ and $\hat{c}$ are presented in Table 9.3. There is again a striking resemblance to the values from Smith's method, which can be explained as follows.

Analogous to equation (9.5) the vector $\hat{\boldsymbol{\theta}}_b$ can be approximated by

$$\hat{\boldsymbol{\theta}}_b \approx \hat{\boldsymbol{\theta}} + \{\quad \nabla^2 L_b(\hat{\boldsymbol{\theta}})\}^{-1} \nabla L_b(\hat{\boldsymbol{\theta}}), \tag{9.15}$$

with $L_b(\boldsymbol{\theta})$ the combined log-likelihood for all annual maxima in the bth bootstrap sample.

For each b the Hessian matrix $-\nabla^2 L_b(\hat{\boldsymbol{\theta}})$ can be approximated by $\mathbf{H}$, giving

$$\mathbf{S}_B \approx \frac{B}{B-1} \mathbf{H}^{-1} \hat{\mathbf{V}}_B \mathbf{H}^{-1}, \tag{9.16}$$

where $\hat{\mathbf{V}}_B$ is the sample covariance matrix of the observed $\nabla L_b(\hat{\boldsymbol{\theta}})$ in the B bootstrap samples. This matrix tends to the true covariance matrix of $\nabla L_b(\hat{\boldsymbol{\theta}})$ as $B \to \infty$. Now each $\nabla L_b(\hat{\boldsymbol{\theta}})$ is the sum of a random sample $\nabla h_1^*(\hat{\boldsymbol{\theta}}), \ldots, \nabla h_n^*(\hat{\boldsymbol{\theta}})$ from the distribution which puts probability mass $1/n$ on each observed $\nabla h_i(\hat{\boldsymbol{\theta}})$. Because $E\{\nabla h_i^*(\hat{\boldsymbol{\theta}})\} = \sum_{i=1}^{n} \nabla h_i(\hat{\boldsymbol{\theta}})/n = \mathbf{0}$, the covariance matrix of $\nabla L_b(\hat{\boldsymbol{\theta}})$ is given by

$$\operatorname{cov}\{\nabla L_b(\hat{\boldsymbol{\theta}})\} = nE\{\nabla h_i^*(\hat{\boldsymbol{\theta}}).\nabla h_i^*(\hat{\boldsymbol{\theta}})'\} = n\hat{\mathbf{V}}_{\nabla h} = \hat{\mathbf{V}}. \tag{9.17}$$

Consequently $\hat{\mathbf{V}}_B$ tends to $\hat{\mathbf{V}}$ as B becomes large and the matrix $\mathbf{S}_B$ is then almost identical to the estimate from Smith's method.

9.3.5 The estimated correlation between $\hat{\tau}$ and $\hat{c}$

For a sample size of 72 the asymptotic formulae (9.3) give a very good approximation to the second-order moments of $\hat{\tau}$ and $\hat{c}$. It turns out that Smith's method, the jack-knife and the bootstrap well reproduce the variances (in particular the variance of $\hat{\tau}$). Some problems are met, however, with the correlation coefficient between $\hat{\tau}$ and $\hat{c}$, see Table 9.4. The theoretical value from the asymptotic formula (9.3) is about 0. Large deviations from 0 are found, however, for $D = 1$ day and $D = 10$ days. This suggests that the correlation estimator can be very variable. To get some insight into the statistical properties of the correlation estimator 100 samples of size 72 were drawn from a Gumbel distribution with parameters $\tau = 0$ and $c = 0.232$. For each sample $\widehat{\text{corr}}\ (\hat{\tau}, \hat{c})$ was estimated by (i) the observed information, (ii) the expected information, substituting $\hat{c}$ for c in equation (9.3), and (iii) Smith's method. Table 9.5 presents a summary of the results. For all three methods the average of $\widehat{\text{corr}}\ (\hat{\tau}, \hat{c})$ is close to the theoretical asymptotic value 0. With respect to the variability of $\widehat{\text{corr}}\ (\hat{\tau}, \hat{c})$ there

Table 9.4 Correlation coefficients between the estimated Gumbel parameters $\hat{\tau}$ and $\hat{c}$ for D-day annual maximum precipitation amounts at De Bilt (1906–77)

D	Expected information, equation (9.3)	Smith's method	Jack-knife	Bootstrap
1	−0.01	0.38	0.37	0.35
2	0.01	0.14	0.13	0.11
4	0.00	−0.11	−0.14	−0.14
7	0.01	−0.08	−0.10	−0.13
10	0.02	−0.23	−0.25	−0.24

Table 9.5 Properties of estimated correlation coefficients between $\hat{\tau}$ and $\hat{c}$ in 100 random samples of size 72 from a Gumbel distribution where the parameters are chosen such that the asymptotic correlation between $\hat{\tau}$ and $\hat{c}$ equals 0

Method of estimation corr$(\hat{\tau}, \hat{c})$	Smallest value	Average	Largest value	Standard deviation
Observed information	−0.07	0.01	0.09	0.03
Expected information	−0.07	0.01	0.08	0.03
Smith's method	−0.38	0.03	0.61	0.18

is hardly any difference between the use of the observed and expected information. For Smith's method, however, the standard deviation of $\widehat{\text{corr}}\,(\hat{\tau}, \hat{c})$ is six times as large as that from the other methods and $\widehat{\text{corr}}\,(\hat{\tau}, \hat{c})$ ranges between -0.38 and 0.61. The values for the annual maxima at De Bilt are all within this range. The same simulation experiment shows that the three methods are almost identical with respect to the standard error of $\hat{\tau}$ and that the standard deviation of the approximated standard error of $\hat{c}$ using Smith's method is about twice as large as that from the other two methods.

9.4 MATHEMATICAL DESCRIPTION OF THE CHANGE OF THE GUMBEL PARAMETERS WITH DURATION

Because $\hat{\tau}$ and $\hat{c}$ are practically uncorrelated, these two parameters can be treated separately. In the first instance the variation of the Gumbel parameters with duration is described by

$$q_j = a + b \ln (D_j/\dot{D}), \; j = 1, \ldots, k \tag{9.18}$$

where q_j represents either τ_j or c_j, D_j denotes the jth duration and $\dot{D}$ is the geometric mean of $D_1, \ldots, D_k$. The standardisation of D_j by $\dot{D}$ facilitates the demonstration of the effect of dependence on the standard errors of the estimated regression coefficients. The regression coefficient a refers to the mean level of $q_1, \ldots, q_k$.

In matrix notation, equation (9.18) can be written:

$$\mathbf{q} = \mathbf{X}\boldsymbol{\beta}, \tag{9.19}$$

with

$$\mathbf{q} = \begin{pmatrix} q_1 \\ \cdot \\ \cdot \\ \cdot \\ q_k \end{pmatrix}, \; \mathbf{X} = \begin{pmatrix} 1 & \ln (D_1/\dot{D}) \\ \cdot & \cdot \\ \cdot & \cdot \\ \cdot & \cdot \\ 1 & \ln (D_k/\dot{D}) \end{pmatrix}, \; \boldsymbol{\beta} = \begin{pmatrix} a \\ b \end{pmatrix}.$$

An estimate of $\boldsymbol{\beta}$ can be obtained from the vector $\hat{\mathbf{q}}$ of the ML estimates at the individual durations. The ordinary least squares (OLS) estimator of $\boldsymbol{\beta}$ reads:

$$\hat{\boldsymbol{\beta}}_{\text{OLS}} = (\mathbf{X}'\mathbf{X})^{-1}\mathbf{X}'\hat{\mathbf{q}}, \tag{9.20}$$

and the covariance matrix of this estimator is given by (Draper and Smith, 1981, p. 110):

$$\text{cov}(\hat{\boldsymbol{\beta}}_{\text{OLS}}) = (\mathbf{X}'\mathbf{X})^{-1}\mathbf{X}'\mathbf{C}\mathbf{X}(\mathbf{X}'\mathbf{X})^{-1}, \tag{9.21}$$

where **C** is the covariance matrix of $\hat{\mathbf{q}}$. The elements of this matrix are the covariances of the ML estimators of a Gumbel parameter (τ or c) for the selected durations.

Because of the correlation between the ML estimates at different durations the OLS estimates can be less appropriate. In this situation it is more obvious to estimate $\boldsymbol{\beta}$ by the generalised least squares (GLS) method, giving (Draper and Smith, 1981, p. 109):

$$\hat{\boldsymbol{\beta}}_{\mathrm{GLS}} = (\mathbf{X}'\mathbf{C}^{-1}\mathbf{X})^{-1}\mathbf{X}'\mathbf{C}^{-1}\hat{\mathbf{q}}, \tag{9.22}$$

with covariance matrix

$$\mathrm{cov}(\hat{\boldsymbol{\beta}}_{\mathrm{GLS}}) = (\mathbf{X}'\mathbf{C}^{-1}\mathbf{X})^{-1}. \tag{9.23}$$

Table 9.6 shows the estimated regression coefficients $\hat{a}$ and $\hat{b}$ and their standard errors. The unknown matrix **C** in equations (9.21), (9.22) and (9.23) has been replaced here by the estimate using Smith's method. The standard errors in Table 9.6 are based on a weighted average of the elements of **C** (or $\mathbf{C}^{-1}$) and are therefore less variable than the estimated covariances themselves. For the OLS method the standard errors are also given for the case where **C** in equation (9.21) is the diagonal matrix with the average of the approximated standard errors of $\hat{\tau}_1, \ldots, \hat{\tau}_k$ (or $\hat{c}_1, \ldots, \hat{c}_k$) on the main diagonal. This negation of correlation leads to a serious underestimation of the standard error of $\hat{a}$. The standard error of $\hat{b}$ is, however, reasonably reproduced.

Table 9.6 also shows that the use of GLS hardly improves the accuracy of $\hat{a}$ and $\hat{b}$. This is in contrast with the studies by Stedinger and Tasker (1985) and Coles and Tawn (1990) for spatially dependent data where much better estimates were obtained by taking spatial dependence into account. The small gain in efficiency with the use of GLS is even questionable because **C** has been estimated

Table 9.6 Estimated coefficients in equation (9.18) with their standard erors (s.e.) and correlation coefficient for the Gumbel parameters τ and c of the annual maximum precipitation amounts at De Bilt (1906–77). The standard errors in parentheses assume independence between the annual maxima at different durations

	$\hat{a}$	s.e. ($\hat{a}$)	$\hat{b}$	s.e. ($\hat{b}$)	$\widehat{\mathrm{corr}}(\hat{a}, \hat{b})$
Parameter τ					
OLS	3.879	0.025 (0.013)	0.403	0.013 (0.015)	−0.01
GLS	3.902	0.024	0.420	0.012	−0.06
Parameter c					
OLS	0.228	0.015 (0.009)	−0.009	0.012 (0.011)	−0.04
GLS	0.226	0.015	−0.012	0.011	−0.13

from the data. It can further be concluded from Table 9.6 that the coefficient b in equation (9.18) is not significant for the Gumbel parameter c.

The adequacy of the model given by equation (9.18) can be tested with the estimated GLS residuals $\hat{\mathbf{e}} = \hat{\mathbf{q}} - \mathbf{X}'\hat{\boldsymbol{\beta}}_{\mathrm{GLS}}$, using the statistic

$$U = \hat{\mathbf{e}}'\mathbf{C}^{-1}\hat{\mathbf{e}}. \tag{9.24}$$

For an adequate model U has a χ^2-distribution with k-2 degrees of freedom. For the Gumbel parameters τ and c the values of U are 18.10 and 0.42, respectively. Because $\mathbf{C}$ has been replaced by an estimate here, the theoretical χ^2_3-distribution does not strictly apply. But nevertheless the large value of U for the parameter τ is a strong indication of lack of fit. Figure 9.2 shows that the GLS line overestimates the parameter values at $D = 4$ days and D = 7 days. The GLS method gives a relatively large weight to the values at $D = 1$ day and D = 10 days, because these values have only one correlated neighbour. It is interesting to compare this regression with the linear regression $y_i = a + bx_i$, $i = 1, \ldots, k$ with autocorrelated disturbances according to a first-order autoregressive process, considered in Johnston (1963, p. 191). When there is no trend or serial correlation in the x_i's, then the asymptotic relative efficiency of the OLS estimate of b is less than 0.35 if the first-order serial correlation coefficient of the disturbances exceeds 0.7. The values y_1 and y_k have also only one

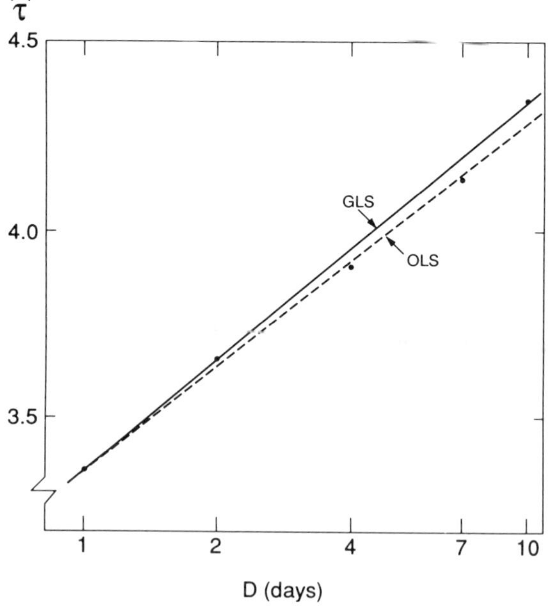

Figure 9.2 Change of the estimated Gumbel parameter τ with duration D

correlated neighbour here, but in contrast to the rainfall depth–duration–frequency curves x_1 and x_k are not the two most extreme values of x. Thus the benefits of the GLS method strongly depend on the configuration of the explanatory variables. The GLS estimate corresponds to the OLS estimate if the columns of the matrix $\mathbf{X}$ are linear combinations of the eigenvectors of $\mathbf{C}$ (Anderson, 1971, pp. 19–20; Chipman, 1979).

Introducing a quadratic term in $\ln(D_j/\dot{D})$ reduces the value of U to 11.43 for the parameter τ. This is still not satisfactory. In fact the OLS line does quite well in Figure 9.2. The largest residual is not more than 0.03. This corresponds to a difference of only 3% in the value of the parameter μ.

9.5 QUANTILE ESTIMATION

The quantile in the annual maximum distribution with exceedance probability $1/T$ is usually called the T-year event x_T in hydrology. For the Gumbel distribution in equation (9.1) x_T is given by

$$x_T = \mu + \sigma y_T = e^{\tau}(1 + c y_T), \tag{9.25}$$

where $y_T = -\ln(-\ln(1 - 1/T))$. An estimate $\hat{x}_T$ of x_T can be obtained by replacing τ and c by their ML estimates for the desired duration (method 1) or by using a rainfall depth–duration–frequency curve (method 2). In this section the rainfall depth–duration–frequency curves are based on the OLS estimates in Table 9.6, assuming c to be the same for all durations. Figure 9.1 at the beginning of this chapter represents one of these curves.

For the annual maximum amounts at De Bilt the correlation between $\hat{\tau}$ and $\hat{c}$ can be neglected. Then the following approximation to the variance of $\hat{x}_T$ is obtained:

$$\operatorname{var}\hat{x}_T \approx x_T^2 \operatorname{var}(\hat{\tau}) + y_T^2 e^{2\tau} \operatorname{var}(\hat{c}). \tag{9.26}$$

In method 1 the variances of $\hat{\tau}$ and $\hat{c}$ are given by equation (9.3). In method 2 the standard error of $\hat{a}$ in Table 9.6 provides an estimate of var $\hat{c}$, whereas an estimate of var $\hat{\tau}$ can be obtained from this table using

$$\operatorname{var}\hat{\tau} \approx \operatorname{var}\hat{a} + [\ln(D/\dot{D})]^2 \operatorname{var}\hat{b}. \tag{9.27}$$

The correlation between $\hat{a}$ and $\hat{b}$ has been neglected here.

For the 4-day annual maximum distribution at De Bilt the two methods of quantile estimation are compared in Table 9.7. From the table it is seen that there is only a small reduction in the standard error of $\hat{x}_T$ if the annual maxima at several durations are considered. The reduction in standard error is even less for the 1-day and 10-day annual maxima because of a larger value of var $(\hat{\tau})$

Table 9.7 Values of the estimated T-year event and their standard errors (s.e.) for the 4-day annual maximum precipitation amounts at De Bilt. The estimate $\hat{x}_{T,1}$ is based on the 4-day annual maxima only, whereas $\hat{x}_{T,2}$ is obtained from the rainfall depth–duration–frequency curves. The quantile estimates and their standard errors are expressed in millimetres

Return period T	$\hat{x}_{T,1}$	s.e.$(\hat{x}_{T,1})$	$\hat{x}_{T,2}$	s.e.$(\hat{x}_{T,2})$	s.e.$(\hat{x}_{T,2})$/s.e.$(\hat{x}_{T,1})$
2	53.5	1.6	55.1	1.4	0.89
10	74.8	3.1	76.9	2.6	0.84
100	101.5	5.4	104.1	4.4	0.82

Table 9.8 Values of the estimated 10-year event from the rainfall depth–duration–frequency curve for De Bilt (Figure 9.1) and their standard errors (s.e.). The standad errors in parentheses assume independence between the annual maxima at different durations. The quantile estimates and their standard errors are expressed in millimetres

D	$\hat{x}_{10}$	s.e.$(\hat{x}_{10})$
1	43.9	1.7 (1.2)
4	76.9	2.6 (1.4)
10	111.2	4.0 (2.7)

from equation (9.27) as a result of the relatively large contribution of the term $[\ln (D/\dot{D})]^2 \operatorname{var} \hat{b}$.

The underestimation of the standard error of $\hat{a}$ in the case of negation of dependence leads also to an underestimation of the standard error of the quantile estimates. Table 9.8 demonstrates this for the estimated 10-year event $\hat{x}_{10}$. The most severe underestimation of the standard error of $\hat{x}_{10}$ occurs at $D = 4$ days, because $[\ln (D/\dot{D})]^2 \operatorname{var} \hat{b}$ in equation (9.27) is relatively unimportant for this duration.

9.6 CONCLUSIONS

Annual maxima of precipitation amounts over different durations show a rather strong dependence. This leads to correlation between the estimated parameters of the assumed Gumbel distribution for the maxima at each duration. A recently proposed procedure by Smith (1992) to assess the effect of spatial dependence on standard errors of parameter estimates can also be applied to approximate the correlation coefficients between the estimated Gumbel parameters in this situation. For a 72-year record at De Bilt the approximated correlation coefficients with this procedure are almost identical to those obtained by using the jack-knife or the bootstrap.

Estimating the change of the Gumbel parameters with duration by generalised least squares (GLS) does not provide an improvement over the use of ordinary least squares (OLS). This is caused by the special configuration of the explanatory variables. The GLS residuals indicate, however, that the relation $\tau = a + b \ln D$ does no sufficiently describe the change of the parameter $\tau = \ln \mu$ with duration D. This lack of fit needs further investigation.

For the Gumbel distribution only a slight improvement can be achieved by estimating a quantile from a rainfall depth–duration–frequency curve. This gain in efficiency might be counterbalanced by small systematic errors in the relations describing the change of the Gumbel parameters with duration. The rainfall–depth–duration–frequency curves should therefore be primarily viewed as a tool to obtain the quantiles at any desired duration.

ACKNOWLEDGMENT

I should like to thank Professor R. L. Smith for a copy of his manuscript 'Regional Estimation from Spatially Dependent Data'.

PART IV

Water Resources

10

Water Resources and Statistics: Past, Present and Future

Vujica Yevjevich
Colorado state University, USA

10.1 INTRODUCTION

Human beings and the randomness of nature

Randomness of the phenomena in the fields of hydrology, meteorology and oceanography must have been evident to the human race thousands of years ago. Chance in the occurrence, magnitude and timing of floods, hurricanes, droughts, heavy snowfall and extreme rainfall would have made people aware of the unpredictability of those of the earth's phenomena that are driven by solar energy. Even the most primitive tribes around the world have the concepts of luck and chance in their languages. The first geophysicists to start instrumental observation of random phenomena would have realised that the best way to extract meaningful information on related random variables from collected data would be through statistical techniques. Water resources, observed as natural hydrologic processes, are random phenomena driven by solar energy through the Earth's water cycle in the atmosphere, oceans and continents. They are periodic–stochastic, spatio-temporal processes.

Periodicity due to astronomic cycles

The common astronomic cycles are the causes of periodicity and near-periodicity in all time series relating to fluids on the Earth which are driven by solar energy.

Statistics for the Environment. Edited by Vic Barnett and K. Feridun Turkman

Their daily and annual cycles create daily and annual periodicities in time series. The cycle of the moon's revolution around the Earth is the cause of almost-periodicity in phenomena such as currents in air and tides in oceans, through its interaction with the daily and annual cycles. The annual and daily cycles have a rate of just over 365. Regardless of the noncommensurability of the cycles of day and year (ratios of cycles are irrational numbers), society and science in practice have accepted them as commensurable or as two superposing cycles. The situation is different for the lunar cycle of the moon's revolution around th Earth. The ratio of lunar to daily cycle is just over 28, and of annual to lunar cycle is about 13. The decimal parts of these ratios are not considered negligible in practice, so the lunar aspects of cyclicity are most often considered as almost-periodicity. That assumption is the reason that tides are considered deterministic but almost-periodic phenomena, which never repeat themselves in the same way at a given place on earth.

Long-range periodicities

In addition to the 'short-range' daily, lunar and annual periodicity (or almost-periodicity), there are the three 'long-range' Milankovitch cyclicities, namely the cycle of the Earth's orbital eccentricity, the cycle of the precession of the equinoxes, and the cycle of the obliquity of the Earth's axis. These three cycles range from about 20 000 to about 100 000 years. They are noncommensurable, so their joint impact is a long-range almost-periodicity in the variation of input of solar energy to any place on Earth. It is considered that glaciations and large climatic fluctuations on the Earth have been induced by this almost periodicity in the last 2 million years.

The above short-range and long-range periodicities or almost-periodicities refer to completely different time-series intervals. They are used differently, the former for the solution of water resources problems and the latter for the study of long-range climatic changes. They are not studied together but separately as two different periodic–stochastic processes.

Three main social and economic cycles

Human civilisations use in their activities and counting three cycles based on day, week and year. The weekly cycle is the creation of these civilisations. As water demands and water supplies from reservoirs demonstrate, the effects of these three cycles are evident in the combined natural and man-affected flows of many rivers and aquifers. Water resources development and use often change the existing daily and annual periodic processes by changing the amplitudes and phases of the harmonics from which these periodicities are composed. They also create a weekly cycle by releasing water according to periodically varying weekly water demands, These are, therefore, human-induced or human-reshaped periodicities in water time series. The seven periodicities or almost-periodicities outlined here are the basis of the three areas of application of statistics in water resources: in studying water supplies, in studying water demands and their interactions with water supplies, and in studying climatic changes.

The origin of stochasticity in water time series

Stochasticity of geophysical periodic–stochastic processes results mainly from turbulence in the atmosphere and the oceans and in water environments on the continental surfaces, apart from other types of random causal processes on the Earth including biological processes. It is best illustrated by the simple passage of solar rays to the Earth through the atmosphere. Just above the atmosphere there is little randomness in the solar 'constant', with the diurnal energy input and the daily total input throughout the year to any place on the Earth's surface being relatively simple periodic functions. However, observations of the input of solar energy at the Earth's surface clearly relate to periodic–stochastic processes. The atmosphere introduces nearly all of the randomness into the process. Geophysics has been able to 'marry stochasticity and periodicity' by producing the observed geophysical periodic–stochastic processes. It should be the role of advanced statistics and probability theory to bring about the mathematical dissociation of periodicity and stochasticity in order to understand, describe and eventually simulate samples of these geophysical processes. Two approaches have been used in treating these processes: studying them with, or without separating stochasticity and periodicity.

Hydrological spatio-temporal processes

All geophysical periodic–stochastic processes can be conceived of as spatio-temporal processes. In the most general case they are conceived as four-dimensional processes: We may study only the temporal process at a particular point, or the spatial variation in one, two or three dimensions along with the temporal variation. Alternatively, the processes may be investigated at a given moment in time along a line, over an area, or throughout a volume. In practice these statistical investigations depend on the particular problem being studied. To observe, investigate and apply these processes in the field of water resources, meaningful but optimal statistical information must be extracted from observed data. Observations result from spatial and temporal statistical sampling. Spatial characteristics are obtained by point sampling, and temporal characteristics by observing random variables in time, either as time point sampling or as continuous recording or accumulation throughout the selected time interval. The sampling concept selected in most cases determines the reliability of optimally extracted information from data. Optimal information is a simple concept, namely that the increase of information also increases the cost, with the economic optimum searched for in practical cases.

10.2 CHARACTERISTICS OF PAST APPLICATIONS OF STATISTICS IN SOLUTIONS OF WATER RESOURCES PROBLEMS

The state of water problems and selection of statistical methods

Applications of statistics and stochastic processes to water resources problems have depended in the past on the character of the problems, on the ongoing approach to their solutions, and on the available statistical techniques. As development, conservation, control and protection of water resources has advanced, and as the problems to be solved have become more complex, so more advanced and sophisticated statistical techniques have become needed and justified. When solutions to water resources problems required only the construction of a single structure, intended for a single purpose with the use of only a single water source, the problems to solve were relatively simple. The design of a well needed information only on the potential maximum water yield and on the level fluctuations of the natural aquifer. For a reservoir the designer mainly or only needed water input and water demand for reservoir sizing, and flood frequency distribution for its spillway design.

Four main statistical methods in hydrology in the past

Until some time between the two world wars, most water resources planners, designers, constructors and operators needed only four simple statistical techniques in order to solve 95% or more of their problems. They were: (1) sampling of observations to describe the properties of particular water resources characteristics; (2) derivation of, or inference of, the probability distribution functions, or simply to fit them to the frequency distributions for the crucial water resources random variables; (3) transfer of information between variables or prediction of future outcomes by correlation and regression; and (4) testing of simple statistical hypotheses, relevant to the extraction or transfer of information on water supply and water demand.

Long-range sampling and the year as a critical subsample

It can be safely stated that water resources problems need two types of sampling. First, they need the longest observed samples, especially time series, which provide information on the basic process with the smallest uncertainty. Second, the standard sampling length of the year (or sometimes the day) is used to empirically produce information on random variables which are functions of

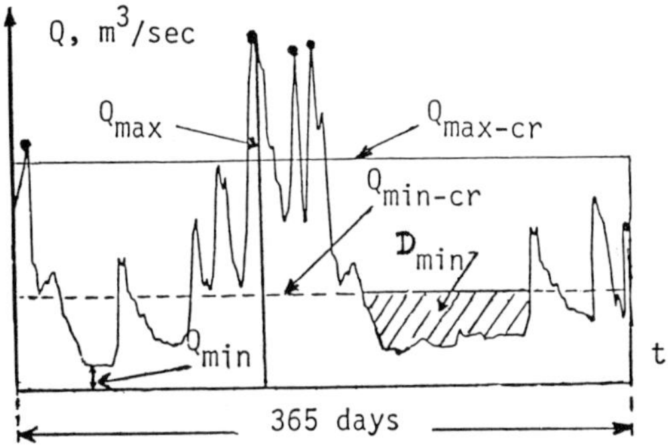

Figure 10.1 Sampling observations within the year

the observed random variable. While the longest available sample leads to the best information on the entire process, the subsample of the year helps produce information which is related to the social time unit needed for economic analyses. Many 'derived variables' referred to this unit have been studied, such as maximum annual flood peak, all peaks above the flood-initiating critical value, flood volume, lowest annual value, maximum total volume of uninterrupted water deficit for a critical demand within the year, and other similar pertinent variables (Figure 10.1). This secondary sampling in essence replaces the often difficult mathematical derivation of statistical characteristics of the 'derived random variables' from the statistical information on the basic process, which is in turn extracted from the longest sample. The new samples are then as long as the number of years of data in the long sample (regardless of whether the original data were collected on an instantaneous, daily, weekly, monthly or any other basis at intervals of less than a year).

Frequency or probability distribution analysis

Techniques for estimating frequencies of hydrological random variables and plotting their positions have enabled water resources specialists to derive freqency distribution curves and to fit the appropriate probability distribution functions (p.d.f.). These sometimes elaborate techniques then permitted the estimation of exceedence or nonexceedence probabilities for variable values outside the range of observed values (Figure 10.2). Problems arise when significantly larger or smaller values (called outliers and inliers respectively) occurred in the historic sample, with these values having in essence much larger or much smaller return periods than the sample size. In some cases there may be a physical reason for the type of tails of the true but unknown p.d.f. They may differ from the tails of the fitted p.d.f. In Figure 10.2, and for the fitted p.d.f. to the annual flows of the River Rhine at Basle, Switzerland out of 150 years of data (1808–1957) only the four lowest values did not closely correspond to

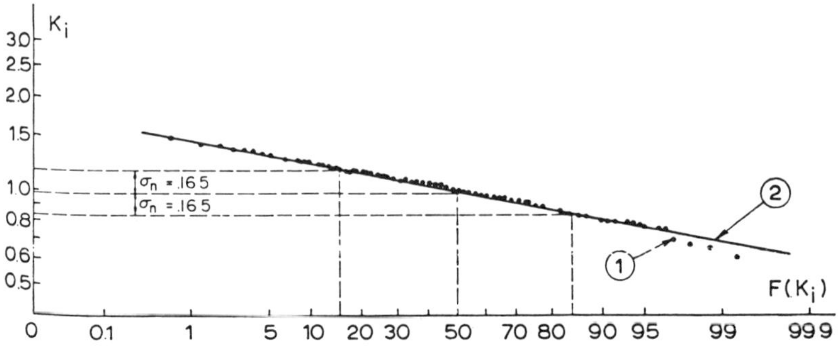

Figure 10.2 Log-normal probability distribution function fitted to frequency distribution of 150 annual flows (1808–1957) of the River Rhine at Basle, Switzerland (flows are given in modular coefficients, or they are divided by the mean)

the fitted log-normal p.d.f. Then, one may speculate that in dry years there was not enough snow and ice melted by high summer temperatures in the Alps, in contrast to the average or humid years. This physical background explains the four extreme inliers of lowest annual flows.

Correlation and regression analysis

Simple or multiple, linear or nonlinear correlation and regression have been important statistical techniques for dealing with hydrology and water resources (Figure 10.3). They have served three basic purposes: filling in missing data in

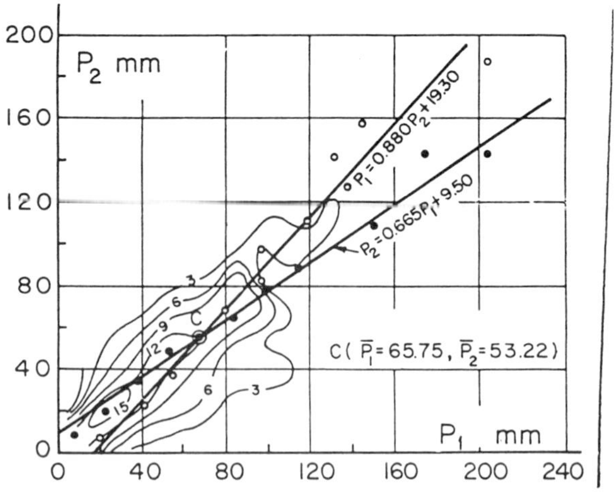

Figure 10.3 Two linear regression lines between monthly precipitation series of two gauging stations, estimated for filling in missing monthly values

observations by correlation between variables of adjacent stations, extracting and/or transferring information, and forecasting future events when physical relationships have permitted a high correlation to exist. The problem in filling in too many missing data, in transfer of information and in forecasting, was always that the resulting new series had smaller estimated variances than they should have. That is, regression equations (Figure 10.3) always give the expected value at a position in the series, thus neglecting the random variation about that value. This fact then meant a smaller estimated variance for a filled in series, or a series with a non-negligible number of values transferred or predicted by correlation and regression. Identification of the variation of the real values about the expected value obtained by regression has recently permitted the generation (simulation) and addition of these generated random values to the values estimated by regression.

Testing statistical hypotheses in hydrology

The fourth significant application of statistics in hydrology and water resources, has been statistical inferential techniques. An example is the inference of the hypothesis that two variables are uncorrelated, or that an annual precipitation or an annual river-flow series was time-dependent (Figure 10.4). If only the first autocorrelation coefficient, estimated for that series, falls outside the confidence limits at a given probability level, drawn around the expected value of that coefficient for an independent series, two hypotheses may be tested: first, that series is time-dependent and second, that the first-order autoregressive model best represents that dependence. Not all inference problems have been as simple as the four examples of Figure 10.4. For example, one complex inference problem in hydrology is the use of stepwise multivariate correlation and regression. In that case, a variable is studied as potentially related to a total of n other variables, from which the stepwise multiple correlation shows only $m (n > m)$ variables to have regression coefficients statistically different from zero. In that case, significant departure from zero of the multiple correlation coefficient obtained is a function of both n and m. In a study by the writer of this paper, the monthly coastal precipitation in California was related to the average

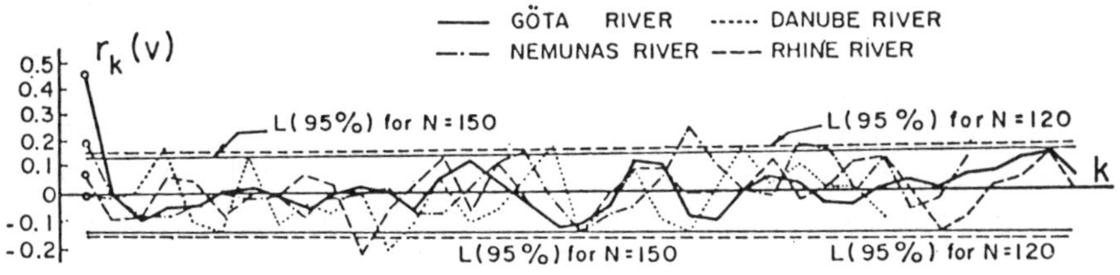

Figure 10.4 Correlograms of annual runoff of four rivers with long records, for which only the first autocorrelation coefficient is likely to be significantly greater than zero

monthly surface temperatures of many quadrants of the Pacific Ocean relatively close to California. There proved to be a multiple correlation coefficient of about $R = 0.40$. Taking into account the number n of sea-temperature quadrants used and the number of significant variables m obtained by the stepwise correlation method, the difference of this R from zero was not statistically significant.

Contributions to statistics by water resources specialists

During the last century of application of statistics to hydrology and water resources, hydrologists and water resources planners and designers not only used the available statistical techniques but also contributed significantly to the development of new methods. An example is the scaling of the cumulative standard normal probability distribution function in such a way that the normal p.d.f. gives a straight line when plotted using Cartesian–normal probablity scales. This probability paper was invented by an American hydrologist–engineer, Allen Hazen, and described in 1914. He plotted the reservoir storage as random variable on the Cartesian scale and used the plotting position $(2m - 1)/2n$ on the probability scale, the plot appearing on p. 1551 of his 1914 paper. Then, log–normal paper followed, with variable values on the logarithmic scale and plotting positions (as the exceedence frequency or probability) on the probability scale, with log-normal p.d.f. plotted on this graph as a straight line (Figure 10.2). Gumbel later introduced the double-exponential scale of so-called 'Gumbel paper', with the double-exponential p.d.f. plotting on this graph as a straight line. Practical people in those times needed an easy graphical fit of p.d.f. to frequency curves, and straight-line fits were most appealing. Unfortunately, no general graph paper could be developed for p.d.f. based on the gamma and beta functions.

Another contribution is the generation of new samples based on information of the observed sample. It was also an American hydrologist–engineer, Charles E. Sudler, who did it first. He stated in his paper (Sudler, 1927, p. 626) that 'an artificial record of, say, 1000 years' length may be built up by dividing the curve [he meant the p.d.f.] into 1000 parts and selecting the values read at the middle of each part.... by some chance method'. Sudler used only 50 values, put them on cards, and after shuffling the deck, he drew all cards one by one (without replacement), to create a 50-year sequence. He used 20 such sequences 'thus providing a 1000-year artificial record'. In the reaction to his paper, two discussants (L. S. Hall and S. Eliassen) pointed out that he should have used sampling with replacement of each drawn card. Sudler countered that he had considered it but decided against because he wanted to compute storages for different series of 50 years, each having the same coefficient of variation. According to V. Klemes' knowledge (pers. comm.), the first synthetic (generated) series made by sampling with replacement was that of Barnes (1954).

A third examples of contribution is the development of the residual mass curve in the design of storage capacities. Sudler (1927) also introduced, apparently for the first time (in North America at least) a river-flow mass curve plotted as the cumulative plot of differences from the mean (this was later called the 'residual mass curve', though the name did not catch on in the American literature). For this, he was congratulated in the discussion of his paper by A. Hazen, who said: 'After having tried this method on an example, the writer wonders that it was not done long ago'. In developing the properties of range, Hurst (1951, 1956, 1957) used the residual mass curve, while Varlet (1923) used the ordinary mass curve, as did Schoklitsch (1923).

A fourth example of the contribution of hydrological engineers to practical advancement in the application of statistics to water resources was the 'stretched-thread method' of mass curves of input and output (supply and demand) of reservoir capacity design, Fig. 10.5. The author who was probably the first to published this method, under the French term 'la règle du fil tendu', was Henri Varlet (1923, reprinted in 1966). If the residual water-supply input mass curve was plotted, and then a parallel line at the distance of the assumed reservoir storage capacity was drawn, the shortest water-demand output mass curve by the 'stretched-thread line' between the two curves is the exact solution in terms of the calculus of variations, to which both linear and dynamic programming are only approximations (Klemes, 1979). Klemes programmed the stretched-thread method as a numerical algorithm (according to him, a relatively trivial task). The algorithm was so fast that a problem of finding an optimal output (finding the 'stretched thread' or 'fil tendu') on a computer for a sample of 30 years of monthly flows was done in one-hundredth of a second, being an exact solution, while at that time (1979)

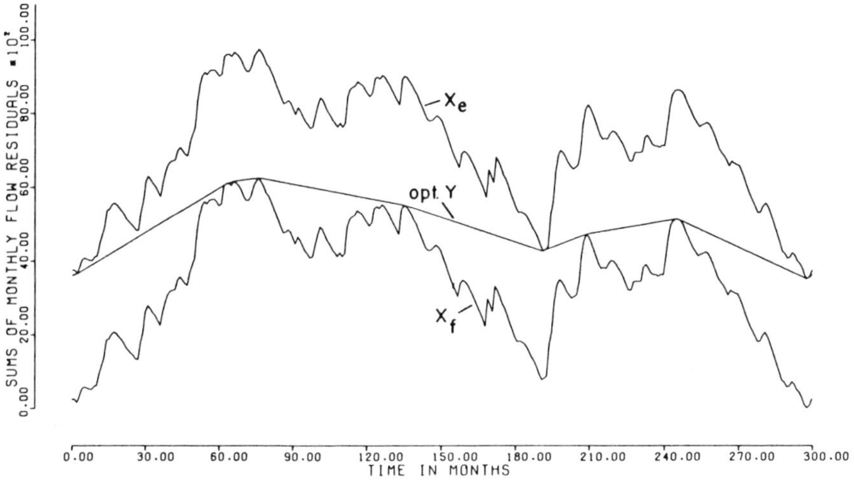

Figure 10.5 The stretched-thread line as the optimal water release from a reservoir with given storage capacity

the solution by dynamic programming took about 10 minutes of computer time and was still only an approximation.

It is characteristic that all the above four examples, as 'firsts', have been made in connection with the analysis of water storage problems. The start of the development of statistics of the range (Hurst, 1951, 1956, 1957; Hurst *et al.* 1965) is due to hydrologist–engineers, H. E. Hurst and his co-authors, also in the study of water storage problems.

10.3 THE PRESENT STATE OF APPLICATION OF ADVANCED STATISTICS IN HYDROLOGY AND WATER RESOURCES

The current complexity of water resources problems

Water resources problems and their solutions have become very complex towards the end of the 20th century. Instead of the single-structure approach, the state of planning is concerned now with systems of various levels, including both subsystems and supersystems. These latter are composed of two or more systems, like the urban water supply, water sewerage and water drainage systems tied together into the urban water supersystem. Instead of one purpose, planners of water resources systems are making every effort to include as many purposes as justified, under the concept of the most economical multi-purpose solution. Instead of using a single source of water for a system (spring, aquifer, lake, reservoir, river, etc.), the modern planning concept is to use all sources of water in the region of a water resources system, thus applying the multi-source concept. These new integrated, comprehensive, multi-structure, multi-purpose and multi-source approaches to development, conservation, control and protection of water resources are having a fundamental influence on the application of statistics and stochastic processes in hydrology and water resources.

A set of reservoirs in a river basin or in any water resources system has both water inputs and water outputs as spatially dependent periodic–stochastic processes. Often, even inputs (supplies) and outputs (demands, uses, or deliveries) are mutually related processes, regardless of whether water supplies are natural, man-affected, or a combination of both types of hydrologic process. A set of wells over an aquifer, or over several adjacent aquifers, may have partial aquifer recharges, which supply individual wells with water, as the spatially dependent processes. Most natural hydrologic periodic–stochastic processes have components, after periodicity in parameters is removed, as the time-dependent (mostly stationary) stochastic processes. In summary, periodicity and trends in parameters, stochastic time-dependence, dependence in space (in these cases, over areas), and dependence between supplies and demands, make water resources input and output processes highly complex. That may be the major reason why solutions of many water resources problems, particularly water storage problems, are still made by empirical and/or sample-generation methods, rather than by mathematically developed water storage theory based on water storage differential equations. The solid bridge between water storage theory and the solution of practical storage problems is, therefore, still missing.

Developments of temporal stochastic models in hydrology

For solutions of complex general-purpose contemporary water resources problems, it was not sufficient to use only classical statistical techniqus, or newly developed or significantly improved statistical techniques. New methods and new developments were still needed. First, hydrologists, and then statisticians working on hydrologic problems, had to investigate the temporal properties of water resources processes in detail (both supply and demand), to the point of developing reliable mathematical-description models. Thus, autoregressive (AR), periodic–autoregressive (PAR), autoregressive–moving-average (ARMA), periodic–autoregressive–moving-average (PARMA), ARIMA, and other models were investigated with various approaches (for a review of many aspects of these investigations, see Salas *et al.*, 1985). While investigations of practical modelling of water-supply time series have been very widespread in the last 30–40 years, the purpose was rarely to mathematically model the water-demand, water-use, or water-delivery time series (for one of the first such cases, see Salas and Yevjevich, 1972). Equation (10.1) gives the simple model, which can be applied either to water-input (supply) and to water-output (demand-delivery) time series, as examples.

$$x_{p,\tau} = \underbrace{\mathrm{Tm}_{p,\tau} + \mathrm{Ts}_{p,\tau}}_{\text{Trend components}} \Big\{ \underbrace{\mu_\tau + \sigma_\tau}_{\text{Periodic components}} \Big[\overbrace{\sum_{j=1}^{m} \alpha_{j,\tau-j}\varepsilon_{p,\tau-j} + \Big(1 - \sum_{i=1}^{m}\sum_{j=1}^{m} \alpha_{i,\tau-i}\alpha_{j,\tau-j}\rho_{|i-j|,\tau-k}\Big)^{1/2}}^{\text{Dependence structure}} \underbrace{\xi_{p,\tau}}_{\substack{\text{Second-order stationary} \\ \text{Independent stochastic component}}} \Big] \Big\} \tag{10.1}$$

This model when applied to water supply series may not have trends in parameters

Periodicity and stochastic modelling

The best models are those which could be supported by the physical characteristics of the process, and not only by the best-fit procedures. Among the various methods of investigation, the following two main approaches have prevailed. In the first approach, periodicity is removed (all significant harmonics of periodic parameters are inferred and removed), then the stochastic dependence is also inferred and removed, resulting in an independent, at least second-order stationary, stochastic process, and whenever feasible also as an independent normally distributed variable. The set of mathematical descriptions thus obtained is conceived of as the mathematical model (see (10.1) for the case of trend–periodic–stochastic time processes). In the second approach, periodicity is not removed, but mathematically modelled along with modelling of the stochastic component.

Main purposes of stochastic modelling of hydrologic time processes

The major hydrologic reasons for modelling time processes for water resources purposes have been: (1) the potential of models for simulation (generation) of sets of new samples, based on the extracted information by the model from observed data, and for the more accurate solution of water resources problems than otherwise feasible; (2) the feasibility of studying various other properties of the time series in question, by simulating new samples, which the direct study did not permit (such as finding experimentally the sampling frequency distributions of hypotheses-testing parameters); (3) the investigation of approximate properties of derived water-resources-related random variables (functions of the basic random variable), with such examples as properties of various ranges and runs, basically by simulating sets of new samples; (4) the investigations which enable explanation of stochastic characteristics of time series by physical processes in nature; and similar reasons.

Input, state and output processes of water systems

To simulate samples of input and output of water resources systems, in general, and of reservoirs, in particular, three basic models are needed: (1) of the (trend)–periodic–stochastic water-supply input process (if it has trends in parameters); (2) of the trend–periodic–stochastic water demand or delivery output process (since outputs nearly always have trends in parameters, because the demand parameters change with time); and (3) of the dependence between independent stochastic components of input and output. Then, planners and designers are able to simulate as many sets of samples of potential future inputs and outputs as a water resources problem may require or justify. This simulation of sets of samples of prescribed sizes (and eventually of the other conditions) enabled the 'experimental determination' of properties of important water-resources-related variables. The best of such examples in hydrology are investigations of ranges and runs, with ranges being-well connected to water storage theory and design, and runs to the solution of flood and drought problems through the theory of crossings. Here also, hydrologist–engineers have contributed their share of advances in statistics and stochastic processes in general, and in their applications in hydrology, in particular.

Ranges and runs as important, derived hydrologic variables

Definitions and properties of ranges, from their means, variances and other parameters to the exact, approximate or asymptotic distributions have been developed over the last 30–40 years. The study was initiated by Hurst (1951), then pursued by statisticians, probabilists and hydrologists, so that by the 1980s significant progress had been made (for a review of that progress up to the late 1970s, see Pegram *et al.*, 1980). It has continued to the present with papers on various aspects of the range appearing in journals occasionally. Investigations of the theory of runs (crossing theory) to droughts have been active from the 1960s on, and still are being intensively pursued at the beginning of the 1990s, mainly by hydrologists and engineers (for a review of results at the earliest stages of investigations, see Guerrero–Salazar and Yevjevich, 1975).

The application of multivariate analysis

These applications in hydrology and water resources have passed through several phases in the last hundred years. Of the four techniques, namely classical multivariate correlation and regression, principal component analysis, factor analysis, and canonical correlation, only the first two have taken root in hydrology. The standard multivariate analysis has sometimes been supplemented by stepwise multiple correlation and regression. For typical recent applications of multivariate techniques in hydrology and water resources, the proceedings book of the 1985 symposium (Shen *et al.*, 1985) is a good example of review. Principal component analysis has been unavoidable in the generation of sets of new multivariate samples, with which a number of mutually dependent multivariate stochastic components of unit variances (or complex periodic–stochastic variables reduced to dependent stationary stochastic components with unit variances) are processed to obtain the same number of mutually independent principal components, though with different variances (classical rotation of the multivariate vector). Several computer algorithms are available for this purpose (some of them designed by hydrologist–engineers). Regardless of several attempts, factor analysis and canonical correlation are not yet standard techniques in hydrology.

Modelling spatial hydrologic processes

Modelling of spatial multi-point hydrologic processes is needed for simulation (generation of samples) of multivariate spatial point-processes. Typical examples are studies of regional drought characteristics by this simulation approach. Three major aspects of the investigation of spatial multivariate processes are: (1) spatial distributions (along a line, across an area, or over a volume) of parameters of models of point time series; (2) relationship between the simple correlation coefficient between the pairs of stochastic components of periodic–

stochastic processes in a region, the characteristics of point time series, and the position and orientation of stations of pairs of correlated series; (3) relationship of correlation coefficients of a point time series with the series of a set of points in the surrounding area.

Areal models of distribution of parameters of the point time series

Models are most often presented in hydrology as the empirically drawn maps of isolines of parameters, estimated from the observed data. Only in relatively flat areas (such as the Great Plains region in the USA) was it feasible to fit the mathematical models to areal distributions of major parameters (examples are Figure 10.6 of monthly average precipitation and (10.2),

$$m = 2.4114 - 0.0991\,Y - 0.1203\,X + 0.0026\,Y^2 - 0.0046\,X^2 + 0.0105\,XY \tag{10.2}$$

where m is the mean monthly precipitation at a station with X its longitude and Y its latitude.

As soon as topography varies significantly and parameter isolines relatively closely follow topography, maps of empirical isolines of parameters necessarily

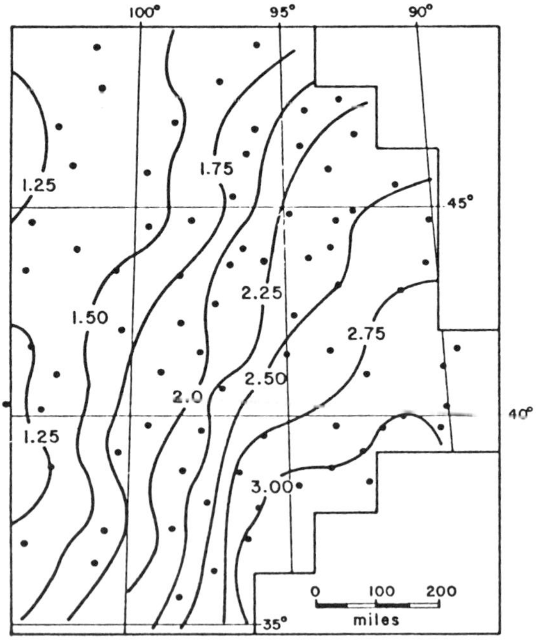

Figure 10.6 Isolines of the mean monthly precipitation in the Great Plains of the USA (with a quadratic function as a good fit)

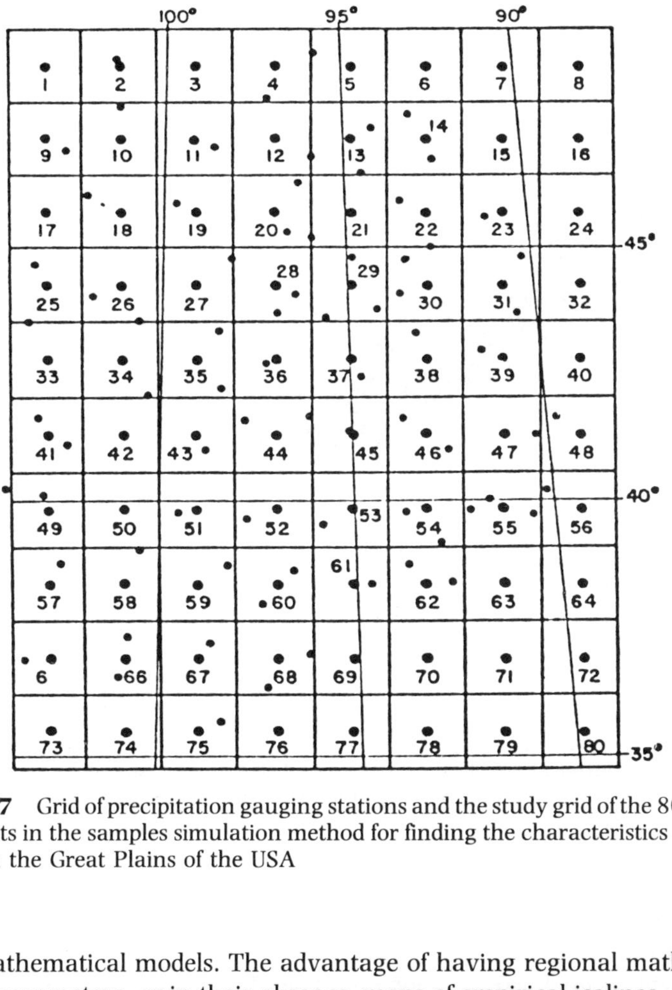

Figure 10.7 Grid of precipitation gauging stations and the study grid of the 80 regularly spaced points in the samples simulation method for finding the characteristics of regional droughts in the Great Plains of the USA

replace mathematical models. The advantage of having regional mathematical models of parameters, or in their absence, maps of empirical isolines, is that for multivariate studies the generation of new sets of samples, say in studing drought or other regional problems, can be related to a regular grid of points instead of using the randomly selected gauging stations of the past. An example of this approach is the study of droughts in the Great Plains (Figure 10.7), where a grid of 80 regular points was used instead of a similar number of precipitation gauging stations (Tase, 1976).

Regional time dependence

Figure 10.8 shows the relationship of the simple correlation coefficient between pairs of independent stochastic components of point periodic–stochastic monthly precipitation time series in the Great Plains of the USA with the distance between

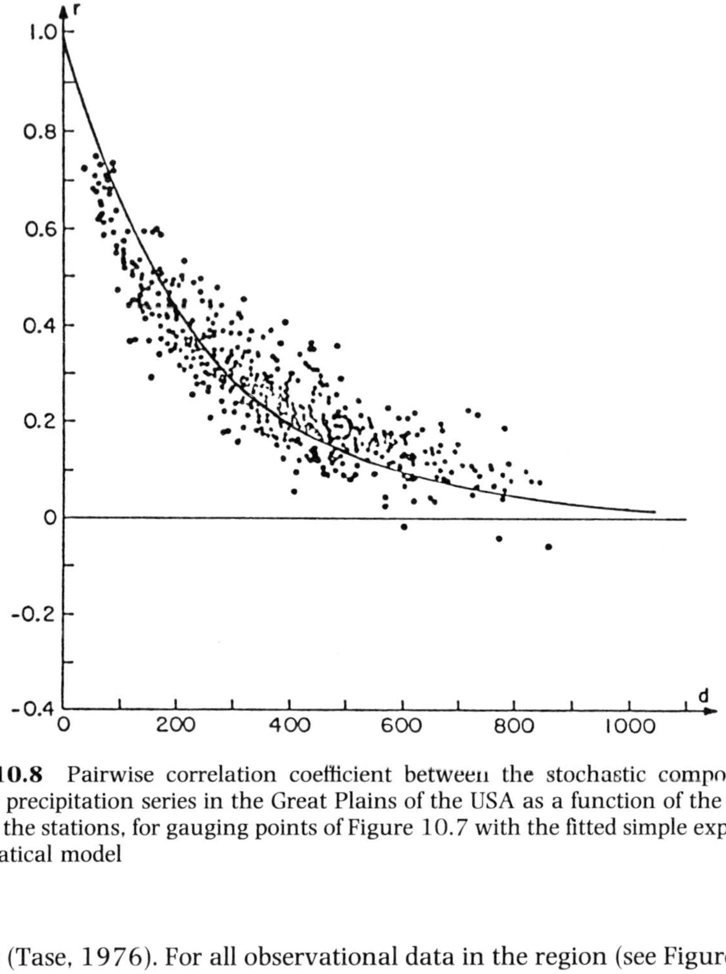

Figure 10.8 Pairwise correlation coefficient between the stochastic components of monthly precipitation series in the Great Plains of the USA as a function of the distance between the stations, for gauging points of Figure 10.7 with the fitted simple exponential mathematical model

stations (Tase, 1976). For all observational data in the region (see Figure 10.7), the independent stochastic components of monthly precipitation were computed. Then taking all the pairs of stations, correlation coefficients are estimated and plotted in Figure 10.8 against the distance, d, between the related pairs of station series. The simple fitted model is

$$r = \exp(-0.00428\, d) \tag{10.3}$$

In general, dependence between r and d of (10.3) may be also a function of latitude and longitude, X and Y, as well as the azimuth of the straight line connecting the two stations. Also, the smaller the time-series interval, the faster r decreases as d increases (or, there is a family of curves in Figure 10.8, each for a given time-series interval, say for the daily time series the fitted function

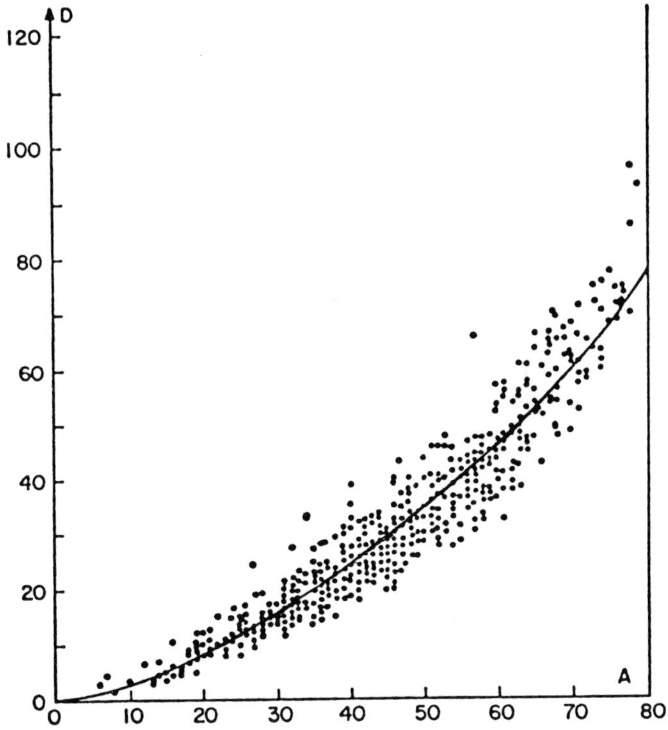

Figure 10.9 The total annual drought water deficit (*D*) in the Great Plains of the USA as related to the annual deficit drought area (*A*), obtained by the samples simulation method, with the fitted quadratic mathematical model

in Figure 10.8 would decrease faster than for the monthly series, and slower for the annual series). The regional dependence models of the type of Figure 10.8 and (10.3) serve then to determine correlation coefficients between the series at the selected grid points (see Figure 10.7) in simulating sets of the potential future multivariate regional samples. In studying droughts of the Great Plains of the USA, Tase (1976) was able to develop approximate relationships between the drought-describing random variables (see example Figure 10.9).

Local spatial dependence

Figure 10.10 shows isolines of the simple correlation coefficient between the normalised stochastic component of monthly precipitation at a particular station and the normalised stochastic components of monthly precipitation around that

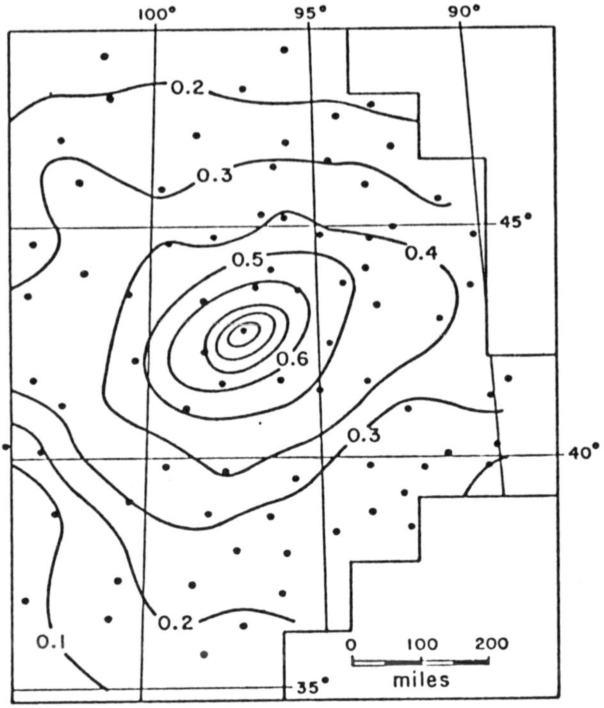

Figure 10.10 Isolines of the correlation coefficient between the stochastic components of monthly precipitation series for a selected station and all the surrounding precipitation gauging stations, showing how that coefficient decreases with the distance and azimuth in the Great Plains of the USA

particular station in the Great Plains of the USA (Tase, 1976). This third type of spatial mathematical model of simply the maps of empirical isolines, as given in Figure 10.10, presents information how for a given time interval of series, the *r-coefficient* decreases in all directions. It is evident that *r* decreases with the distance from the particular station, and its value may also depend on the azimuth and the position of the station. The application of this information is useful in simulation of samples which are generated for the purpose of integrating the new subsystems into the existing systems (say, a new hydro-electric power plant or reservoir into the system of hydro-electric power plants or reservoirs).

While it was easy to demonstrate in the preceding text that applications of statistics in hydrology and water resources have been determined by the ever-more complex problems to be solved, it is much less easy to prove it that it will continue to be so in the future. A reasonable expectation, however, can be derived from the analyses of likely future complex water resources problems, requiring advanced applications of statistics.

10.4 EXPECTED FUTURE APPLICATIONS OF STATISTICS IN WATER RESOURCES

Statistics and future water resources problems

Twelve topics of potential future water resources problems are now selected, with the purpose of showing some insights into the potentially required future applications of statistics and stochastic processes in hydrology and water resources development, conservation, control and protection. Because there are many areas in the world in which water resources are still at their initial stages of development, the present applications will continue at many locations along with the new problems requiring either improved techniques or the invention of new methods.

Reallocation of water rights

As society's needs change, the reallocation of water rights already granted or sold will become an ever more important aspect of water resources development. Sensitive aspects of this reallocation will demand the most accurate information not only on distribution of water quantity and quality variables in space and time, but also on social and environmental impacts of the existing but especially of the future reallocated water rights. To accomplish this task, sharpening of existing and inventing new statistical techniques will be desirable.

Pollution of environment

Societies worldwide have become very much aware of the danger of a continual increase in pollution of all the Earth's environments in the last decade of the 20th century. Pollution has several aspects. Its two major problems are to stop or to significantly decrease future pollution, and to clean the already polluted water, soil, land, air, and subterranean regions. To stop water pollution will require strict control of toxic and other polluting substances which may be reaching water environments. To do that safely and economically, vastly detailed information is needed. For that purpose, pollution materials and processes must be well understood. The most cost-effective sampling techniques should be selected, improved, applied, or developed for monitoring purposes. In these efforts, the most advanced bridges should be built between the physical, chemical and biological aspects of water pollution and the old and new statistical sampling and information extracting methods. This may be the biggest challenge to hydrologists, statisticians and the other water-quality specialists.

Cleaning polluted water environments

Experience has already shown that it is much easier to replace polluted water in an environment by clean water than to clean the polluted environment itself.

In this latter case, particles of an alluvial aquifer, fissures and granular porosity surfaces of rocks, or walls and bottoms of large karstic channels and caves may be so polluted that cleaning water environments may require as long a time as that over which the pollutants were deposited. As most pollution-related phenomena are predominantly either stochastic, periodic–stochastic, or trend–periodic–stochastic processes, again the scientific progress in cleaning polluted water environments (and also other polluted environments) will depend on the quality and the quantity of statistical information to be obtained by the most advanced methods of monitoring and information procurement.

Controversies between water resources and environmental protection

Controversies have increased in recent times between the proposed solutions to water resources problems and the prevailing concepts of how to protect the environment. Unfortunately, most of these 'battles' are fought, lost or won, with relatively meagre information, regardless of the fact that often large and very expensive reports on the interaction between water resources solutions and the environment have been produced. What are the risks of further failures in the future? What are the uncertainties in evaluating risks? What are the chances that the cost will not exceed the budget and the solution will not fail? These and similar questions will need more reliable answers in the future than are available now. Society is likely to need a new breed of statisticians and hydrologists in the future. They should not only be versed in all the ramifications of obtaining the needed statistical information but should also well understand the physics, chemistry and biology of environmental problems, so to be able to build the most advanced bridges between the water resources and environmental sides of the problems.

Climatic changes, water resources and environment

The ongoing concerns for climatic changes and their expected impacts are often based on claims which are supported either by a limited amount of data and information, or on change-prediction models that do not yet encompass many factors that might generate, affect, reverse, or mitigate the changes. Essentially, the first problem of the complex set of questions related to climatic changes is the reliability of selected change-discrimination techniques. Because all climatic variables are random processes in space and time, only the most advanced and the new yet-to-be-developed change-discrimination methods could help society determine what should be done to meet the challenges of eventual climatic changes. It is surprising that the ongoing international debate on climatic changes does not involve dozens of the best statisticians specialised in the discrimination of changes in stochastic processes.

Ageing and renovation of water resources systems and structures

Water resources systems have limited physical lives. Most of the world's urban water resources systems, especially of old and large cities, are in a dilapidated state. They need renovation or complete replacement. A similar situation applies to other old structures, such as large dams. Some assessments predict that renovation of urban water systems will require investment in the USA of somewhere between 100 and 150 billion dollars in the 21st century. Ageing is not uniform because of many chance causes. Therefore, assessments of the state of deterioration of systems and structures must be conducive to the various types of statistical sampling and the corresponding statistical interpretation. In some cases the statistical techniques have yet to be developed.

Improvement of the existing water resources systems

It is a sound policy to review important water resources systems occasionally (say, every five years or so). It should be ascertained whether a relatively small investment could significantly increase the benefit from the system. The justification is simple. New data, new social values or demands, new technologies or discoveries, new methods of solving old or new problems, and similar innovations or factors, may lead reviewers to produce an increase of the benefit/cost ratio of a system by the relatively modest investment. In that approach, more reliable analyses of old, but particularly of new, statistical data, including data on performance of the system in the immediate past, may play a crucial role in this line of future activities.

Changing risks in function of new information and demands

In general, one can safely claim for hydrology and water resources problems that the smaller a risk to be accepted, the less accurately it need be estimated. The estimated 100-year flood or drought usually should be less uncertain than the estimated 10 000-year flood or drought. A similar situation exists with the risk of pollution impacts. As societies become less and less tolerant of risks (because of the increased stakes), or even request the impossible of the risk-free society, there is a tendency to pressure for reduction of risks. One should be reminded of recent studies of risks of floods overtopping and destroying large dams. The assessment was that the increase of flood spillway capacities at about the 5000 largest dams in the USA, to meet the risk reduction would require an additional investment of about 20 billion dollars. Similar conditions hold with danger from radioactivity-related disasters, and of various pollution accident risks. To meet these demands and expectations of society, new approaches and new methods for assessment and reliable computation of general types of risks, and their combinations, are needed. Risk assessment and abatement has become a new, very responsible speciality of the application of statistics, in general, and of its application in water resources and hydrology, in particular.

Conflicts of interests in water resources development

Conflicts have increased with time. Litigations in courts or via arbitration have multiplied manyfold in the last 50 years. In most cases, the crucial aspects in litigation have become the availability of reliable data and acceptable decision-making information of a statistical nature to all parties involved. The precise determination of available water and reliable estimation of variation, impacts, damages, costs, benefits and risks, usually dominate the conflict-resolution discussions. Again, progress in the application of statistics to hydrology and water resources will become a decisive factor in many litigation outcomes.

Marketing water and water rights

Water and water rights have usually been allocated by political and administrative decisions. There is a tendency at present, in the general worldwide push toward market economies, to consider water and water rights as market commodities. Water or water rights may be available for sale at the market price rather than appropriated by governmental decisions. In that trend, the reliability of statistical information on water and water-related economic parameters may affect both the acceptance of market mechanisms and the price of water or water rights.

Changes in priority of use of water sources

In ancient times, because of the abundance of their water and the reliability of flow and location, springs were the prime source supplying cities and small communities with water. The alluvial groundwater aquifers came second. The third priority was given to water from lakes, the fourth to rivers, and the last to recycling with treatment of used water or of saline water. Mainly because of pollution, it seems that these priorities will change in the future. The water supply reservoirs at the upper parts of small river basins may become the first priority as the water source. The reason is that the only future pollution of the waters of these reservoirs may be airborne, if their catchments are protected against surface and underground pollution. The other sources (springs, aquifers, lakes, rivers and recycled waters) are likely to be in the second priority, by virtue of their water quality, pollution danger, feasibility of cleaning and final chemical composition. Monitoring of water quality and statistical analyses of the resulting observations will play the key role in deciding on the priority use of the various sources of water for human uses.

Drinking water industrially produced

If emissions of pollutants into the air, water and soil continue to increase at the present pace or faster, it is likely that the best drinking water will be industrially processed as a 'manufactured products'. In that task, all minerals and toxins will be removed and required dissolved chemicals will be added according to nutritional and medical criteria. By producing a variety of types of water (say, types A, B, C, D, E, etc.) the market may satisfy the needs and tastes of people, like the variety of mineral waters at present on the market fit the tastes and needs of individuals. These waters would be used only for cooking and drinking purposes, or in a limited quantity. Here, the sophisticated statistical analyses of the qualities of the best drinking waters may lead to standards for future industrially processed types of marketed drinking waters.

Application of advanced statistics and stochastic processes in solutions of future water resources problems

As the above short discussions of the chosen examples of water resources problems has demonstrated, several future applications of statistics need to be upgraded, improved or developed. Some of them are:

1. The advanced statistical application for the most accurate discrimination of changes in stochastic processes, especially of climatic changes, of changes induced by human activities in the environment, of intended changes by water development, etc.
2. The increase in accuracy of extraction and/or transfer of information in hydrology and water resources in the case of complex trend–periodic–stochastic processes, especially for basic parameters, since the process mean determines benefit, the required storage capacity is proportional to the standard deviation of input – output, as well as depending on the measure of autocorrelation of this difference.
3. The further development of feasibility to find how much of existing information in observed data, measured by the entropy coefficient, becomes *de facto* extracted or transferred by a given statistical technique.
4. The improvement of existing methods and development of new methods of estimating the risks related to highly complex natural and/or man-induced processes, whenever feasible by combining statistical and physical methods of investigating complex risk problems.
5. The advancement of joint application of deterministic and stochastic methods of information extraction and/or transfer, in which, application the physical characteristics determine the type of mathematical model (p.d.f., structural time model, spatial model, etc.) while the statistical methods serve to estimate model parameters from observed data.
6. A renewed effort to improve model inputs and outputs in planning, design

and operation of reservoirs, by building the missing bridge between water storage theory and decision making in sizing of storage capacities.

7. Substantial improvement in monitoring water quality, water-quality-related accidents, impacts on the environment, the complex interactions of environmental processes, etc., with the improvement and development of new monitoring techniques which match the complexities.
8. The development of new methods of sampling and information extraction for cases in which every square metre of a catchment, or every cubic metre of an aquifer, has variable but spatially dependent hydrologic properties, with the use of the best statistical and physical estimation of a limited number of representative model parameters for an entire catchment or aquifer.

11

On the Estimation of Parameters of Rare Events in Environmental Time Series

M. Ivette Gomes,
University of Lisbon, Portugal

11.1 INTRODUCTION AND PRELIMINARIES

Whenever extreme-value analysis is required there are usually a few parameters of particular interest, among which are: (a) an *index of 'regular variation'* γ, which is the shape parameter in the generalized extreme-value (GEV) model with distribution function

$$G_\gamma(x) = \begin{cases} \exp[-(-1-\gamma x)^{1/\gamma}] & \text{for } 1-\gamma x > 0 \text{ and } \gamma \neq 0, \\ \exp[-\exp(-x)] & x \in \mathbb{R} \text{ if } \gamma = 0; \end{cases} \quad (11.1)$$

(b) *high quantiles* of the underlying model, i.e. high levels χ_p such that $F(\chi_p) = 1 - p$, with p small; (c) the *return period* for a level, i.e. the expected waiting time between independent exceedances of a specified high level; (d) the *extremal index* θ, which may roughly be defined as the reciprocal of the expectation of the duration of extreme events.

Statistical analysis of extreme values giving particular emphasis to (a), (b) and (c) has been carried out by several authors (e.g. Smith, 1989, and Davison and Smith, 1990, among other recent papers). However, up to now, not much

Statistics for the Environment. Edited by Vic Barnett and K. Feridun Turkman

attention has been paid to (d), although this parameter θ, of real interest only for initial underlying *dependent* samples (the usual situation in practice), is important both in its own right and because of its effect on (a), (b) and (c).

In Section 11.2 we review the asymptotic behaviour of exceedances of high levels and of upper order statistics. In Section 11.3 we deal with the estimation of high quantiles and the influence of an extremal index on such estimation. Different approaches to statistical extremes will be considered. In Section 11.4 a simulation has been developed to study the properties of the different estimators described in Section 11.3, and in Section 11.5 the estimation of the extremal index will be re-examined. The procedures developed have been applied to a few sets of hydrological data, and an example of such application to the level of the river Douro at Fraga hydrometric station is given in Section 11.6. For some of the sets of data considered, simple Markovian models may be used to represent the physical process behind the data. The estimation of the extremal index is then less complex, and when dealing with exceedances over chosen thresholds, the influence of all excesses (instead of only the independent excesses) on the estimation of the parameters of the generalized Pareto distribution may be easily studied.

11.2 THE ASYMPTOTIC BEHAVIOUR OF EXCEEDANCES OF HIGH LEVELS AND OF UPPER ORDER STATISTICS

Let $\{X_n\}_{n\geqslant 1}$ denote a stationary dependent sequence of random variables (r.v.'s), under general local and asymptotic dependence restrictions, usually valid in most practical situations (Leadbetter and Nandagopalan, 1989). Then, whereas in an independent, identically distributed (i.i.d) structure, any point process limit for the time-normalized exceedances of high levels is a Poisson process, in this case such process limit is a compound Poisson process. This means that there is a clustering of high exceedances, where the underlying Poisson points represent cluster positions, and the multiplicities correspond to cluster sizes.

For such a class of stationary sequences, there exists a well-defined extremal index θ, $0\leqslant\theta\leqslant 1$, which plays a key role in determining the intensity of cluster positions. It may be proved that for many interesting situations the extremal index is the reciprocal of the limiting mean cluster size. We have $\theta=1$ for i.i.d. sequences, i.e. high exceedances appear individually, and $\theta>0$ for 'almost all' cases of interest.

We shall illustrate the notion of extremal index with sample paths from Markovian max-autoregressive sequences

$$X_n=\alpha \max(X_{n-1}, Y_n) \qquad n\geqslant 1,\ 0<\alpha<1, \tag{11.2}$$

where X_0 is a r.v. with distribution function (d.f.) H_0, and $\{Y_n\}_{n\geqslant 1}$ is a sequence

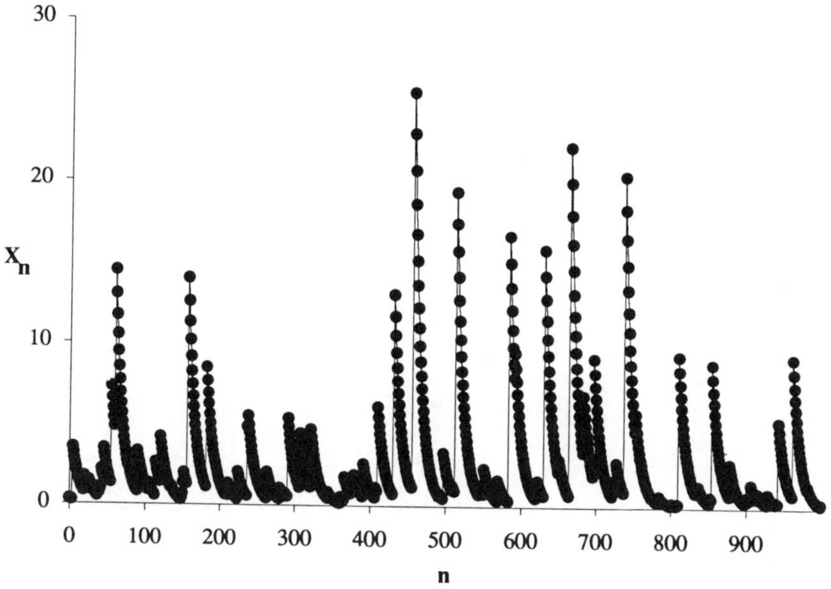

Figure 11.1 Sample path from a Markovian structure (11.2), with stationary GEV d.f. $G_{-1}(x-1)$, and with extremal index $\theta = 0.1$

of i.i.d. r.v.'s, independent of X_0, with d.f. H. These models and related ones have been revealed to be of great importance in hydrology, and inference for this model has been developed in Gomes (1990). It is a very rich model in respect of clustering of high exceedances. Under stationarity, with stationary d.f. F in the domain of attraction for maximum values of a GEV (γ) r.v. with $\gamma < 0$ ($F(\alpha x) = F(x)H(x)$ is the stationarity equation), it has been shown by Alpuim (1989) that we get an extremal index $\theta = 1 - \alpha^{-1/\gamma}$. Figures 11.1 and 11.2 show the differences between sample paths associated with the structure (11.2) for the same GEV stationary d.f., $G_{-1}(x-1)$, and for $\theta = 0.1$ and $\theta = 0.9$, respectively.

We note the large size for the clusters of high exceedances in Figure 11.1 (the limiting mean cluster size should be 10). We also note the shrinkage of maximum values for $\theta = 0.1$, compared to $\theta = 0.9$.

Let us now consider the influence of the extremal index on the limiting behaviour of upper order statistics. Let $\{X_n\}_{n\geq 1}$ be a stationary dependent sequence with stationary d.f. F, under the general dependence restrictions that enable us to guarantee the existence of an extremal index θ, and let $\{X_n\}_{n\geq 1}$ be the associated sequence of i.i.d. r.v.'s from $F(\cdot)$. Let us put $M_n^X = \max_{1\leq i\leq n} X_i$, $M_n^Y = \max_{1\leq i\leq n} Y_i$. For any $\tau > 0$, let $u_n = u_n(\tau)$, $n \geq 1$, be a level such that

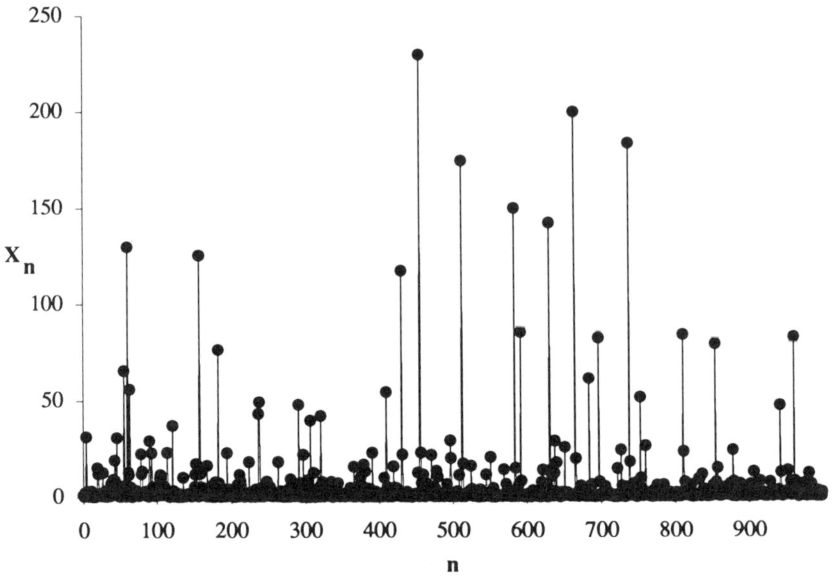

Figure 11.2 Sample path from a Markovian structure (11.2), with stationary GEV d.f. $G_{-1}(x-1)$, and with extremal index $\theta = 0.9$

$n(1-F(u_n)) \to \tau$, as $n \to \infty$, i.e. u_n is the $(1-\tau/n)$-th quantile of $F(\cdot)$. (This holds for every d.f. such that $(1-F(x-))/(1-F(x)) \to 1$, as $x \to \infty$, which will be assumed throughout.) Then $S_n^Y = \sum_{j=1}^n I_{[Y_j > u_n]}$ is a binomial r.v. that approaches, as $n \to \infty$, a Poisson r.v., whereas $S_n^X = \sum_{j=1}^n I_{[X_j > u_n]}$ goes, as $n \to \infty$, to a compound Poisson r.v. Consequently, $P[M_n^Y \leqslant u_n] = P[S_n^Y = 0] \to \exp(-\tau)$, as $n \to \infty$, whereas $P[M_n^X \leqslant u_n] = P[S_n^X = 0] \to \exp(-\theta\tau)$, as $n \to \infty$, i.e. τ, the intensity of the Poisson limiting process becomes $\theta\tau$.

This means that if there exist $\{a_n\}_{n \geqslant 1} (a_n > 0)$, $\{b_n\}_{n \geqslant 1}$, and a nondegenerate r.v. L such that $(M_n^Y - b_n)/a_n \to L$, as $n \to \infty$, where L has a d.f. $G_\gamma(x)$, in (11.1), then $(M_n^X - b_n)/a_n \to L^*$, with d.f. $G_\gamma^\theta(x)$. We thus have, for these dependent structures, a shrinkage of maximum values—$P[M_n^X \leqslant x] \approx F^{n\theta}(x)$—but the type of the limiting law is not modified, since $G_\gamma(x)$ is stable for maxima, and consequently, $G_\gamma^\theta(x) = G_\gamma[(x-\lambda)/\delta]$, $\lambda = (1-\theta^{-\gamma})/\gamma$, $\delta = \theta^{-\gamma}$.

However, the same does not happen for other upper order statistics. For instance the limiting d.f. of $M_n^{(k)}$, the kth largest order statistic among $(X_1, X_2, \ldots, X_n)$, may be expressed as a mixture of the limiting d.f.'s of the i top order statistics, $1 \leqslant i \leqslant k$, in an i.i.d. structure. The difference is even greater if we consider the joint limiting distribution of $(M_n^{(1)}, \ldots, M_n^{(k)})$ in both situations (Hsing, 1988), i.e. the presence of clustering does not affect the distribution of the maximum, but it does affect the distribution of other order statistics once, for instance, the second largest value may occur in the same cluster as the maximum. This means that the more recent approaches to statistical extremes may eventually be more troublesome than Gumbel's classical approach, once they take into a account, even if implicitly, the joint behaviour of upper order statistics. So we need to examine precisely how the statistical theory of extremes is affected and what to do.

11.3 THE INFLUENCE OF THE EXTREMAL INDEX ON THE ESTIMATION OF UPPER QUANTILES

Suppose that we have access to a sample of size n, not necessarily i.i.d., but stationary from a model F, that may be split into k subsamples of size r, i.e. $r = n/k$. Assume also that we want to estimate a value χ_p that is exceeded with a small probability p, i.e. we want to estimate χ_p such that $F(\chi_p) = 1 - p$ with p small.

Under an i.i.d set-up, with underlying model F, and denoting by M_r the maximum of r observations, we know that $P[M_r \leqslant a_r x + b_r] = F^r(a_r x + b_r) \underset{r\to\infty}{\longrightarrow} G_\gamma(x)$, where $G_\gamma(x)$ is the GEV distribution. We thus have, asymptotically, a GEV model for the M_r's, i.e. $F^r(x) \approx G_\gamma[(x - \lambda_r)/\delta_r]$, if and only if $F(x) \approx G_\gamma^{k/n}[(x - \lambda_r)/\delta_r]$, and consequently, for small p, $\chi_p \approx \lambda_{n/k} + \delta_{n/k}[1 - (np/k)^\gamma]/\gamma$.

On the other hand if we consider excesses over a high level u, we have asymptotically a Pareto model for the independent excesses, i.e. $P[X - u \leqslant t | X > u] \approx P_\gamma(t/\beta) = 1 - (1 - \gamma t/\beta)^{1/\gamma}$, $1 - \gamma t/\beta > 0$, $t > 0$, and consequently, also for small p, $\chi_p \approx u + \beta[1 - (p/(1 - F(u)))]^\gamma/\gamma$.

We shall now consider the main approaches to statistical extremes. We shall adopt a practical point of view, with the consideration of suitable estimates of χ_p, and without the specification of the theoretical conditions, on the tail behaviour of F and on the way p and k, as eventual functions of n, the original sample size, behave as $n \to \infty$, that imply consistency of the corresponding estimators.

(a) Under a 'peaks over threshold (POT)' approach, and under an i.i.d set-up, the consideration of k excesses, or equivalently, from a practical point of view, the consideration of a level $u = x_{n-k:n}$, where $x_{n-k+1:n}$ denotes the kth largest observation in the sample $1 \leqslant k \leqslant n$, gives rise to the estimate $\hat{\chi}_p = x_{n-k:n} + \hat{\beta}[1 - (np/k)^{\hat{\gamma}}]/\hat{\gamma}$, where $\hat{\beta}$ and $\hat{\gamma}$ are, for instance, maximum likelihood estimates of β and γ, respectively.

If we are working with a stationary sample, with extremal index θ, then in order to avoid the joint distribution of all excesses over a high level (difficult to obtain in general and strongly related to the type of dependence) we need to consider $k^* \approx k/\theta$ upper observations. Consequently, the consideration of k independent excesses is equivalent, from a practical point of view, to the choice of a level $u = x_{n-k^*:n}$, $k^* \geqslant k$, which gives rise to an estimate

$$\hat{\chi}_p^{(1)} = x_{n-k^*:n} + \hat{\beta}_{(1)} \frac{1 - \left(\frac{np}{k^*}\right)^{\hat{\gamma}(1)}}{\hat{\gamma}_{(1)}}, \tag{11.3}$$

based on the maximum likelihood estimation of β and γ. Notice that the estimation of the extremal index is implicitly effected in the choice of the sample of independent excesses, i.e. we consider the estimate $\tilde{\theta}_n = k/k^*$.

If a dependent model is valid, like the max-autoregressive model, we may take into account all k excesses together with the structure of dependence between clusters, which is straightforwardly obtained from (11.2). The behaviour of an estimate of χ_p equivalent to (11.3) compares with the behaviour of $\hat{\chi}_p^{(1)}$, whenever it is possible to choose optimally, in a way specified later, the k independent excesses that provide the estimate in (11.3).

The consideration of the *multivariate dependent structure* (Weissman, 1978; Gomes, 1981; Smith, 1986; Tawn, 1988) $(Z_1 > Z_2 > \cdots > Z_k)$ from an extremal process with p.d.f.

$$\frac{1}{\delta_k} g_\gamma\left(\frac{z_k - \lambda}{\delta}\right) \prod_{j=1}^{k-1} \frac{g_\gamma\left(\frac{z_j - \lambda}{\delta}\right)}{G_\gamma\left(\frac{z_j - \lambda}{\delta}\right)}, \text{ if } z_1 > z_2 > \cdots > z_k,$$

is equivalent to the POT approach, bringing exactly the same difficulties. Indeed, as was mentioned before, this model is obviously seriously affected if the underlying structure is dependent. For a comparison between this model and a generalization of Gumbel's model, see Gomes (1985).

(b) Under *Gumbel's classical approach*, inference on the tail behaviour of F is based on the sample of the k maximum values of the subsamples of r observations, which are assumed to come from a GEV model, $G_\gamma[(x - \lambda_r)/\delta_r]$, $\lambda_r \in \mathbb{R}$, $\delta_r \in \mathbb{R}^+$ and $\gamma \in \mathbb{R}$, with parameters to be estimated from the sample.

Once again, if we are working with a stationary sample, with extremal index θ, eventually smaller than one, then, with M_r denoting the maximum of r consecutive observations, we have $P[M_r \leqslant x] \approx G_\gamma^\theta[(x - \lambda_r)/\delta_r] = G_\gamma[(x - \lambda'_r)/\delta'_r]$, where $\lambda'_r = \lambda_r + (\delta_r/\gamma)(1 - \theta^{-\gamma})$, $\delta'_r = \delta_r \theta^{-\gamma}$, and this implies the approximation $\chi_p \approx \lambda'_{n/k} + \delta'_{n/k}[1 - (np\theta/k)^\gamma]/\gamma$. The estimation of χ_p depends then on the estimation of $\lambda'_{n/k}$, $\delta'_{n/k}$, γ and θ. Once again $\lambda'_{n/k}$, $\delta'_{n/k}$ and γ may be estimated from maximum likelihood, and θ from a suitable method of estimation of the extremal index.

If we use the estimate of θ suggested in (a), we obtain

$$\hat{\chi}_p^{(2)} = \hat{\lambda}'_{n/k} + \hat{\delta}'_{n/k}[1 - (np/k^*)^{\gamma(2)}]/\hat{\gamma}_{(2)}, \tag{11.4}$$

where k^* may be again the number of observations needed in order to get k independent excesses.

(c) Under a semi-parametric approach (de Haan, 1990), assuming a strong mixing stationary sequence with extremal index θ, we employ, in the estimation of F, the consistency of F_n^*, the empirical d.f. of the sample. Thus, if we want to estimate the location and scale parameters under a semi-parametric approach using a few upper observations from the original sample, $x_{n-j+1:n}$, $1 \leqslant j \leqslant s$, it seems sensible to consider the estimates $\tilde{\lambda}'_{n/k} = x_{n-k^*+1:n}$, $\tilde{\delta}'_{n/k} = (x_{n-k^*+1:n} - x_{n-2k^*+1:n})[\hat{\gamma}/(2^{\hat{\gamma}} - 1)]$ where k^* should be equal to k/θ, i.e. k^* may again be regarded as the number of observations needed to obtain k independent excesses of the level $x_{n-k^*:n}$. This gives rise to the estimate

$$\hat{\chi}_p^{(3)} = x_{n-k^*+1:n} + (x_{n-k^*+1:n} - x_{n-2k^*+1:n})[1 - (np/k^*)^{\hat{\gamma}(3)}]/(2^{\hat{\gamma}(3)} - 1), \tag{11.5}$$

where $\hat{\gamma}(3)$ is a semi-parametric estimate of γ. In the simulations presented below, and for a max-autoregressive sequence (11.2), whose maximum is in the domain of attraction of a G_γ d.f., $\gamma < 0$, we have considered Hill's estimator for γ; see Hill (1975).

Finally we should note that the classical methodology, that takes $\theta = 1$, obviously provides serious under estimation of the upper quantiles, whether we use (11.3), (11.4), or (11.5).

11.4 COMPARISON OF ESTIMATORS OF UPPER QUANTILES

We shall now consider max-autoregressive stationary structures (11.2) with different extremal indices and from a stationary $G_{-0.5}(2(x-1))$ model, with G_γ given in (11.1), i.e. a Fréchet model with shape parameter equal to 2. We consider the values $n = 1000$ and $k = 4$, 5, 8, 10, 20, 25, 40, 50, 100, 125, 200 and 250. Although the increasing of the sample size k under approches (a) and (c) is less problematic, to fix k appears to be the fairest situation.

In order to see the behaviour of the different estimators of γ, and the influence of the existence of an extremal index different from 1, we consider, in Figures 11.3 and 11.4, the mean values of the estimators of γ obtained for max-autoregressive sequences with extremal indices $\theta = 0.9$ and $\theta = 0.5$, respectively. Linear interpolation is used between integer values of k. We consider approaches (a), (b) and (c), for which we use the notations POT, GUMBEL and HILL, respectively. The POT estimator of γ is based on Pareto excesses, maximum likelihood estimation, and an optimal procedure of selection of the independent excesses, only possible here because of our knowledge of the simulated model. The consideration of a suboptimal, *ad hoc* practical procedure gives similar results

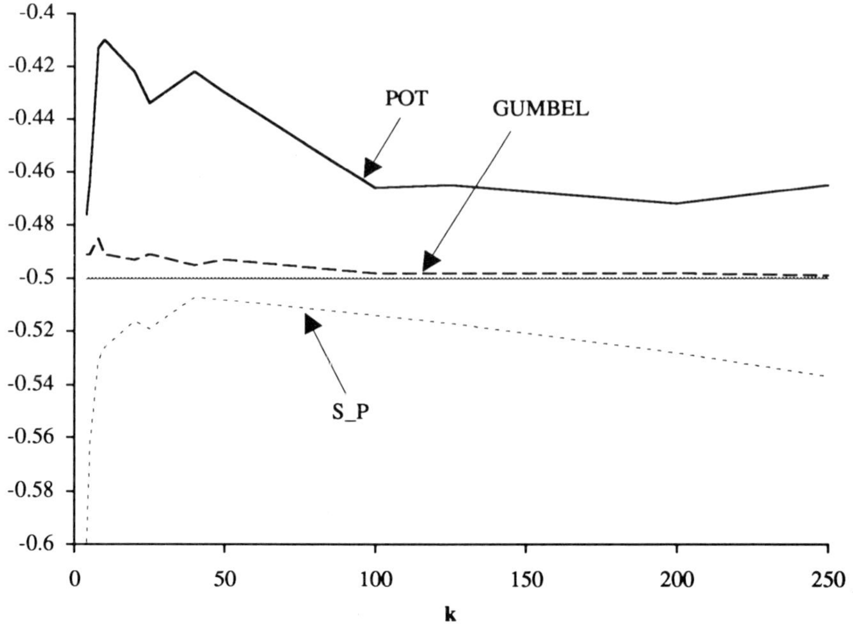

Figure 11.3 Mean values of POT, Gumbel and Hill estimators of γ, for $n = 1000$, and for max-autoregressive samples (11.2), with extremal index $\theta = 0.9$, and from a GEV stationary d.f. $G_{-0.5}(2(x-1))$, with index of regular variation $\gamma = -0.5$

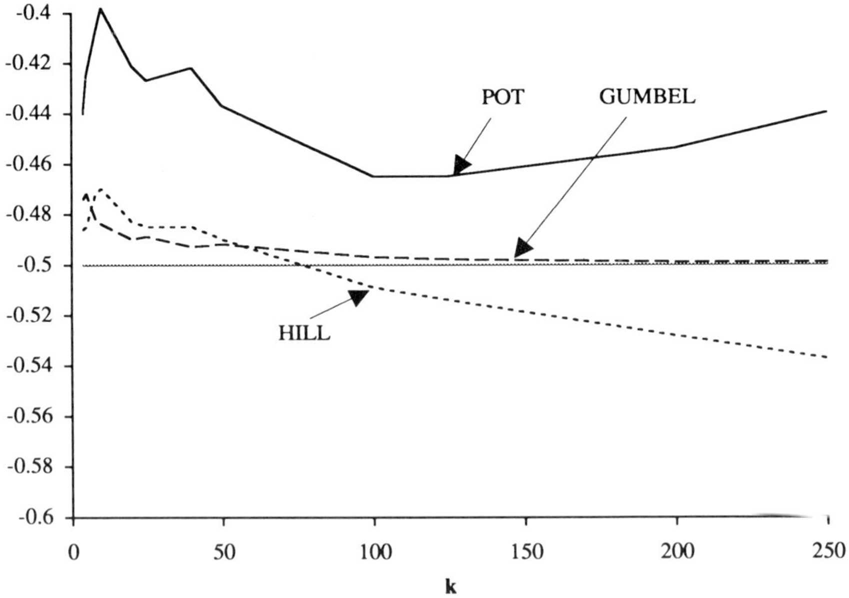

Figure 11.4 Mean values of POT, Gumbel and Hill estimators of γ, for $n = 1000$, and for max-autoregressive samples (11.2), with extremal index $\theta = 0.5$, and from a GEV stationary d.f. $G_{-0.5}(2(x-1))$, with index of regular variation $\gamma = -0.5$

for $k \leqslant 150$, but produces high positive bias and large variance for larger values of k. We should remark that convergence of the Newton–Raphson algorithm to solve the maximum likelihood equations was not always achieved for the small values of k (those samples have obviously been discarded from the whole simulation). The Gumbel estimator of γ is based on a GEV model for the k maximum values. The Hill estimator of γ is $\tilde{\gamma}(3) = \sum_{i=1}^{k-1} \log(X_{n-k+1:n}/X_{n-i+1:n})/(k-1)$, which showed in the simulations a better behaviour than the equivalent estimator based on the 'independent' k upper order statistics.

The results presented are based on 500 replicated samples. The Hill estimator of γ turns out to be more sensitive to the existence of a small extremal index. Although not clear from the figures, where we have decided not to represent (for legibility) the behaviour of the POT estimator of γ under a suboptimal choice of the independent excesses, it is worth mentioning that the pot estimator of γ is quite sensitive, for large values of k, to the choice of the independent excesses.

Since there is, in practice, no difference in the mean square errors of the different estimators considered, when instead of $\theta = 0.9$ we have $\theta = 0.5$, we present, in Figure 11.5, the mean square errors of the estimators of γ considered before, for a max-autoregressive sequence with extremal index $\theta = 0.9$.

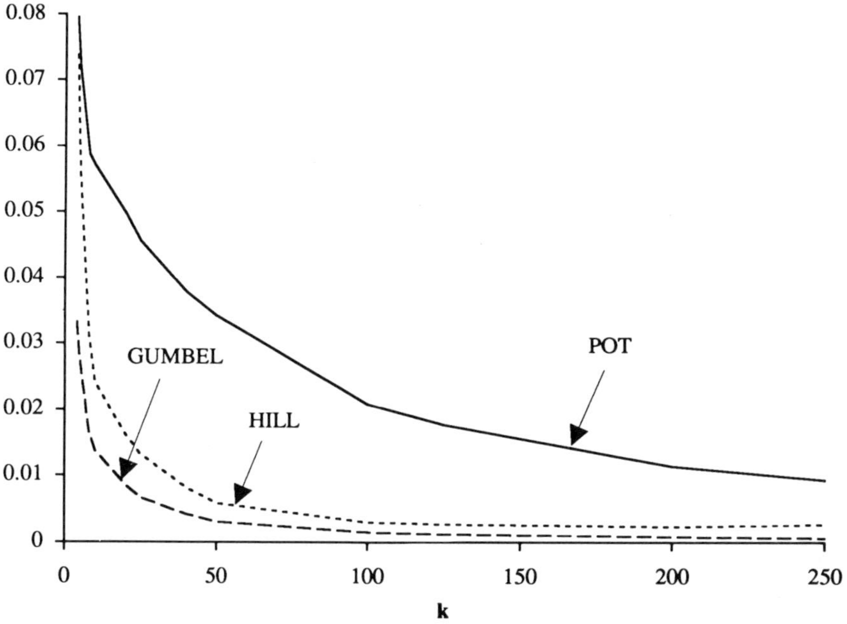

Figure 11.5 Mean square errors of POT, Gumbel and Hill estimators of γ, for $n = 1000$, and for max-autoregressive samples (11.2), with extremal index $\theta = 0.9$, and from a GEV stationary d.f. $G_{-0.5}(2(x-1))$.

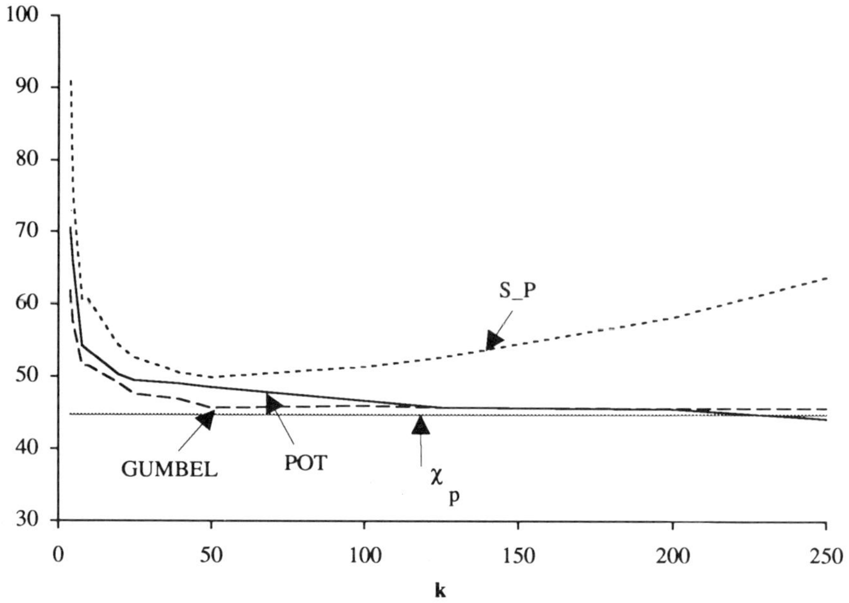

Figure 11.6 Mean values of POT, Gumbel and S_P estimators of χ_p, for $n = 1000$, for $p = 0.0005$, and for max-autoregressive samples (11.2), with extremal index $\theta = 0.9$, from a GEV stationary d.f. $G_{-0.5}(2(x-1))$

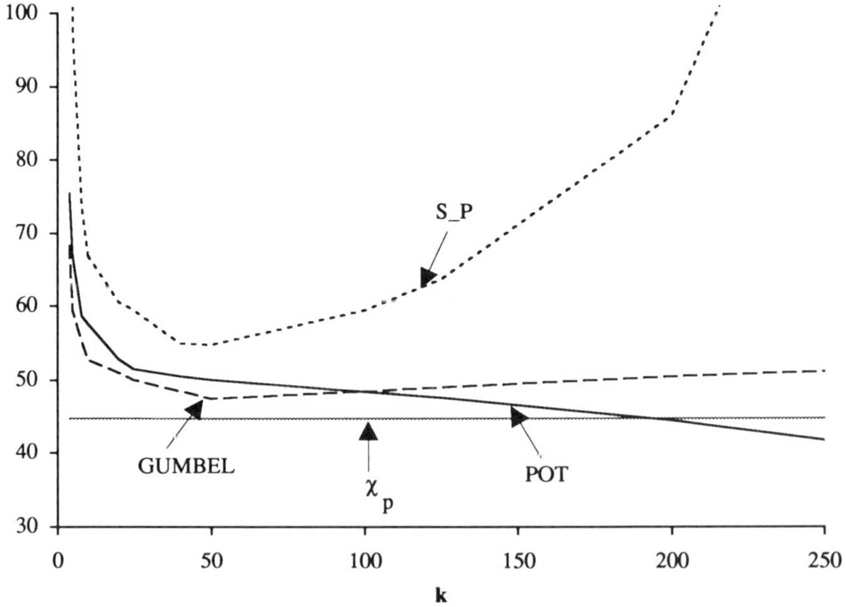

Figure 11.7 Mean values of POT, Gumbel and S_P estimators of χ_p, for $n = 1000$, for $p = 0.0005$, and for max-autoregressive samples (11.2), with extremal index $\theta = 0.5$, from a GEV stationary d.f. $G_{-0.5}(2(x-1))$

Thus it appears that the POT estimator of γ is the one with worst behaviour among the estimators considered. Similar results have been obtained for other stationary d.f.'s, for other values of n, and for other simulated models.

Let us now consider the behaviour of the estimators of χ_p, corresponding to the estimates in (11.3), (11.4) and (11.5), and denoted by POT, GUMBEL and S_P (standing for semi-parametric), respectively. We have selected for $n = 1000$ the value $p = 0.0005$. In Figures 11.6 and 11.7 we present the mean values of the referred estimators of χ_p, for max-autoregressive sequences with extremal index $\theta = 0.9$ and $\theta = 0.5$, respectively. The true value of χ_p, represented in the figures, is $\chi_p = 44.72$.

Notice that, apart from the problematic choice of k^* (the choice here is optimal, but a suboptimal choice brings the same problems mentioned before), the POT estimator of χ_p seems to be less sensitive to the existence of a small extremal index, for small values of p, provided we choose a suitable k. From a mean-square-error point of view, Gumbel's estimator has better behaviour, as is suggested by Figures 11.8 and 11.9, where we present the square root of the mean square errors of the estimators of χ_p considered before, once again for max-autoregressive sequences with extremal indices $\theta = 0.9$ and $\theta = 0.5$, respectively.

Note also, that for the same number of observations, and with the same estimator of the extremal index, Gumbel's approach turns out to be less sensitive

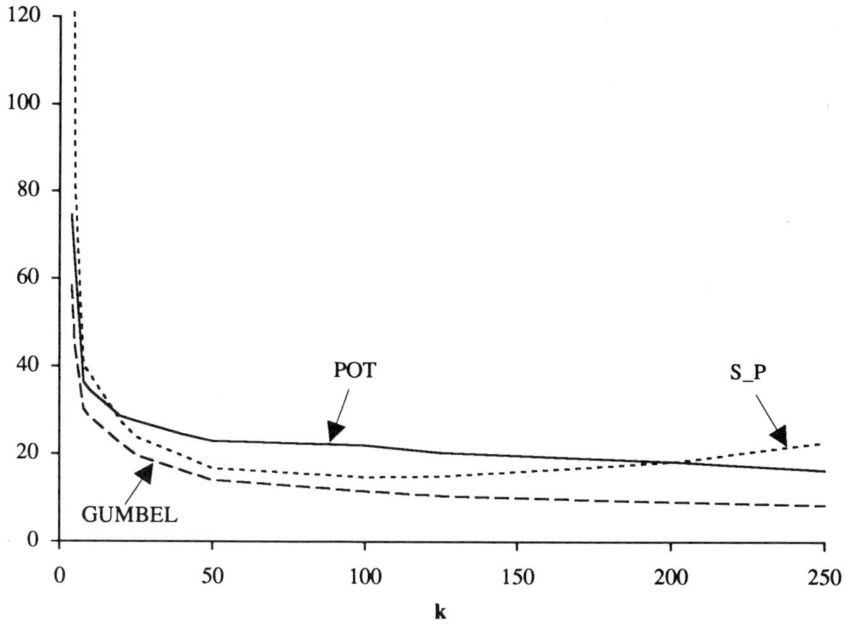

Figure 11.8 Mean square errors of POT, Gumbel and S_P estimators of χ_p, for $n = 1000$, for $p = 0.0005$, and for max-autoregressive samples (11.2) with extremal index $\theta = 0.9$, from a GEV stationary d.f. $G_{-0.5}(2(x-1))$

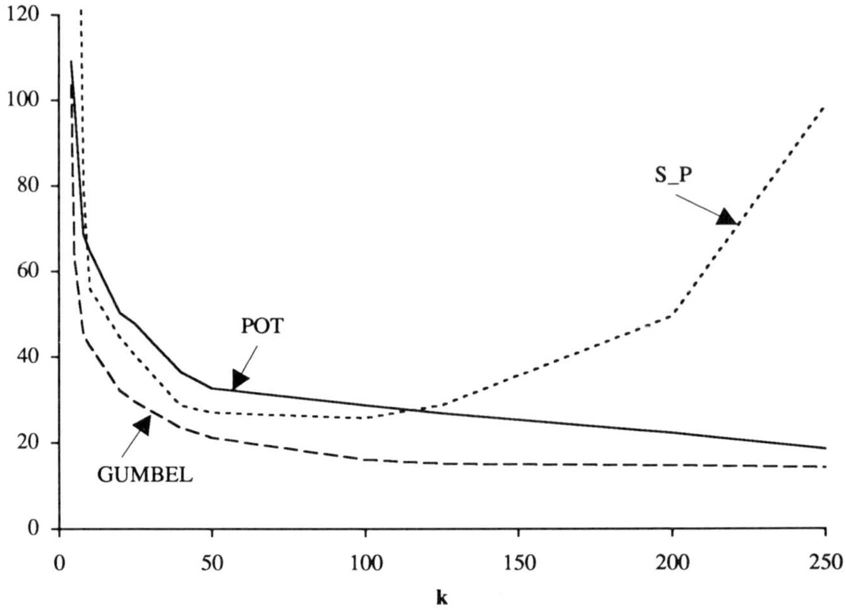

Figure 11.9 Mean square errors of POT, Gumbel and S_P estimators of χ_p, for $n = 1000$, for $p = 0.0005$, and for max-autoregresive samples (11.2), with extremal index $\theta = 0.5$, from a GEV stationary d.f. $G_{-0.5}(2(x-1))$

to the existence of a small extremal index. This is cnhanced when we consider smaller values of θ. Of course, good estimates of the extremal index are always necessary.

11.5 THE ESTIMATION OF THE EXTREMAL INDEX

For the class of stationary dependent processes with d.f. F we are considering, we know that *clusters of exceedances* may be identified asymptotically as *runs of consecutive exceedances* and *cluster sizes* as *run lengths*, i.e. for every $\tau > 0$, $n(1 - F(u_n)) \to \tau$ as $n \to \infty$ if and only if $P[M_n^X \leqslant u_n] \to \exp(-\theta\tau)$ as $n \to \infty$ and this happens if and only if $P[X_2 \leqslant u_n | X_1 > u_n] \to \theta$ as $n \to \infty$. Moreover, the distribution of the number of events in a cluster, given that there is at least one, (i.e. with r_n suitably chosen) takes the form

$$\pi_n(j) = P\left[\sum_{i=1}^{r_n} I_{[X_i > u_n]} = j \,\middle|\, \sum_{i=1}^{r_n} I_{[X_i > u_n]} > 0\right]$$

$$\cong P[X_2 > u_n, \ldots, X_{j+1} > u_n, X_{j+2} \leqslant u_n | X_1 \leqslant u_n < X_2] \to \pi(j),\ j \geqslant 1,$$

as $n \to \infty$, where I_A denotes, as usual, the indicator function of A.

Given a sample $(X_1, X_2, \ldots, X_n)$, an obvious nonparametric estimator of θ is suggested: namely that for a suitable threshold u, we put

$$\hat{\theta}_n^{(1)} = \left(\sum_{j=1}^{n-1} I_{[X_j \leqslant u < X_{j+1}]} \right) \Big/ \left(\sum_{j=1}^{n} I_{[X_j > u]} \right).$$

If we have R clusters of sizes $C_1, C_2, \ldots, C_R$, we can use the estimator used in Leadbetter *et al.* (1989),

$$\hat{\theta} = R/(C_1 + \cdots + C_R) = R / \sum_{j=1}^{n} I_{[X_j > u]} = \hat{\theta}_n^{(1)},$$

under an 'exceedance run' definition of clusters. Blocks may also be considered (Hsing, 1989), but it seems we do not gain anything, although we avoid the ambiguities of block size choice and of threshold choice.

But how should we choose the threshold $u = u_n$? Since $n[1 - F(u_n)]$ should go to τ, as $n \to \infty$, the obvious choice seems to be $X_{n-\tau:n}$. However, it may be proved that consistency of $\hat{\theta}_n^{(1)}$ may only be achieved if $1 - F(u_n) = \gamma_n \to 0$, as $n \to \infty$, with $n\gamma_n \to \infty$, i.e. we should take a percentage of high values equal for instance to $\tau/\sqrt{n}$ (instead of τ/n); see Nandagopalan (1990). However, from a practical point of view things do not differ much, and the empirical study of the influence of u on the estimation of θ (Gomes, 1990) has shown that such an estimator of θ is not as insensitive to the value of u as has been suggested by several authors. Estimates of θ obviously increase with increasing level and do not show a definite sign of stabilising. We also note an obvious conclusion: the level should be proportional to $1/\theta$. We should mention (among nonparametric methods of estimation of θ aiming to select an 'optimal' high level u) the method which seems to have provided the best results, among the methods simulated in Gomes (1990). Since the mean cluster size may be identified with $1/\theta$ a decreasing function of θ and we noticed in different simulations that the ideal percentage of high values chosen was directly proportional to $1/\theta$, we have implemented an iterative procedure to estimate θ, beginning with an arbitrary value, for instance $\tau = [\sqrt{n}/2]$. Choosing the level $u = X_{n-\tau:n}$, and according to the estimate θ_0 of θ obtained, we consider recursively $\tau = [\sqrt{n}/(2\theta_0)]$, whenever $\theta_0 \geqslant 1/\sqrt{n}$. When we get a difference between successive iterates smaller than a prescribed small $\varepsilon > 0$, we take that iterate as the estimator $\hat{\theta}_n^{(2)}$ of θ.

Canto e Castro (1992) suggests another estimator, denoted here by $\hat{\theta}_n^{(3)}$, that provides an alternative candidate to the estimation of θ. The choice of u is in this case not problematic, once u is chosen on the basis of a mean number of up-crossings equal to one in each of the k_n blocks of size r_n with $k_n r_n = n$ (the second local maximum in each block being then a suitable candidate for the level).

Table 11.1 Mean values and standard errors of estimators of the extremal index

θ		$n = 1000$	$n = 3000$	$n = 5000$
	$\hat{\theta}^{(1)}$	0.880 (0.058)	0.883 (0.047)	0.887 (0038)
	$\hat{\theta}^{(2)}$	0.878 (0.057)	0.883 (0.044)	0.891 (0.037)
0.9	$\hat{\theta}^{(3)}$	0.853 (0.058)	0.871 (0.040)	0.876 (0.034)
	$\hat{\theta}^{(4)}$	0.901 (0.021)	0.900 (0.014)	0.900 (0.013)
	$\hat{\theta}^{(5)}$	0.907 (0.056)	0.899 (0.044)	0.899 (0.033)
	$\hat{\theta}^{(1)}$	0.509 (0.092)	0.500 (0.069)	0.500 (0.059)
	$\hat{\theta}^{(2)}$	0.487 (0.073)	0.493 (0.051)	0.495 (0.042)
0.5	$\hat{\theta}^{(3)}$	0.486 (0.064)	0.494 (0.047)	0.498 (0.041)
	$\hat{\theta}^{(4)}$	0.510 (0.059)	0.510 (0.039)	0.500 (0.032)
	$\hat{\theta}^{(5)}$	0.518 (0.072)	0.505 (0.051)	0.503 (0.044)
	$\hat{\theta}^{(1)}$	0.035 (0.055)	0.120 (0.049)	0.110 (0.043)
	$\hat{\theta}^{(2)}$	0.080 (0.026)	0.089 (0.018)	0.090 (0.014)
0.1	$\hat{\theta}^{(3)}$	0.113 (0.021)	0.108 (0.015)	0.106 (0.013)
	$\hat{\theta}^{(4)}$	0.110 (0.022)	0.110 (0.015)	0.105 (0.013)
	$\hat{\theta}^{(5)}$	0.107 (0.022)	0.103 (0.014)	0.102 (0.012)

The use of bootstrap methodology may also be interesting for the estimation of the extremal index. Suppose X_i, $1 \leqslant i \leqslant n$, arise from a stationary dependent structure, with underlying d.f. $F(\cdot)$. If we can properly smoothly bootstrap the sample, generating an i.i.d. sample from $F(\cdot)$, the fact that the location and scale parameters (λ', δ') and (λ, δ) of r maximum values of the stationary sequence and of the i.i.d. associated sequence, respectively, are related by $\theta^\gamma = \delta/\delta'$, $\gamma = (\delta - \delta')/(\lambda' - \lambda)$, enables us to get the estimator $\hat{\theta}_n^{(4)}$ from the parametric estimation of δ, δ' and γ. We have here used such a method, which has provided good results, as may be seen in Table 11.1 where we present the mean values and standard errors of the different estimators of θ considered, for $n = 1000(2000)5000$, based on 500 replications of max-autoregressive samples, again with stationary d.f. $G_{-0.5}(2(x-1))$, and with extremal indices $\theta = 0.9$, 0.5 and 0.1. The estimator $\hat{\theta}_n^{(5)}$ is the estimator of θ considered in Section 11.3, for $k = [\sqrt{n}]$. In $\hat{\theta}_n^{(1)}$ we have also chosen $\tau = [\sqrt{n}]$, and this was also the starting value for $\hat{\theta}_n^{(2)}$. For $\hat{\theta}_n^{(3)}$ we have taken $k = [\sqrt{n}]$ groups and in $\hat{\theta}_n^{(4)}$ we have taken samples of maximum values of size $k_n = [\sqrt{n}]$.

The use of $\hat{\theta}_n^{(4)}$ in the estimation of quantiles has not modified the patterns presented before but has slightly improved the quality of the estimators (mainly in respect of standard erors). The replacement of θ by its true value (without estimation!) greatly improves the quality of the estimators. This means that under all approaches to statistical extremes, the extremal index needs to be properly estimated. It even seems sensible to estimate θ first, and then to choose the k independent excesses on the basis of the $k^* \approx k/\theta$ largest order statistics in the sample, choosing only the representatives of the clusters or otherwise considering all excesses if we can model the type of dependence between the members of the clusters.

11.6 AN EXAMPLE OF APPLICATION—LEVELS OF THE RIVER DOURO AT FRAGA HYDROMETRIC STATION

Available data are daily mean levels, from the early 1940s to 1990, for several hydrometric stations at different places on Portuguese rivers. We present here results only for the hydrometric station at Fraga, for which available data runs from 1946/47 to 1989/90. Stationarity of observations was achieved by considering only the period November–February. The choice of this period has been done on the basis of graphical evidence, together with information on the tail behaviour for the different months.

The use of Gumbel's classical methodology in Section 11.3(b) leads us to (184.1, 113.5, −0.32) as estimates of $(\lambda', \delta', \gamma)$ in the GEV model $G_\gamma[(x-\lambda')/\delta']$ in (11.1), fitted to the 44 selected maximum values. The standard errors associated to those point estimates are (19.4, 16.6, 0.13), respectively; see Prescott and Walden (1980) for the asymptotic covariance matrix of the maximum likelihood estimators of $(\lambda', \delta', \gamma)$ in the GEV model.

For clusters we obtain the value 0.37 for the estimate of the extremal index θ using the method of Section 11.5. An asymptotic standard error of 0.12 may be associated with this estimate. This means that we have an estimate of 2.7 days (with a standard error equal to 0.9) for the mean time that the river level spends above high levels, and consequently the classical methodology that takes $\theta = 1$ may provide serious under estimation of the high quantiles of the underlying model.

Suppose we use the method of Section 11.3(a) for excesses over a high threshold, to infer the other parameters of rare events. In order to choose 44 (the number of years registered) independent excesses, we have considered the 118 upper observations associated with a level $u = 212\,\mathrm{m^3 s^{-1}}$ and have chosen by means of *ad hoc* procedures the 44 representatives of the clusters, and associated excesses. In fact, estimation of the index of regular variation is highly sensitive to the choice of the threshold u. For this set of data and for the level u referred to above, we obtained a positive estimate of γ equal to 0.23, with an associated standard error of 0.12. The estimate of the scale parameter β in the generalized Pareto model was then 178.8 with an associated standard error equal to 33.5 (see Smith, 1984, for the asymptotic covariance matrix of the maximum likelihood estimators of β and γ in the generalized Pareto model).

For the semi-parametric approach of Section 11.3(c) we obtained results more similar to those obtained through Gumbel's approach. Thus having obtained

Table 11.2 Estimates of p-quantiles of the Douro data, obtained through the different statistical approaches considered

p	$N=1/p$	GUMBEL	POT	S_P	POT_ALL
0.004	250	444.5	466.0	435.4	422.6
0.005	200	402.1	438.5	398.5	394.3
0.008	125	322.2	375.6	329.5	338.5
0.010	100	288.2	343.3	300.4	313.7
0.020	50	196.9	231.6	222.8	243.0
0.04	25	123.8	100.6	161.5	181.1

through Pickand's estimate a negative value for γ (Pickands, 1975), we then worked with Hill's estimator, which provided an estimate of γ equal to -0.34. The second level required here for the estimation of δ' is $u' = 153\,\mathrm{m^3 s^{-1}}$, the 238th order statistic of the whole sample.

Notice that if we were interested in a test of the null hypothesis $\gamma = 0$ (for instance by means of a likelihood ratio test or any other form of statistical test (see, for instance, Tiago de Oliveira and Gomes, 1984, or Gomes and Alpuim, 1986) there would be no reason to reject the hypothesis at the usual significance levels. That does not happen generally for the sets of data analysed, where the

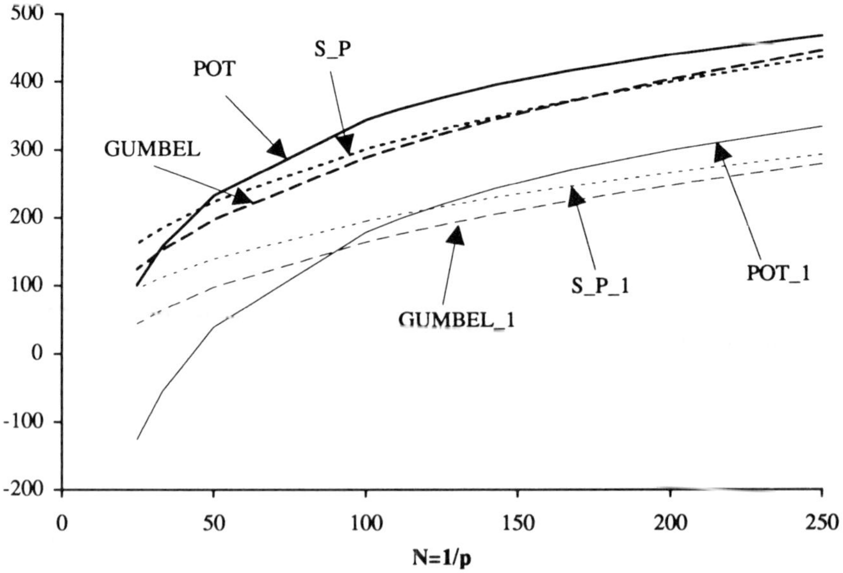

Figure 11.10 Estimates of high quantiles χ_p of the Douro data versus $N = 1/p = 25$, 50, 100, 125, 200, 250, and estimates that would have been obtained with $\theta = 1$

tails seem to be reasonably heavy ($\gamma \ll 0$). Almost all sets of hydrological data analysed have shown the existence of a negative index of 'regular variation' and of an extremal index smaller than unity.

In spite of the great variability in the estimation of γ, the estimation of moderately high quantiles has been shown to be largely insensitive to the approach to statistical extremes that is adopted. However, under all approaches to statistical extremes, the extremal index θ needs to be properly estimated, because of its high influence on the estimation of high quantiles. We should notice that for this hydrometric station we could also fit a max-autoregressive model (11.2), with $\alpha = 0.91$, corresponding to an estimate of γ equal to -0.19. In Table 11.2 we present for some values of p the estimated value of χ_p, obtained from the different methods, denoted POT [(11.3)] GUMBEL [(11.4)], S_P[(11.5)] and POT_ALL (all excesses together with (11.2)).

Figure 11.10 is the graphical representation of Table 11.2, apart from POT_All whose graph does not differ much from S_P. The estimates that would have been obtained if we had considered $\theta = 1$ are also indicated and denoted by POT_1, GUMBEL_1 and S_P_1, in line with the notation used before.

ACKNOWLEDGMENTS

The author thanks the Direcção-Geral dos Recursos e Aproveitamentos Hidráulicos for having provided the data.

12

Extreme Sea Levels

Jonathan A. Tawn
University of Lancaster, UK

12.1 INTRODUCTION

Worldwide public speculation about possible changes in mean sea levels, and climate variability, together with the increasing requirement to exploit coastal areas, and protect existing coastal developments, have led to a re-evaluation of all aspects of the design of coastal flood-prevention schemes. This engineering design problem requires statistical input in the form of information about the likelihood of extreme sea levels for the site of interest. This statistical information is used to determine the height of the sea wall (or 'dike') so that the risk of flooding meets certain design criteria.

In north-west Europe much low-lying coastal land is densely populated with communities relying on flood-prevention schemes to reduce significantly the risk of flooding. In past centuries flood defences were designed on an *ad hoc* basis and consequently were often breached, causing large death tools in well-documented storm events, see Lamb (1991). This position remained largely unchanged until the events of 31 January to 1 February 1953, when an extreme meteorologically induced strom surge (varying spatially between 2 and 3 metres over the southern North Sea) coincided with a large astronomical high tide to cause severe flooding along the coasts of the Netherlands and eastern England, resulting in the deaths of approximately 1600 and 350 people respectively. In response to this event the Dutch government appointed the Delta Committee (1960) to re-evaluate design levels on the basis of existing data, whereas the UK government set up the Storm-Tide Warning Service to provide short-term forecasts of extreme events, and revised design levels were produced (Lennon, 1963; Suthons, 1963). In both countries considerable upgrading of flood defences subsequently took place.

Statistics for the Environment. Edited by Vic Barnett and K. Feridun Turkman

Since 1953 many additional tide gauges have been installed, so extensive data in the form of hourly sea-level observations are now available for a network of sites. Using these extended data, together with more advanced statistical modelling, the Dutch have recently completed an evaluation of the design levels recommended by the Delta Committee, see van der Made *et al.* (1989) and de Haan (1990). In the UK since 1953, considerable damage has resulted from flooding in a number of notable events: on 11–12 January 1978 and 1 February 1983 areas of eastern England were flooded; on 13 December 1981 coastal regions along the Bristol Channel were flooded (almost seriously affecting Hinkley Point nuclear power station); on 16 October 1987 some limited flooding occurred on the south coast of England; and on 25–7 February 1990 a small town, Towyn, on the North Wales coast was severely flooded. Some revision of design levels have been made, through extensions of data sets (Graff, 1981) and the development of new methodology (Pugh and Vassie, 1980). However, due to the continuing problems and the increased access to data it was clear a new initiative was required to improve the statistical procedures for design level estimation and to apply these methods to all UK coastal sites. This initiative took the form of first funding a studentship and subsequently the UK Ministry of Agriculture Fisheries and Food funding the Estimates of Extreme Sea Conditions project to be undertaken by the author in collaboration with the Proudman Oceanographic Laboratory. In this chapter the present position of the study is reviewed and the strategy for future work outlined. Throughout, connections will be drawn with the Dutch approach.

First, details of the physical processes involved are given. For any site the observed hourly sea level Z_t, after averaging out surface waves, comprises three unobserved additive components: mean sea level MSL_t, tidal level T_t, and surge level S_t, so that

$$Z_t = \mathrm{MSL}_t + T_t + S_t. \tag{12.1}$$

Mean sea level, a measure of variability in the data of frequency longer than a year, varies as a result of changes in land and global water levels. The nonstochastic astronomical tidal component, generated by changing forces on the ocean produced by planetary motion, can be written as

$$T_t = \sum_{i=1}^{m} h_i \cos(w_i t + \phi_i). \tag{12.2}$$

Here $\{w_i\}$ and m are global constants, and the amplitudes and phase lags, $\{h_i, \phi_i\}$, depend on the position of the site (Pugh, 1987). The stochastic surge component, generated by short-term climatic behaviour, is identified as the residual series $\{Z_t - \mathrm{MSL}_t - T_t\}$. Around the UK, MSL_t appears to be approximately linearly increasing (Woodworth, 1987, 1990), the dominant cycle in the tide has a period of 12 h 26 min, and the autocorrelations of the surge are negligible for lags greater than approximately 50 hours (roughly the progression time for a weather system). Extreme sea levels typically arise in storms which happen to produce large surges at or around the time of a high tide.

In Section 12.2 the extreme-value-theory models which are exploited in this chapter are described. For each site of interest the usual summary parameter for the design of sea flood defences is the return level, $z_i(p)$, defined as the level exceeded by the annual maximum hourly sea level in year i with probability p. Typically the complete relationship between $z_i(p)$ and p is of interest, though in some cases attention is restricted to $p^{-1} = 50$, 100, 250 and 1000. In Section 12.3 various proposals for the estimation of $z_i(p)$ using data from only the site of interest are reviewed. Since these procedures are independently applied to each site they fail to exploit the known spatial coherence of trends and surges, and the propagation of tides. Consequently, these marginal methods for estimating design levels are inefficient, and it is unclear how estimates derived from them should be interpolated to other sites of interest. In Section 12.4 a spatial approach, developed by Coles and Tawn (1990), which overcomes these problems is reviewed. In Section 12.5, application of this spatial model is illustrated with extensive data from Australia and the UK. In Section 12.6 the question of how the methods can be extended to incorporate surface-wave variability is discussed. Finally, in Section 7 the proposed strategy for the remainder of the study is outlined.

12.2 BACKGROUND TO EXTREME-VALUE MODELLING

This section summarises the classes of asymptotically justified models for extreme values of stationary sequences used in this chapter. The initial discussion concentrates on results for independent and identically distributed random variables, from classical extreme-value theory through to a point process methodology. Then, some relevant results are discussed for the case in which the independence and stationarity assumptions are relaxed.

Suppose $X_1, \ldots, X_n$ is a sequence of independent and identically distributed (i.i.d.) random variables with distribution function F, and $M_n = \max(X_1, \ldots, X_n)$. The classical result of extreme-value theory is the characterisation of possible limiting distributions of the sequence M_n, suitably normalised. Specifically, suppose there exist normalising sequences $a_n > 0$ and b_n such that, as $n \to \infty$,

$$\Pr\{(M_n - b_n)/a_n \leqslant x\} = F^n(a_n x + b_n) \to G(x), \tag{12.3}$$

where G is a nondegenerate distribution function; then the characterisation is that

$$G(x) = \exp\{-[1 - k(x - \mu)/\sigma]^{1/k}\}, \tag{12.4}$$

subject to $1 - k(x - \mu)/\sigma > 0$ and $\sigma > 0$. Here μ and k may be any real numbers and G is called the generalised extreme-value distribution, GEV(μ, σ, k), with location, scale and shape parameters given by μ, σ and k respectively. The case $k = 0$, interpreted as the limit as $k \to 0$, gives $G(x) = \exp\{-\exp[-(x - \mu)/\sigma]\}$, for $x \in \mathbb{R}$ and $\sigma > 0$. In these cases F is said to be in the domain of attraction of G.

Though the distribution of the maximum of the sequence is usually the objective of an extreme-value analysis of a particular process, the characterisation given by equations (12.3) and (12.4) may not be the most helpful in practice. This is principally because estimation is forced to be based on the observed series of maxima only, whereas much more extensive data—which also contain information about the extremal behaviour of the process—may also be available.

A point process approach to the problem can be used to exploit these additional extremal data (Pickands, 1971; Smith, 1989). Suppose equations (12.3) and (12.4) hold for some normalising sequences a_n and b_n, and we define a sequence of point processes on $\mathbb{R}^2$ by

$$P_n = \{(i/(n+1), (X_i - b_n)/a_n): \quad i = 1, \ldots, n\}.$$

The effect of the norming in the second component is to 'collapse' all but the extreme values to the lower end-point of the rescaled distribution, while away from this boundary the process behaves like a nonhomogeneous Poisson process. More specifically, the process P_n converges weakly to a Poisson process on sets which exclude the lower boundary in the second component. From equation (12.4) the intensity of the limiting process must be

$$\Lambda\{(t_1, t_2) \times (x, \infty)\} = (t_2 - t_1)[1 - k(x - \mu)/\sigma]^{1/k}$$

for $0 \leqslant t_1 \leqslant t_2 \leqslant 1$ and $1 - k(x - \mu)/\sigma > 0$. From this limiting Poisson process characterisation of all extreme values of the series, various widely exploited special cases can be derived. For instance, the limiting joint distribution of the r largest (r fixed) order statistics (Weissman, 1978) is easily derived. Similarly, the generalised Pareto distribution as a model for the distribution of exceedances over a high threshold is obtained, since as $n \to \infty$

$$\Pr\{(X_i - b_n)/a_n > u + x \mid (X_i - b_n)/a_n > u\} \to \frac{\Lambda\{(0,1) \times (u + x, \infty)\}}{\Lambda\{(0,1) \times (u, \infty)\}}$$

$$= [1 - kx/(\sigma - ku + k\mu)]^{1/k}, \qquad (12.5)$$

see Pickands (1975) and Davison and Smith (1990).

The theory of extreme values of dependent stochastic processes has been extensively developed to cover various theoretical models for dependence and nonstationarity (Leadbetter *et al.*, 1983). For a stationary process $X_1, \ldots, X_n$, with marginal distribution function F, a natural generalisation of property (12.3), provided a mild mixing condition that ensures sufficiently spaced extremes of the process are independent holds, is that as $n \to \infty$

$$\Pr\{(M_n - b_n)/a_n \leqslant x\} \approx F^{n\theta}(a_n x + b_n) \to G^{\theta}(x) = G_*(x). \tag{12.6}$$

Here G, a_n and b_n are as in equations (12.3) and (12.4), and $0 \leqslant \theta \leqslant 1$ is a parameter, called the extremal index for the process, which is determined by the amount of clustering amongst the extreme values of the process: θ^{-1} is the limiting mean cluster size. The limit G_* is a generalised extreme-value distribution with the same shape parameter as G but with different location and scale parameters. The point process characterisation has also been extended to cover the stationary case (Hsing, 1987), and the distribution of extreme order statistics (Hsing *et al.*, 1988), with property (12.5) continuing to hold.

For nonstationary sequences a general theory of extreme values is much less helpful since there are many ways in which a series may exhibit nonstationarity. For this chapter the only specific result needed concerns the existence of a generalisation of the extremal index. For X_i, with distribution function F_i Husler (1986) and Smith (1988) show that for large n

$$\Pr\{M_n \leqslant x\} \approx \left\{\prod_{i=1}^{n} F_i(x)\right\}^{\theta}, \tag{12.7}$$

where θ has the same interpretation as in the stationary case.

12.3 SEPARATE SITE METHODS

In this section the various proposals for estimation of the marginal behaviour of extreme sea levels at a particular site are reviewed. These methods fall into two categories, direct and indirect, according to whether or not the constituent processes are separately modelled.

12.3.1 Direct methods

The essence of the direct methods is to analyse the extremes of the sea-level series, $\{Z_t\}$. The most fundamental direct approach is the traditional annual maxima method (Gumbel, 1958), which has been extensively applied to the UK sea-level data by Lennon (1963), Suthons (1963) and Graff (1981). Here the method is first presented in the stationary case and then extended to series with trends.

In the stationary case, the annual maximum sea level in year i, $M_{Z,i} = \max_{t \in \tau_i} Z_t$ where τ_i is the indexing set of observations in year i, is taken to follow a generalized extreme-value distribution. The justification for this model is that the asymptotic limit (12.4, 12.6) is assumed to be an adequate approximation. The quality of the approximation depends on the form of $\Pr\{Z_t \leqslant z\}$ for large z and the number of observations in a year, $|\tau|$ (this number is the same for each year, so $|\tau_i| = |\tau|$ for all i). Although a_n, b_n and θ are unknown they can be incorporated into the location and scale parameters which need to be estimated anyway: consequently

$$\Pr\{M_{Z,i} \leqslant x\} \approx \exp\{-[1 - k(x-\mu)/\sigma]^{1/k}\}. \qquad (12.8)$$

Under this model the first two moments are easily shown to be

$$E(M_{Z,i}) = \mu + \sigma k^{-1}[1 - \Gamma(k+1)] \qquad (12.9)$$

$$\operatorname{var}(M_{Z,i}) = \sigma^2 k^{-2}[\Gamma(2k+1) - \Gamma(k+1)^2] \qquad (12.10)$$

where Γ is the gamma function.

For many sites the series of annual maxima appear to have nonstationary features such as trends. Examination of the data shows that it is usually adequate to consider a simple case of nonstationarity, namely a linear trend with identically distributed errors. Thus model (12.8) is extended to $M_{Z,i} \sim \mathrm{GEV}(\mu_i, \sigma, k)$, with $\mu_i = a + ib$. Replacing μ with μ_i in equations (12.9) and (12.10) it can be seen that $E(M_{Z,i})$ varies linearly with time, but $\operatorname{var}(M_{Z,i})$ is independent of time. Thus a linear model with identical GEV error structure through time is a suitable model for $M_{Z,i}$.

The level $z_i(p)$ used for design of flood defences satisfies $\Pr\{M_{Z,i} \leqslant z_i(p)\} = 1 - p$. Therefore, from equation (12.8), with μ_i, replacing μ,

$$z_i(p) = \mu_i + \sigma k^{-1}\{1 - [-\log(1-p)]^k\}. \qquad (12.11)$$

Typically maximum likelihood is used to fit the GEV model to the series of annual maxima, under the assumption of independence of $M_{Z,i}$ from year to year. Substitution of the maximum likelihood estimates of μ_i, σ and k into equation (12.11) gives the maximum likelihood estimate of $z_i(p)$, denoted by $\hat{z}_i(p)$. The estimated standard error of $\hat{z}_i(p)$, $\mathrm{se}(\hat{z}_i(p))$, can be obtained using the delta method and from this pointwise confidence intervals of $z_i(p)$ derived. An alternative approach, which is computationally slightly more expensive, is to derive profile-likelihood intervals. Cox and Hinkley (1974) describe these methods in general and Davison and Smith (1990) discuss their merits in extreme-value applications.

Various proposals have been made to extend the annual maxima method to incorporate all independent large sea-level observations in the estimation of the annual maximum distribution. The model underlying each proposal is the nonhomogeneous Poisson process limit for independent and identically distributed random variables. Typically, as in the annual maximum approach, the asymptotic limit is assumed to be an adequate approximation, and estimation is via maximum likelihood. In practice, since large values of the process tend to cluster in the form of storms, some declustering is necessary to remove temporal dependence. Declustering takes the form of identifying independent clusters and extracting cluster maxima only. The properties of the resulting process of cluster maxima are consistent with the limiting model (Hsing, 1987).

Special cases of the point process approach have been applied. Smith (1986) and Tawn (1988) each develop methods based on the limiting joint distribution of the top r (r fixed) independent order statistics, whereas the Delta Committee (1960) models exceedances over a high threshold using an exponential distribution, which is a special case ($k = 0$) of the generalised Pareto distribution, given by expression (12.5). The nonhomogeneous Poisson process likelihood for points exceeding a threshold, u, is

$$\prod_i \left\{ \exp\{-[1 - k(u - \mu_i)\sigma]^{1/k}\} \prod_{m=1}^{N_i} (\sigma^{-1}[1 - k(y_{i,m} - \mu_i)/\sigma]^{1/k-1}) \right\} \quad (12.12)$$

(Smith, 1989), where $\prod_i$ is the product over years, and $y_{i,m}$ is the mth cluster maximum of N_i, which exceeds u in year i. Extensions to this likelihood to account for seasonality and missing data are given by Smith (1989). Depending on whether or not u is taken to be an order statistic determines whether the likelihood reduces to that for the joint distribution of top order statistics or for the generalised Pareto distribution.

In each of the foregoing approaches the limiting models are assumed to apply exactly in the finite approximation. The more recent Dutch approach, described by Dekkers and de Haan (1989), Dekkers *et al.* (1989) and de Haan (1990), differs in this respect by assuming the limit to exist but have not yet been reached. The key implication of this is that estimation is no longer by maximum likelihood. Instead, two schemes depending on the properties of distributions in the domain of attraction of the limiting models are used. To the author's knowledge the merits of this penultimate approach relative to maximum likelihood estimation have not yet been assessed: however, it appears de Haan's (1990) scheme is at present restricted to single samples from a common distribution and is not suitable for the more complicated modelling of trends and covariate models described here.

There are two major deficiencies in the direct form of approach applied to sea levels. First, no allowance is made for the fact that the tidal component of the sea level is deterministically induced by astronomical forcing. Consequently sea levels are a nonstationary sequence, so the validity of the limiting models is questionable. More important is that since the tides vary considerably as they propagate along the coastline, estimates for $z_i(p)$ obtained from direct methods are highly site-specific and so are of limited use for neighbouring sites. To take account of these factors an understanding of the spatial properties of sea levels is vital. Methods to overcome the two deficiencies are discussed in Sections 12.3.2 and 12.4 respectively.

12.3.2 Indirect methods

An alternative approach, driven by knowledge of the underlying physical processes, is to model separately the constituent parts of the process. Here the UK and Dutch approaches differ in physical assumptions made, choice of observational data to analyse, and the statistical method of estimation.

First consider the UK approach, which uses all hourly observations. The first version of this method was developed by Pugh and Vassie (1980) under the name 'joint probabilities method'. Although the tidal constituent is correctly accounted for by this version of the method, the statistical components of the procedure are crude nonparametric methods which lead to significant bias in estimates of $z_l(p)$. Tawn and Vassie (1989, 1991) and Tawn (1992) propose a series of amendments to the method, terming the resulting procedure the revised joint probabilities method (RJPM). A summary of Tawn's (1992) account of the RJPM is given here for the case where the surge distribution does not depend on either the tidal state or time, so the surge sequence is taken to be stationary with F_s the distribution function of hourly surges.

The initial step of the RJPM is the estimation of the distribution function of hourly surges. From the asymptotic property (12.6) the distribution of the annual maximum surge is assumed to be

$$F_s^{|\tau|\theta_s}(x) \approx \exp\{-[1 - k_s(x - \mu_s)/\sigma_s]^{1/k_s}\} \quad \text{for } x > u_s, \qquad (12.13)$$

subject to $1 - k_s(x - \mu_s)/\sigma_s > 0$ and $\sigma_s > 0$. This requires both the number of observations per year, $|\tau|$, and the threshold, u_s, to be sufficiently large for the asymptotic limit (12.6) to be an adequate approximation. Here $\theta_s, 0 \leqslant \theta_s \leqslant 1$, is the extremal index for the hourly surge sequence. Under the assumptions that motivate approximation (12.13)

$$F_s(x) = \exp\{-(|\tau|\theta_s)^{-1}[1 - k_s(x - \mu_s)/\sigma_s]^{1/k_s}\} \quad \text{for } x > u_s, \qquad (12.14)$$

hence equation (12.14) provides a parametric family of tail behaviours, that has some theoretical justification, for the distribution of hourly surge levels. Different estimation methods are used for F_s above and below the threshold, u_s, the estimation of μ_s, σ_s and k_s is via maximum likelihood applied to the point process model, described in Section 12.2, with likelihood (12.12). The estimator of θ_s^{-1} is the sample mean size of independent clusters in the surge sequence, where the cluster maxima exceeds the threshold u_s (a slightly adapted version of this proposal is given by Tawn (1992)). Below the threshold the data will be relatively dense, so the empirical distribution function of hourly surges is a sufficiently accurate estimate there.

The second stage of the RJPM involves obtaining the distribution of annual maximum sea levels. From the process representation (12.1) and property (12.7) it follows that

$$\Pr\{M_{Z,i} \leqslant z\} = \Pr\{S_t \leqslant z - \mathrm{MSL}_t - T_t:\ \text{for } t \in \tau_i\}$$

$$\approx \left\{ \prod_{t \in \tau_i} F_s(z - \mathrm{MSL}_t - T_t) \right\}^{\theta}.$$

Here θ, $\theta_s \leqslant \theta \leqslant 1$, is the hourly sea-level extermal index. Estimation of θ is via the same approach as for θ_s but based on the hourly sea-level series. This parameter is of independent interest as θ^{-1} is the mean overtopping duration of extreme sea levels.

The final stage of the RJPM concerns the estimation of the design level $z_i(p)$. By the definition of $z_i(p)$,

$$\left\{ \prod_{t \in \tau_i} F_s(z_i(p) - \mathrm{MSL}_t - T_t) \right\}^{\theta} = 1 - p. \qquad (12.15)$$

An estimate of $z_i(p)$ can be obtained by solving (12.15). A numerical solution is required if $z_i(p) < u_s + \max_{t \in \tau_i} (\mathrm{MSL}_t + T_t) = u$, as at least one term in the product will involve the empirical distribution function for hourly surges. However, for $z_i(p) \geqslant u$ all terms in the product correspond to surge levels greater than u_s so, using the parametric form (12.14) $z_i(p)$ is found to satisfy

$$-\theta^{-1}\theta_s \log(1-p) = |\tau|^{-1} \sum_{t \in \tau_i} \{1 - k_s(z_i(p) - \mu_s - \mathrm{MSL}_t - T_t)/\sigma_s\}^{1/k_s}. \qquad (12.16)$$

An estimate of $z_i(p)$ can be obtained by solving equation (12.16), having replaced the parameters $\theta, \theta_s, \mu_s, \sigma_s$ and k_s by their estimates. The standard error of this estimate may be calculated by the delta method. Within this framework there is scope for the model to be extended to take account of seasonality and trends in extreme surges. Tawn (1992) illustrates the RJPM with data from various UK sites which exhibit widely differing physical characteristics.

Now consider the Dutch approach, as described by de Haan (1990), in which the data used are observed high-water levels, $\{Z_t^* = \mathrm{MSL}_t^* + T_t^* + S_t^*\}$ (i.e. approximately one observation every 12 h 26 min). The method reduces to the same three steps as the RJPM, except an extreme-value analysis is also applied to the tidal series. First, the direct approach to extreme-value modelling, described in Section 12.3.1, is separately applied to both the $\{T_t^*\}$ and $\{S_t^*\}$

series. Estimation is again by the de Haan (1990) penultimate approach producing estimates $\hat{k}_{T^*}$ and $\hat{k}_{S^*}$, for the shape parameters of the respective series. The second stage, obtaining the distribution of the annual maximum sea level, relies on the use of a property about sums of independent variables with distributions that have regularly varying upper tails. Specifically, this result is used to give an estimated shape parameter for the annual maximum distribution of high waters, namely $(\hat{k}_{T^*}^{-1} + \hat{k}_{S^*}^{-1})^{-1}$: this is valid only when both $\hat{k}_{T^*}$ and $\hat{k}_{S^*}$ are positive. The location and scale parameters of the annual maximum high-water distribution are then estimated using de Haan's (1990) penultimate approach applied to the high-water data that are observed at the same times as the declustered series of extreme surges. The annual maximum distribution is of GEV form, therefore completely specified, so estimation of $z_i(p)$ is immediate.

The decision of the Dutch to use high-water levels only is motivated by the physical knowledge that for shallow-water sites the tide and surge interact (Prandle and Wolf, 1978), leading to the surge series being nonstationary over the tidal cycle. The RJPM overcomes tide–surge interaction by allowing some, or all, of the parameters of the extreme surge process to depend on the tidal level (Tawn, 1992, Section 4.4), so all observations can be used. In this respect the RJPM makes more efficient use of data in situations where the tidal range is small relative to surge variability, or when some of the parameters of the extreme surge process can be taken to be homogeneous over the tidal cycle. On the other hand, the RJPM does require adequate modelling of the nontrivial tide–surge interaction. Now consider the handling of the tide. The Dutch approach treats this deterministic constituent as a stochastic component, but since the behaviour of the tidal series can be predicted exactly, the restriction to a tail estimator for the tide is approximate and highly inefficient. If this tail estimator is to be used the question of whether a deterministic tidal function can really be *declustered* should be addressed. Furthermore, in some regions of the UK there is significant evidence that $k_s < 0$ so de Haan's (1990) tail convolution approach will not be applicable.

12.4 SPATIAL MODELLING

The methodology presented in Section 12.3 can only be applied independently to each individual site, which means that the known spatial coherence of trends, surges and the propagation of tides are not accounted for. Extreme sea levels along a coastline are typically generated by the same physical mechanisms, so the parameters which describe the distribution will be spatially coherent. Models which describe the separate constituents are best suited to exploit this spatial coherence as the individual parameters should change smoothly along a coastline.

A multivariate RJPM is therefore required. Consequently a model for the joint distribution, over all sites, of all surge-cluster maxima exceeding site-based thresholds is needed. This multivariate extreme-value problem is of the kind discussed by Coles and Tawn (1991) and Joe *et al* (1992): but the high dimensionality of the problem currently restricts applicability. A simpler approach, proposed by Coles and Tawn (1990), is to model the joint distribution of the sea-level annual maxima over all data sites accounting for dependence between sites, and to model the changes in each of the parameters of the distribution over sites in a way which is consistent with the properties of the underlying generating process. A brief outline of this approach is given here.

For the case when the tide and surge are independent and $k_s = 0$ then, for $z_i(p) > u$, equation (12.16) reduces to

$$z_i(p) = \mathrm{MSL}(i) + \mu_s + \sigma_s \log I(\sigma_s) - \sigma_s \log\{-\theta^{-1}\theta_s \log(1-p)\}$$

where

$$I(\sigma_s) = |\tau|^{-1} \sum_{t \in \tau_i} \exp\{T_t/\sigma_s\} \quad \text{and} \quad \mathrm{MSL}(i) = \mathrm{MSL}_t \quad \forall t \in \tau_i,$$

with T_t defined in (12.2). The tidal series directly influences the $I(\sigma_s)$ and θ terms, for all $t \in \tau_i$, each term being approximately invariant to year i. Thus a good approximation to the distribution of the annual maximum sea level in year i is a GEV $(\mu^*_{s,i} + \sigma_s \log I(\sigma_s), \sigma_s, 0)$. Coles and Tawn (1990) argue that when $k_s \neq 0$, or if there is tide–surge interaction, this structural dependence on the tide continues to hold: thus

$$M_{Z,i} \sim \mathrm{GEV}\,(\mu^*_{s,i} + \sigma_s \log I(\sigma_s), \sigma_s, k_s).$$

Here $\mu^*_{s,i}$ primarily depends on properties of extreme surges and mean sea level through $(\mu_s, \sigma_s, \theta_s)$ and MSL(i) respectively. Thus the principal influence of the tide on the annual maximum sea-level distribution is through the location parameter only.

Now let $M_{Z,i,j}$ define the annual maximum sea level at site j in year i, so the marginal distribution for the spatial model for annual maxima is

$$M_{Z,i,j} \sim \mathrm{GEV}(\mu^*_{s,i,j} + \sigma_{s,j} \log I_j(\sigma_{s,j}), \sigma_{s,j}, k_{s,j}),$$

where the additional subscript indicates that the surge parameters and tidal series can vary over sites. From annual maximum sea-level data and tidal information only the parameters $\mu^*_{s,i,j}$, $\sigma_{s,j}$ and k_s are identifiable, so these become the focus of concern. The $\mu^*_{s,i,j}$ parameter is a combination of mean sea-level and surge parameters. Experience with annual maximum data suggests that a suitable model is of the form $\mu^*_{s,i,j} = \alpha_j + i\beta_j$. Since sea-level data are recorded using tide gauges, which measure the height of the ocean surface relative to the adjacent land, estimated values for β_j result from a combination of vertical land movements and actual sea-level variations. Thus the spatial variation in the β_j parameters will be complex, but smooth. Since the α_j parameters depend on trends prior to the base year, in the absence of information on mean sea level at site j these parameters are not spatially modelled. The parameters σ_s and k_s are characteristics of the extreme surges that occur near high tides and so should reflect the spatial continuity of the surge process. By identifying the underlying relationship between these parameters over sites, and relating them to a physically based covariate, increased precision of estimates of the GEV parameters and return levels will be obtained, and the estimates will be robust to erroneous values which are evident in historical records. Furthermore, a physical basis for prediction of return levels for sites with insufficient data for any form of extreme sea-level analysis will be made possible.

The estimation of the relationships between the GEV parameters over sites in a coastal region requires the simultaneous use of all the annual maximum data for that region when estimating the associated parameters. The method of estimation must account for the spatial dependence in the annual maxima from site to site. Coles and Tawn (1990, Section 3(d)) attempt to model the progression of extreme surges along the coastline using a first-order Markov model for spatial dependence between sea-level annual maxima. Results in Davies and Flather (1978) and Provis and Radok (1979) suggest such a model captures the key features of the observed spatial dependence between sites for the UK and Australian data. Violations of the assumptions of this model are unlikely to be critical as primary interest is in the model for marginal distributions, not the dependence structure itself.

The corresponding likelihood function is given by Coles and Tawn (1990, equation 3.1), as are various diagnostic tests to aid model selection. It is important to appreciate that accounting for spatial dependence in the likelihood analysis allows a transfer of information between any two sites that have nonzero dependence. Thus, if one site has a longer series of data than another, then there is an increase in the precision of estimates for the second site. This is particularly beneficial for the estimation of separate trends at each site. Furthermore, censored observations for one site can be treated as missing without introducing bias.

12.5 APPLICATION OF THE SPATIAL MODEL

12.5.1 UK data analysis

The data set is an updated subset, edited for erroneous data, of that collected and analysed by Graff (1981), consisting of annual maximum sea levels for 61 mainland UK coastal sites, which are measured relative to a common datum. The positions of these sites are shown in Figure 12.1. The number of annual maxima available for each site ranges from 10 to 136, with a mean of 43 per site. Details of the analyses of these data, by Coles and Tawn (1990) and Dixon and Tawn (1992), are summarised here.

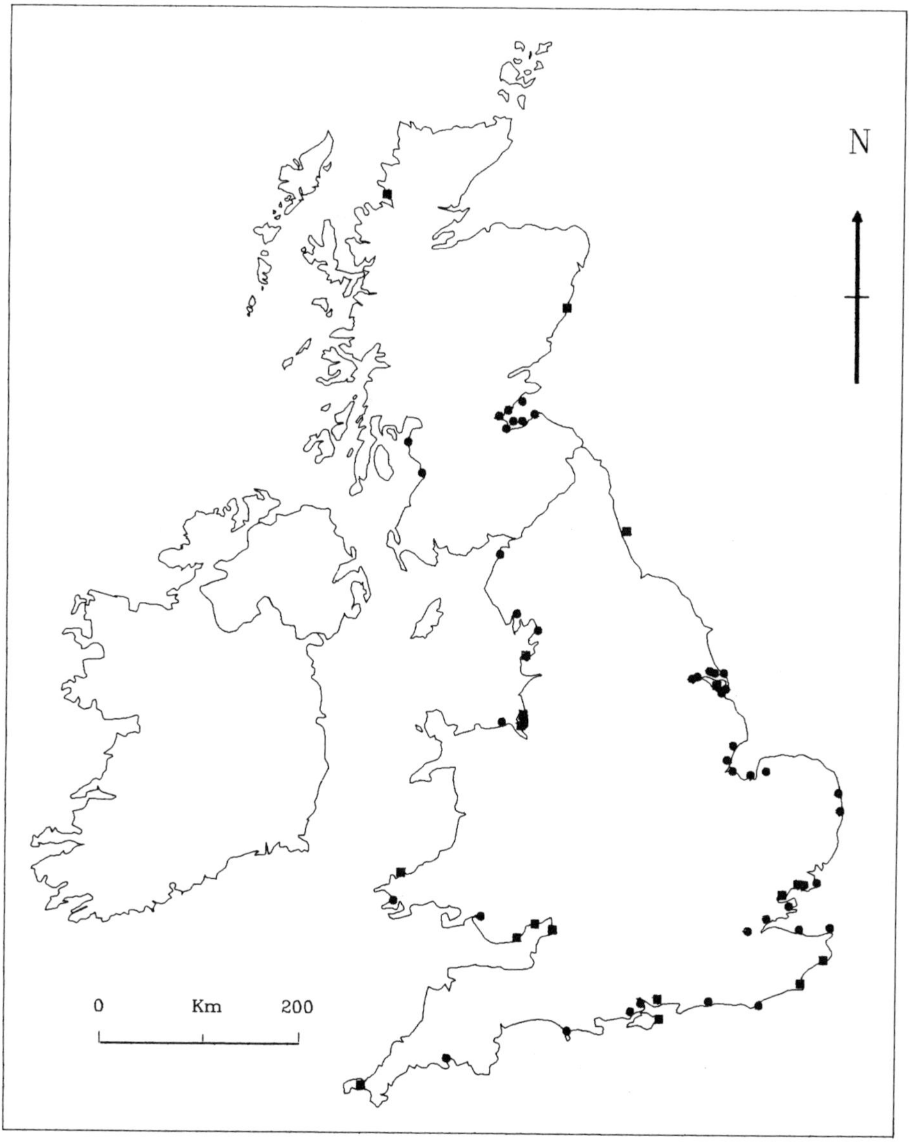

Figure 12.1 Map showing coverage of the UK sites, with annual maximum data, that are considered in this study

Since for some sites no information is currently available about the mean sea level, no attempt has been made to spatially model the α_j parameters. Coles and Tawn (1990), however, do show that the location parameter of the annual maximum distribution is highly related linearly to the tidal function $\sigma_{s,j} \log I(\sigma_{s,j})$ over data sites. Only $\sigma_{s,j}$, $k_{s,j}$ and β_j are spatially modelled. Because of the close proximity of the sites and the simplicity of changes in the extreme surge process along coastal stretches, Coles and Tawn (1990) found that a linear model, on the covariate—distance from a reference point, is suitable for both log $\sigma_{s,j}$ and $k_{s,j}$ on separate coastal stretches. The main findings were that: on many coastal stretches a homogeneous model for $k_{s,j}$ is suitable; estuaries have slightly different behaviour from neighbouring open coastlines; along the south coast $\sigma_{s,j}$ increases with distance east; and along the east coast $\sigma_{s,j}$ increases with distance south until Lowestoft and subsequently decreases slightly until the Thames estuary. The latter is consistent with knowledge of surge amplification and tide–surge interaction along the east coast (Prandle and Wolf, 1978).

To reduce the computational complexity of the problem in the Coles and Tawn (1990) analysis the trend is estimated using the marginal annual-maxima method, described in Section 12.3.1. The resulting β_j estimates did not display enough spatial stability to be sensibly modelled spatially. With the use of increased computing power, Dixon and Tawn (1992) simultaneously estimate the trend parameters and the spatial model within a joint likelihood analysis. As with the marginal trend estimation, this gives a separate estimate of the trend for each site; however, in this case the GEV error distribution, modelled

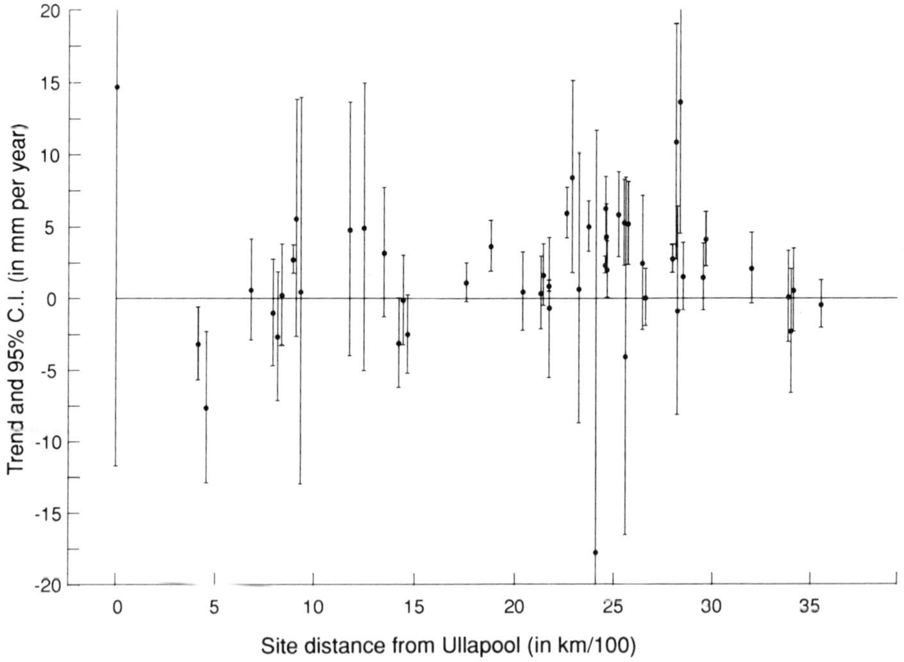

Figure 12.2 Trend estimates with 95% confidence intervals obtained by the spatial annual-maxima method described in Section 12.5. These estimates are plotted against anticlockwise coastal distance from Ullapool

spatially, accounts for both the spatial dependence and the coherence of the scale and shape parameters. In other words, smooth changes in the variability of the error distribution are modelled. The resulting trend estimates, plotted in Figure 12.2, have greater spatial stability than the corresponding trend estimates obtained using a separate site analysis.

To accommodate both the observed variation in the trend estimates over data sites and the expected coherence in the underlying trend, an estimator which is both smooth and flexible is required. As no simple parametric form fits these trend estimates Dixon and Tawn (1992) use a weighted smooth nonparametric kernel regression estimator to obtain the estimate displayed in Figure 12.3. This figure shows significant spatial variation in trends around the coast. Large trends are observed in the south-east, and small or negative trends can be seen around the Scottish coast. This general picture is consistent with knowledge about north–south tilting of the UK owing to glacial iostatic uplift in the north and land subsidence in the south and south-east.

Dixon and Tawn (1992) subsequently examine the influence of vertical land movements on the spatial variation of extreme sea-level trends. A smooth spatial estimate of trends in land levels, obtained from Shennan (1989), is removed from the spatial estimate of extreme sea-level trend (Figure 12.3) to give the

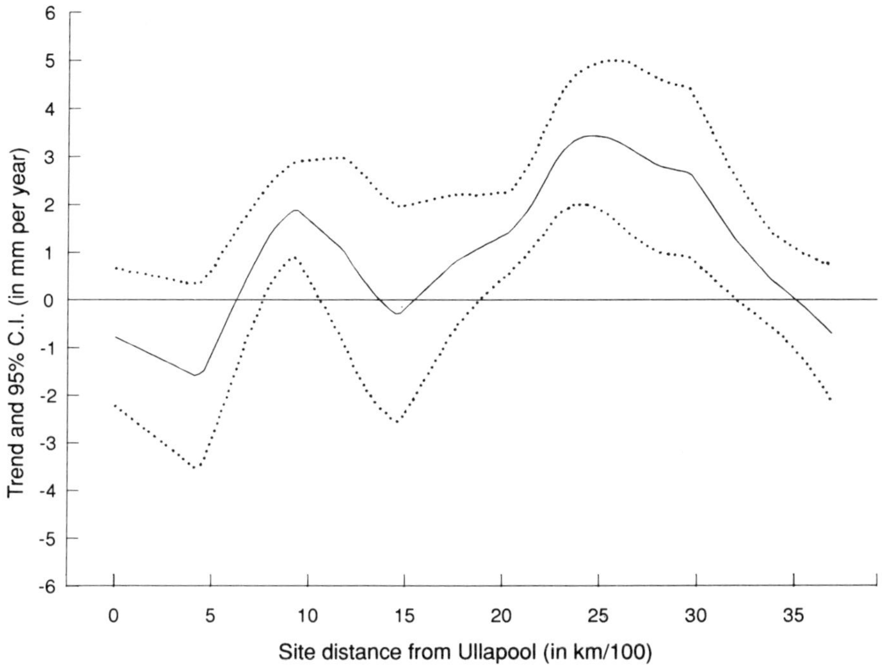

Figure 12.3 Trend estimates, obtained from the kernel smoothed regression estimates, as described in Section 12.5.1. The abscissa is the same as in Figure 12.2

spatial behaviour of actual extreme sea level trends. The resulting difference is approximately constant, 1.0 to 1.2 mm per year, over the entire coastline showing that the spatial variation in UK-observed extreme sea-level trends is almost entirely due to land-movement trends. These findings are similar to those for UK eustatic mean sea-level trends (Shennan and Woodworth, 1992) and are consistent with estimates for global mean sea-level rise (Barnett, 1984; Gornitz and Lebedeff, 1987), suggesting that there is no trend in extreme surges.

12.5.2 Australian data analysis

This section summarises the analysis of Tawn and Mitchell (1992). The principal dataset analysed consists of annual maximum and annual mean sea levels for 30 Australian coastal sites, measured relative to local site-specific datum. The positions of these sites are shown in Figure 12.4. The number of annual maxima available for each site ranges from 15 to 86, with a mean of 28 per site. For each site a full series of tidal predictions is also available.

There are a number of aspects of the Coles and Tawn (1990) models which need to be extended, as there are some features of the Australian data which

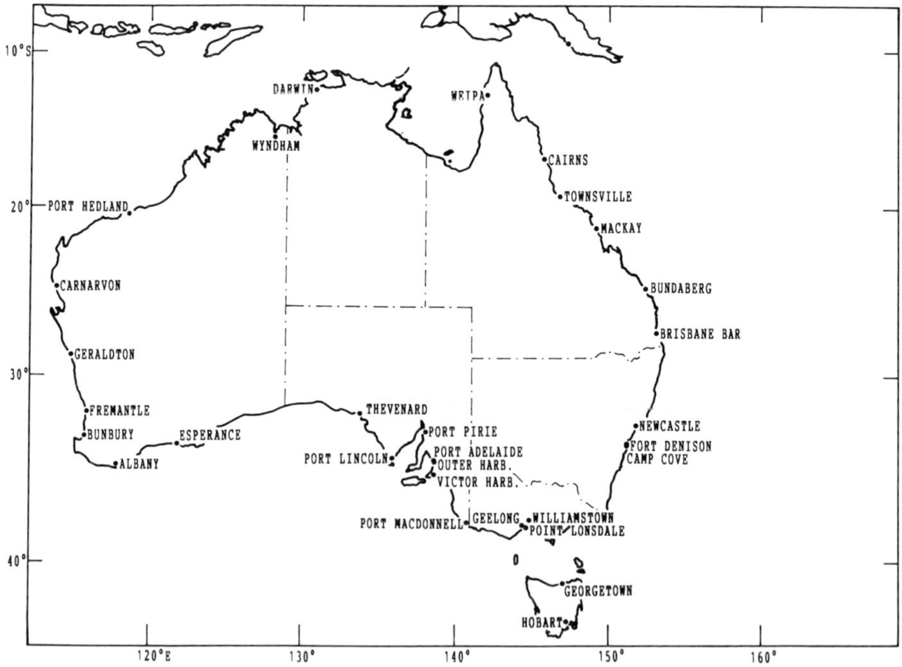

Figure 12.4 Map showing the positions of the 30 primary Australian sites

require special attention. These are:

(i) the poor coverage of the coastline by data sites with long records, a consequence of which is that the simple coastal distance–covariance models Coles and Tawn (1990) use are not appropriate;
(ii) the presence of inter-annual variation due to the Southern Oscillation (Pariwono *et al.*, 1986), which means that the nonstationarity can no longer be modelled by a simple linear trend.

With regard to the latter, the series $\{M_{Z,i} - \mathrm{MSL}(i)\}$ is found to be approximately invariant to Southern Oscillation effects. This standardised annual maximum variable is also found to be approximately stationary through time, which parallels for Australia the results of Section 12.5.1 that suggest trends in extreme sea levels are entirely due to trends in mean sea levels. A further advantage of studying the standardised annual maximum variable is that it is invariant to the local datum, and so can be spatially modelled. Thus, models of the form

$$M_{Z,i,j} - \mathrm{MSL}_j(i) \sim \mathrm{GEV}\,(\mu_j, \sigma_{s,j}, k_{s,j})$$

were considered. Here $\mathrm{MSL}_j(i)$ is the mean sea level in year i at site j, $\mu_j = \alpha_j + \sigma_{s,j} \log I_j(\sigma_{s,j})$, and the parameters α_j, $\sigma_{s,j}$ and $k_{s,j}$ are to be spatially modelled.

For the standardised annual maxima strong statistical evidence was found in favour of taking the $k_{s,j}$ parameters to be homogeneous along separate coastal stretches. These regions were empirically determined but are in independent agreement with existing oceanographic and meteorological knowledge (Linacre and Hobbs, 1977). These k-homogeneous coastal regions, together with corresponding estimates of common shape parameters, are given in Table 12.1.

Figure 12.5 plots the estimated location parameter μ_j against the estimated tidal function, $\sigma_{s,j} \log I_j(\sigma_{s,j})$, for all sites. The parameter estimates used here were obtained under the k-homogeneous model for the separate coastal regions.

Table 12.1 Homogeneous coastal regions for the shape-parameter. The numbers in parentheses are the corresponding standard errors for the common shape-parameter estimates. The Brisbane–Point Lonsdale region includes Tasmania

Region index for sites	Homogeneous coastal region	Common shape parameter, k (associated standard error)
A	Wyndham–Weipa	0.232 (0.092)
B	Cairns–Bundaberg	−0.107 (0.084)
C	Brisbane–Point Lonsdale	0.184 (0.048)
D	Port Macdonnell–Thevenard	0.045 (0.050)
E	Esperance–Albany	0.327 (0.131)
F	Bunbury–Port Hedland	−0.044 (0.051)

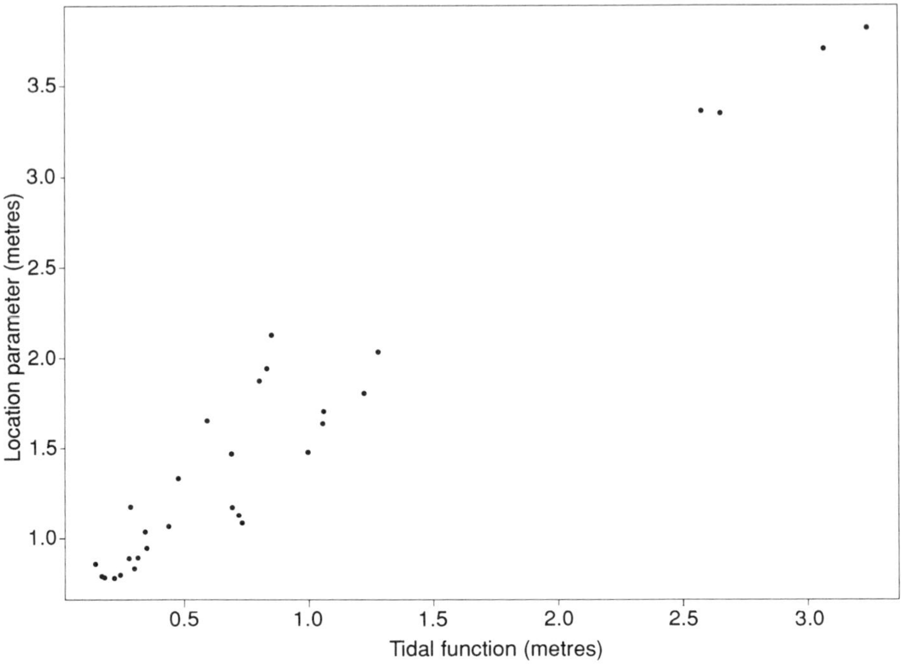

Figure 12.5 Plot of estimated GEV location parameter, μ_j, against the tidal function $\sigma_{s,j} \log I_j(\sigma_{s,j})$, for all sites

Although there appear to be two linear relationships in Figure 12.5, of most importance is the good linear relationship observed, with gradient consistent with the relationship predicted by the RJPM model. This leaves only the $\{\alpha_j, \sigma_{s,j}\}$ parameters to be spatially modelled. Justified by an argument that only the surge variability varies over sites, Tawn and Mitchell (1992) suggest that a suitable model for α_j is

$$\alpha_j = c + d\sigma_{s,j}.$$

Figure 6 shows this relationship. Despite the variability in each component due to the relatively large estimation error associated with each parameter, it is clear the parameters are related and the relationship varies from region to region. Finally, Tawn and Mitchell (1992) suggest that $\sigma_{s,j}$ could be linearly modelled on surge standard deviation.

Once the basic spatial model is identified, data from 48 other sites, which have limited data, are incorporated into the analysis. The spatial model is then re-estimated, with and without the covariance model for $\sigma_{s,j}$, and return-level estimates compared with those obtained using the basic annual-maxima method

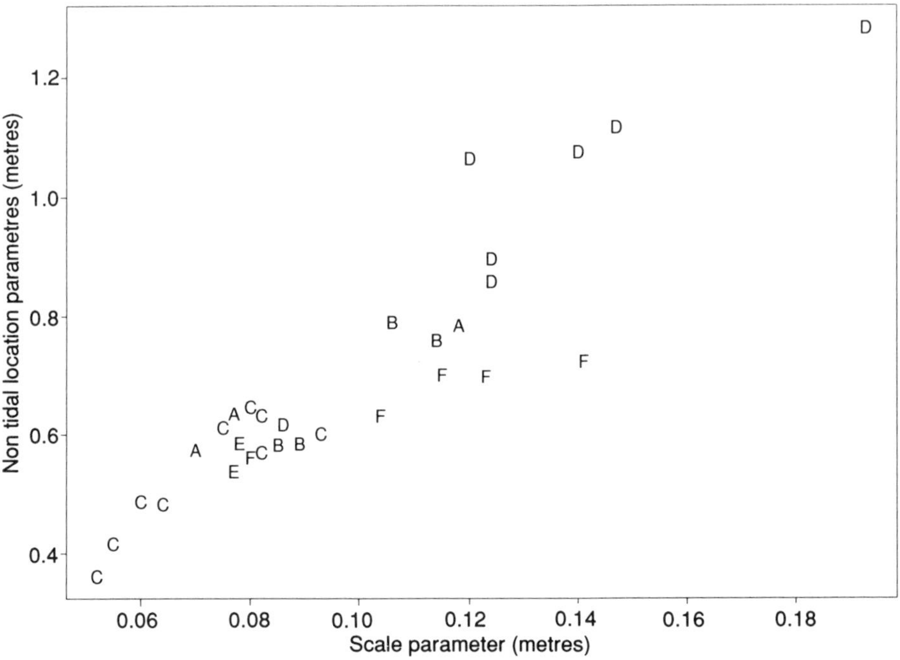

Figure 12.6 Plot, for all sites, of the scale parameter, $\sigma_{s,j}$, against the location parameter after the tidal contribution is removed, $\alpha_j = \mu_j - \sigma_{s,j} \log I_j(\sigma_{s,j})$. The letters on the plot refer to the coastal region the site is in. These regions are defined in Table 12.1

for those sites studied in the initial analysis. Table 12.2 shows the corresponding results for the South Australian sites. For the longer-record sites the agreement is generally good, with the spatial method producing more precise estimates. For the shorter-record sites the estimates seem to be consistent with neighbouring sites which have longer series and with known physical changes in this region.

12.6 INCORPORATING WAVES IN THE ANALYSIS

Presented in Sections 12.3 to 12.5 has been a series of solutions to the design problem of estimating a suitable height for a coastal flood-prevention scheme ignoring the presence of surface waves. As posed this problem is over simplified. Waves can have a considerable influence on the degree of flooding, since flooding can occur when the sea level remains below the sea wall but waves overtop the wall. Current practice is that waves are accounted for by an *ad hoc* additive adjustment to the design level. However, since the most important function of the sea wall is to reduce the volume of water overtopping within a storm, the

Table 12.2 South Australian data sites, data spans and estimates and standard errors (s.e.) for the $p = 0.01$ return level of the standardised annual maximum, in metres with respect to mean sea level. Estimates were obtained using: (A) separate site analyses, (B) spatial model with σ_i free from site to site, (C) spatial model with σ_i linearly related to the surge standard deviation at the site

Site, span of data and number of observations	Return level and standard error (A)	(B)	(C)
Port Macdonnell 1963–85 (22)	1.19 (1.11)	1.24 (0.05)	1.27 (0.05)
Cape Jaffa 1980–82 (3)		1.43 (0.06)	1.61 (0.06)
Victor Harbor 1965–89 (24)	1.83 (0.37)	1.70 (0.06)	1.80 (0.07)
Port Noarlunga 1958–73 (6)		2.36 (0.12)	2.42 (0.08)
Outer Harbour 1942–89 (46)	2.40 (0.10)	2.47 (0.06)	2.59 (0.08)
Port Adelaide 1933–89 (34)	2.48 (0.12)	2.59 (0.07)	2.65 (0.08)
Port Lincoln 1965–89 (25)	1.80 (0.15)	1.84 (0.05)	1.80 (0.06)
Wallaroo 1983–89 (7)		2.17 (0.10)	2.17 (0.08)
Port Pirie 1923–89 (65)	3.01 (0.16)	2.83 (0.06)	2.94 (0.09)
Port Augusta 1981–83 (3)		2.91 (0.19)	3.22 (0.10)
Thevenard 1966–89 (22)	2.21 (0.30)	2.32 (0.09)	2.21 (0.08)
Fowlers Bay 1980–82 (3)		1.79 (0.10)	1.67 (0.06)
Eucla 1966–82 (6)		1.66 (0.09)	1.76 (0.06)

design should be based on extremes of this variable. Clearly, knowledge of $z_i(p)$ and θ^{-1} give only part of the required information.

Extensive wave-tank modelling (Hydraulics (Research Station, 1980) has been used to determine an empirical relationship between the expected overtopping discharge rate, Q_t, sea level Z_t, inshore wave height H_t (measured as significant wave height, which is 4 times the standard deviation of sea surface elevations recorded every 0.5 seconds over a 20-minute period, at time t), and inshore wave period P_t (also measured over the 20-minute sampling period). This relationship takes the form

$$Q_t = b(Z_t, H_t, P_t),$$

where the function b depends on characteristics of the sea-wall design. In essence, the problem reduces to estimating quantiles of the distribution of the annual maximum overtopping discharge rate in year i: i.e. to estimate $q_i(p)$, such that

$$\Pr\{\max_{t \in \tau_i} Q_t \leqslant q_i(p)\} = 1 - p.$$

As with the problem of estimating $z_i(p)$ there are two possible approaches–direct and indirect methods–depending on whether or not the constituent processes are modelled. As in Section 12.3, the indirect approach has certain advantages as the complete physical process is modelled, see Coles and Tawn (1993) and Delft Hydraulics (1991). However, in this case there is more than one stochastic component to be modelled, so the key problem is of a multivariate extreme-value form involving the estimation of the dependence between extremes of the constituent processes. Coles and Tawn (1993) discuss this problem further and illustrate the methodology with an example.

12.7 FURTHER WORK

Sections 12.3 and 12.6 have described the current status of the statistical methodology for extreme sea levels. The primary objective of the UK study is now to apply these methods rigorously, the RJPM in particular, in a consistent way to all UK data sites. This will involve a thorough examination of the seasonality of extreme surges; the spatial properties of tide–surge interaction; and the sensitivity of the methods to both the choice of the threshold, u, and the declustering scheme.

Once these separate site analyses are complete, then other multivariate extensions of the RJPM will be examined. One possibility involves developing a spatial point process model for extreme surges, with covariate models for the surge parameters. Following Smith (1992), the dependence between surges at different sites could be ignored in the likelihood analysis, but incorporated in the standard error calculation. Consequently, estimates of return levels will be produced for the entire coastline together with improved estimates at data sites.

The statistical aspects of dependence modelling in the extreme sea levels study are only partially developed, leaving scope for important statistical and practical work still to be completed for problems involving coastal and offshore structures.

ACKNOWLEDGMENTS

I thank C. W. Anderson, S. G. Coles, M. J. Dixon and J. M. Vassie for helpful comments on an earlier version. The work was partly supported by the Ministry of Agriculture Fisheries and Food, and grants from the Nuffield Foundation and the Royal Society. The work used equipment provided by SERC under the complex stochastic systems initiative.

PART V

Dynamics of Fish Populations

13

Modelling and Managing a Salmon Population

Terry Speed
University of California, Berkeley, USA

13.1 INTRODUCTION

This chapter is a review of some collaborative research in which we have been involved concerning the chinook salmon (*Oncorhynchus tshawytscha*) population of the San Joaquin river system in California, USA. The size of this population has fluctuated widely, and generally declined over the last few decades, and regulatory agencies are now showing a lot of interest in encouraging or enforcing measures aimed at restoring the population to its former strength. There are many factors which are believed to be contributing to the decline in the population size, and there is disagreement as to their relative importance. It is therefore not surprising that statistical and other kinds of models have played a role in discussions concerning the measures which should be adopted to 'save the salmon'. Our initial involvement came about when we were asked by a consulting firm working for an irrigation district to comment on the appropriateness of a linear regression model being proposed by state government agency. Not long after this we become acquainted with a complex simulation model of the salmon population being used by the same consulting firm, and for different reasons we found both modelling approaches unsatisfactory. Both of these models are mentioned briefly below, and the core of this chapter reviews our not altogether successful attempt to do better, highlighting some of the questions that people would like to answer using the model.

In order for this chapter to be self-contained, it will be necessary to give some background to the problem concerning the region, the life cycle of chinook

Statistics for the Environment. Edited by Vic Barnett and K. Feridun Turkman

salmon, the suspected causes of the decline in population size, and the particular questions of interest. After doing this, we will give the brief discussion of the alternative approaches mentioned above, and turn to our model, developed in cooperation with Frank Ligon of EA Engineering, Science and Technology of Lafayette, California. As we are publishing a technical article on this work for a fisheries journal, we will not give full details of the model develpment, but rather concentrate on general issues associated with its nature and use.

13.2 BACKGROUND

Chinook salmon are commercially the most important species of salmon in California, and the majority of them originate in California's Central Valley. The Sacramento and the San Joaquin are the two main rivers draining this Central Valley and they meet in the Delta area in the north-east of San Francisco Bay. Salmon which return to spawn in the fall (autumn) are known as fall-run salmon, and fall-run chinook are the only salmon found in the San Joaquin system above the Delta. These salmon spawn in three tributaries of the San Joaquin: the Stanislaus, the Tuolomne and the Merced. Each of these rivers has dams on it, and the salmon are restricted to areas below the lowest dam in each case, as none of the dams has a fish ladder or other means of passage for the salmon.

The problem that government agencies, irrigation districts, farmers, fisherman, environmentalists and many others are concerned with is the decline in the size of the chinook salmon population over the last few decades, a problem that is seen by most parties as being directly related to the issue of water resource control. The following statement by the California Department of Fish and Game to a public meeting in July 1987 sets the issue out clearly:

> Chinook salmon, a very significant fishery, may be affected by this action of the State Water Resources Control Board. The mortality of juvenile salmon is a major limiting factor in the production of fish in the San Joaquin drainage. Spawning runs in the San Joaquin drainage are 70% or less of historic levels due largely to inadequate streamflows in the San Joaquin River, its tributaries and the Delta when young migrate to sea.

To back up this statement, we plot in Figure 13.1 the annual escapement in the San Joaquin system for the years 1953–90. 'Escapement' is the term used for the estimated number of fish returning to spawn in the river in any given year. The decline is very clear. The statement by the California Department of Fish and Game just given not only sets out the problem, if offers an explanation for it, and one of the issues we address below is the extent to which the available data support this explanation.

In order to examine possible causes of the decline in fish numbers, we need to know something about the life cycle of the chinook salmon. As is well known,

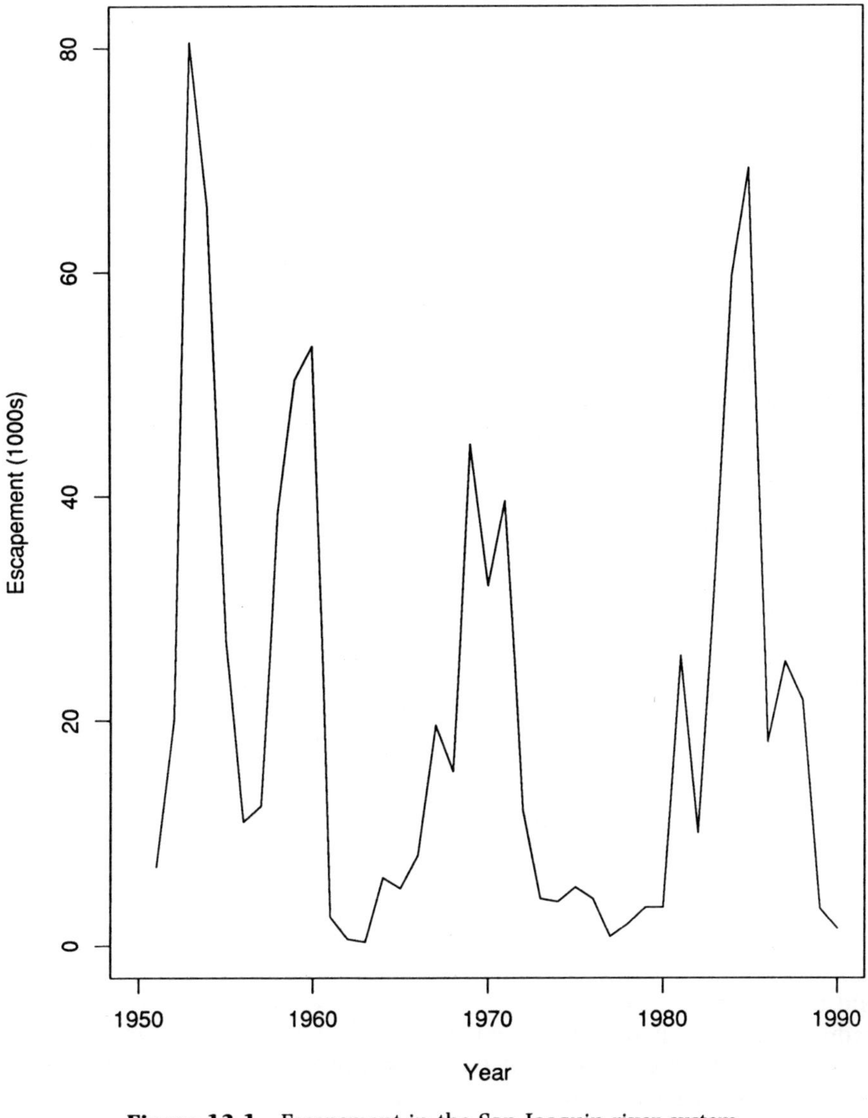

Figure 13.1 Escapement in the San Joaquin river system

salmon are unusual fish in that they spend the initial part of their life in freshwater, then, late in their first year or early in their second year, migrating out to sea where they spend the majority of their lives, later returning (if they have not perished before then) to the freshwater to spawn. In our particular case, fall-run chinook salmon return to the upper parts of the tributaries of the San Joaquin River, just below the lowest dams, in mid-to-late fall. The females make nests called redds, which are cleared regions in gravels in suitable shallow parts of the rivers, and then lay their eggs in the redds. After the males have fertilised the eggs, the latter are covered over with gravels and left to incubate. The adult fish then die. What are known as alevins hatch from the eggs in late winter and early spring after incubating for approximately 2 months and emerge from the gravels after 3-4 weeks ready to live as fry. Fry mature in the river near where they emerged, or further downstream, perhaps in the mainstream San Joaquin River and sometimes as far as the Delta. After a few months of growth these juveniles undergo a physiological transformation known as 'smoltification', and they become known as smolts. Once this process begins, smolts start to migrate downstream to grow in salt water, although some juveniles remain in tributaries for more than a year before smolting and migrating out to sea. During the smoltification and migration stage, imprinting of the 'home stream' water occurs in natural smolts, ensuring that with high probability they can find their way back to their home stream to spawn as adults at 2, 3, or 4 years of age. Once they reach the ocean, the greatest natural mortality of young salmon occurs during the first year of life, when they are still small enough to be eaten by a range of predators. Otherwise, they grow quickly into sizeable fish which are prized by sport and commercial fisherman alike.

Following this brief summary of the salmon's life cycle, we can now list a number of factors which may be contributing to the decline of the population size. As we will see, many are related to the size of spring river flows. It must be remembered that under natural conditions, winter snows would melt in the spring and lead to substantial river flows. Nowadays, this water is collected in dams and diverted or released as required for irrigation, the generation of hydroelectric power, or other purposes. A key issue in this debate concerns the extent to which spring river flows above a specified level should be required to improve the habitat and facilitate the downstream migration of juvenile salmon.

The first class of factors which can limit the population size concern the spawning gravels and the redds. The availability of an adequate supply of high-quality gravels is important to ensure that eggs which are laid survive and hatch. Gravels can become clogged with fine material, and lose the vital hydrodynamic properties which ensure an adequate supply of oxygen for the eggs, and there is evidence that this is occurring. The occasional high flow resulting from a good snow season may well have served in the past to flush away fine material, effectively cleaning the gravels, but such high flows are now quite rare due to the control exercised over the water. The area of the gravels available for spawning is probably an important factor in determining population size, although not one of recent origin, because there have been dams on these rivers for many decades now. Later-spawning fish tend to superimpose their redds on existing redds, with resulting loss of eggs, and this effect is exaggerated by the fact that offspring of later-spawning salmon tend to have lower viability. The reason for this is that such progeny will only become ready to smolt in late spring or early summe, when the sources of mortality mentioned below operate most strongly. This phenomenon constitutes an example of so-called density-dependent mortality, about which we will say more below. Put briefly, the problem of redd superimposition suggests that beyond a certain point the population may not be able to benefit from further spawning fish; indeed it may become worse off due to the lower viability of juveniles from later spawners.

As soon as they emerge from the gravels, juvenile salmon are susceptible to predation by other fish in the streams. It is believed that the level of predation is related to flow, because higher flows tend to cause turbidity in the stream and the reduced visibility tends to protect the young salmon from predators. There is also evidence that higher water temperatures reduce the survival of juvenile salmon, and there is a clear inverse relationship between stream flow and water temperature. Furthermore, it is fairly well established that smolts tend to migrate out to sea during periods of high flow, with the consequence that reduced or very regular low-to-moderate spring flows may delay this downstream migration.

There are further sources of juvenile salmon mortality which are to some extent related to streamflow. Associated with the Sacramento–San Joaquin river system is two large canals or aqueducts transporting water from near the Delta to central and southern California. These 'export facilities', as they are termed, have pumping stations diverting water from the lower reaches of the San Joaquin and the Delta. Not only do the export operations take a larger fraction of the available flow at the times when it is lowest, and needed most to help preserve a favourable environment for the salmon, there is also evidence that substantial numbers of smolts become entrained and perish in the pumping facilities.

A final possible contributor to the decline in the salmon population to add to this no doubt incomplete list, is increased commecial and sport fishing. In the next section we develop a model which may prove useful in assessing the relative contributions of some of these factors to the decline in numbers.

13.3 MODELS AND METHODS

In the previous section we listed several possible density-independent sources of juvenile salmon mortality, mostly related in some way to spring flow, and one possible density-dependent source, redd superimposition. Here the term 'density' refers to the population size, for in the case of redd superimposition, we saw the possibility that more spawners would not necessarily lead to more juvenile salmon, because of the reduced viability of later emerging young. On the other hand, there are no obvious reasons why most of the flow-related sources of mortality or reduced survival, such as water temperature, delayed downstream migration, and entrainment in export facilities, should be density-dependent, although this may well be the case with predation. For this reason, a first division of sources of juvenile salmon mortality might well be into the density-independent ones, which are related to spring flow, and density-dependent ones, which are related to an index of population abundance. Before proceeding further with these ideas, let us see that there seems to be a strong prima-facie case for the involvement of spring flow in the declining numbers.

To connect spring flow with population abundance, as indicated by annual escapement, we need to consider the appropriate alignment for the two time series. Young salmon hatching from eggs laid in the fall of year t (1960, say) emerge and mature in the following spring, and so it is the streamflows in the spring of year $t+1$ (1961) which affect the survival of this cohort. Our indications of the size of this cohort come from the escapements in the fall of years $t+2$, $t+3$ and $t+4$ (1962, 1963 and 1964) when the fish return to spawn as 2, 3, and 4-year-olds. The majority of fish that spawn now return to spawn at 2 or 3 years of age—we shall see more precise numbers below—and so we can get a rough indication of the possible impact of flow on abundance by comparing the fall-escapement time series (E_t) with the average of two spring-flow time series, that lagged by one year (Q_{t-1}) and that lagged by two years (Q_{t-2}), corresponding to an average of the flow conditions $1\frac{1}{2}$ years earlier, which affected the 2-year-old spawners, and those $2\frac{1}{2}$ years earlier, which affected the 3-year-old spawners in the escapement. A plot along these lines is given in Figure 13.2, where it is clear that the two series co-vary to a considerable extent. Also included in Figure 13.2 is an estimate of the annual number of fish caught which come from the San Joaquin river system. The basis of this estimate is explained in Speed and Ligon (1993).

Where should we go from here? Before turning to the model, we mention two other approaches which were—and still are—in use when we first became involved in this issue. In a sense they are two extremes: a linear regression model and an extremely complex discrete-time simulation model. Our aim was to find a statistical model which, on the one hand, was less simplistic then the linear regression model, and on the other hand, was based on population biology, but was more modest than the simulation model in the extent to which it involved unobserved states, variables and parameters.

The linear regression model is easy to describe: it simply involved regressing escapement E_{t+2} in year $t+2$ against spring flow Q_t in year t, i.e. in fitting the model $E_{t+2} = a + bQ_t$ to the two time series. Just such a fitted line is displayed along with the 1955–84 data in Figure 13.3; this is Figure 12 of California Department of Fish and Game (1987). What is wrong with such a model? Of course the answer to this question depends on the use to which the model is put, so let us point out two uses which we think would be inappropriate: relying upon the model to 'predict' escapement two years later, on the basis of a given year's spring flow; and using the model to fix a minimum level of flow, with the hope that if spring flows were manipulated to exceed this minimum level, then escapement two years later would exceed the value corresponding to this level on the regression line. It seems likely that the authors of this model do intend it to be used in these two ways. Rather than give a lengthy explanation

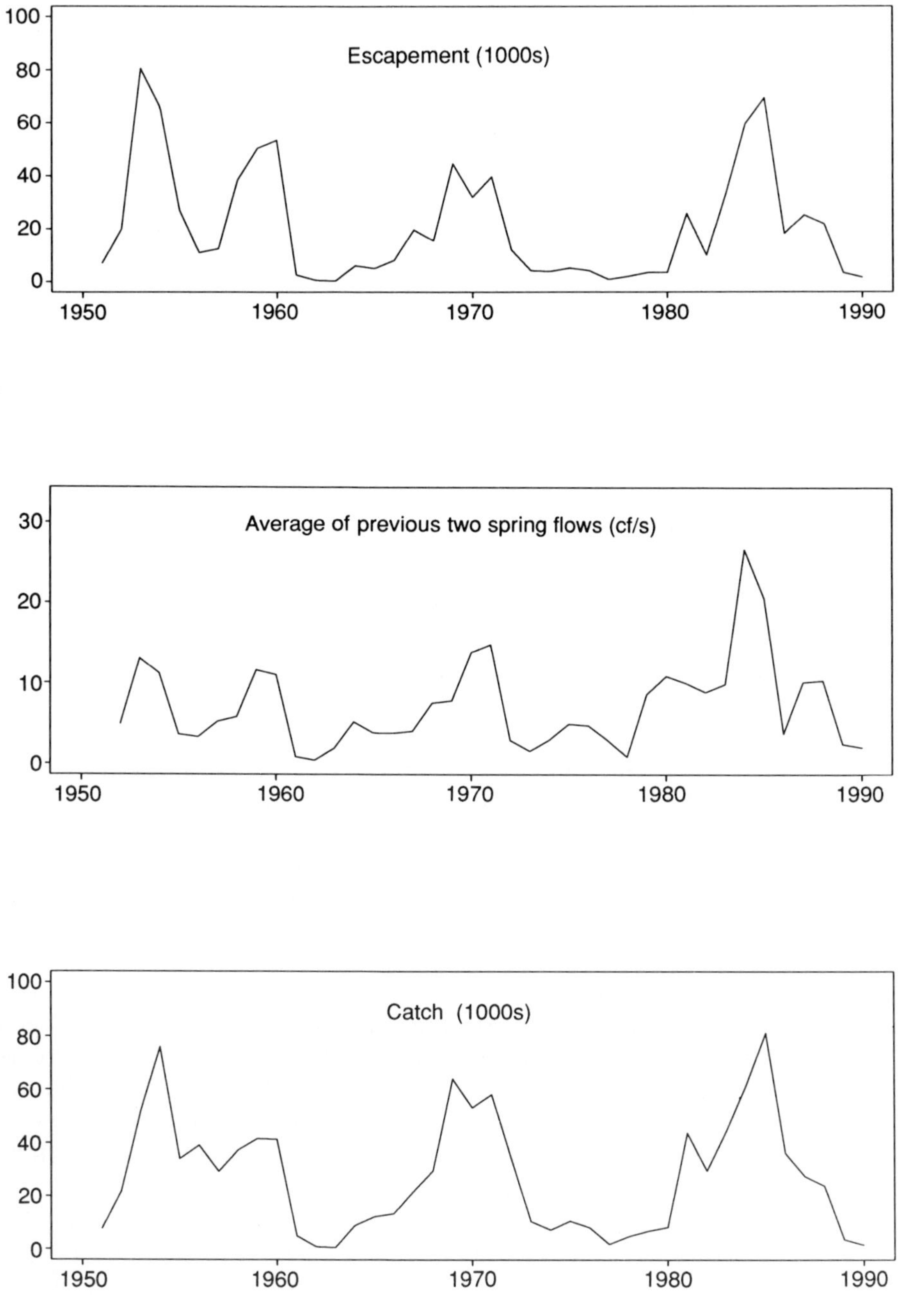

Figure 13.2 Escapement, average lagged flow and catch in the San Joaquin river system. (Average lagged flow is the average of the spring flows at Vernalis $1\frac{1}{2}$ and $2\frac{1}{2}$ years earlier.)

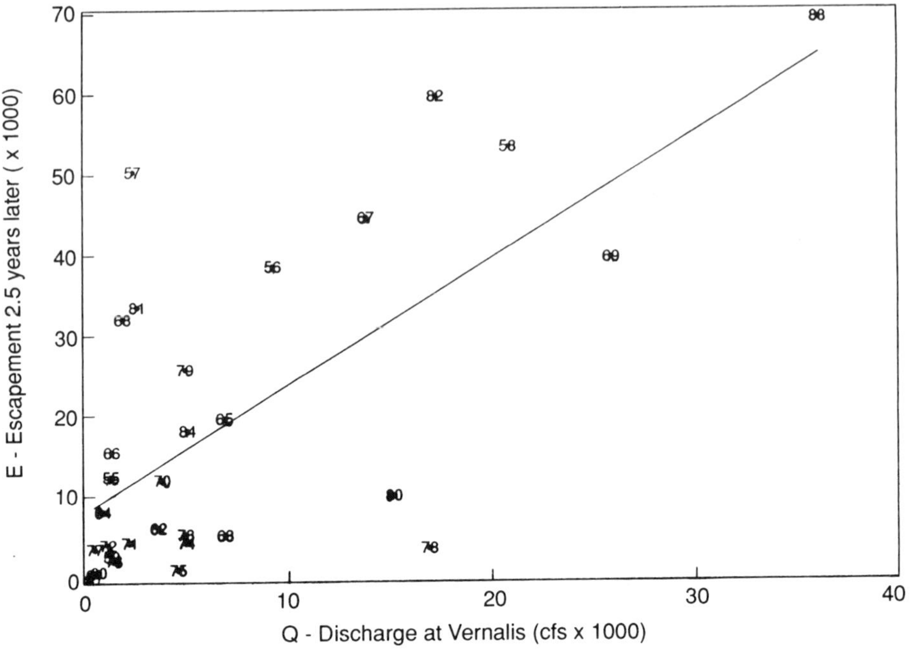

Figure 13.3 Regression of escapement on spring flow $2\frac{1}{2}$ years earlier

of our reasons for being sceptical about these uses of the linear regression model, reasons which will come up later, as they also apply to our own elaboration of the model, let us point out some of its other obvious deficiencies. Firstly, any give year's escapement contains a mixture of 2-, 3- and 4-year-olds, and the proportions of each age-class varies with the relative size of that cohort. It is cohort size which would be affected by the magnitude of the spring flow when members of that cohort are maturing, and attempts to relate flow to mixtures of cohorts must run the risk of missing or mistaking the relationship of interest. A second problem concerns the use of the formulae of simple linear regression with a highly autocorrelated series such as (E_t). Even if the relationship between (E_t) and (Q_{t-2}) was what we are interested in, and was linear, we would need to modify the standard errors associated with the estimates of a and b, and perhaps even the estimation method itself, to take the autocorrelation into account.

We need not pursue this line much further, however, because the decomposing of a series such as (E_t) into separate components from different cohorts is already a common technique in fisheries research. But before turning to our model, let us briefly outline the other extreme kind of model. The linear regression model just described is not only somewhat unsatisfactory from a statistical viewpoint, it also fails to incorporate fisheries biology, even the fact that each escapement is a mixture of age-classes. At the other extreme we have San Joaquin River System Chinook Salmon Population Model (EACH), described in EA Engineering, Science and Technology (1991), which is a discrete-time deterministic simulation model incorporating very detailed assumptions concerning the biology of salmon. This model provides a weekly time-step account of eight life stages of salmon over five years, moving with flow through five geographic regions in the drainage basin. At each time step, salmon life stages are subjected to intrinsic factors such as redd superimposition and natural mortality, anthropogenic factors such as fishing and flow diversion to exports, and environmental factors such as the amount of spawning habitat or predation. The model has an extremely large number of unobservable states, and must include a similarly large number of assumed empirical relationships and assumed values for key population parameters. In Figure 13.4 we display one of the five geographic regions of the model. The model is then 'tuned' to 'track' a given body of data, in this case escapements from 1973 to 1986, and subsequently 'validated' on an independent set of data. This is an activity which bears some resemblance to the fitting of statistical models, but it is important to realise that no explicit modelling of errors or variation is included: it is simply a deterministic model with many more parameters than observations.

How can one criticise such a model? Even if it really has correctly incorporated the biology, which seems extremely unlikely given the number of life-history parameters and relationships that need to be assumed, there remains the problem of fitting the available data, estimating further parameters and assessing the overall goodness-of-fit. It seems that we can only carry out this exercise using familiar statistical methods and thinking when the complexity of the model is reduced considerably. We must oversimplify the biological relationships assumed, and we take this to be the price that must be paid in order to use statistics. In our view, the linear regression model was 'too statistical', and 'not biological enough', whereas the simulation model, even putting aside the fact that many details were probably wrong, was 'too biological', and 'not statistical enough'. Another frequent problem with such simulation models is their sensitivity to changes in the values of key parameters. Sometimes it can be too great; at other times too small. The model we will now consider has many obvious defects, but it does incorporate *some* biology, and it *is* amenable to *statistical analysis.* If statistical analysis is to help at all in this sort of problem, then it must be through the use of such models, intermediate in complexity between the traditional regression-type black-box statistical model, and the most highly detailed model which could be developed in the given context.

We will now describe our model and how it was fitted to the available data. It is a state-space model of the kind pioneered by Kalman (1960), and widely used in the environmental sciences by Young (1984), and in fisheries by Mendelssohn (1988), Schnute (1991) and Sullivan (1992), see also Brillinger *et al.* (1980). The idea is that the numbers of fish R_t^2, R_t^3 and R_t^4 in the salmon population aged 2, 3 and 4 years are unobserved variables which develop in

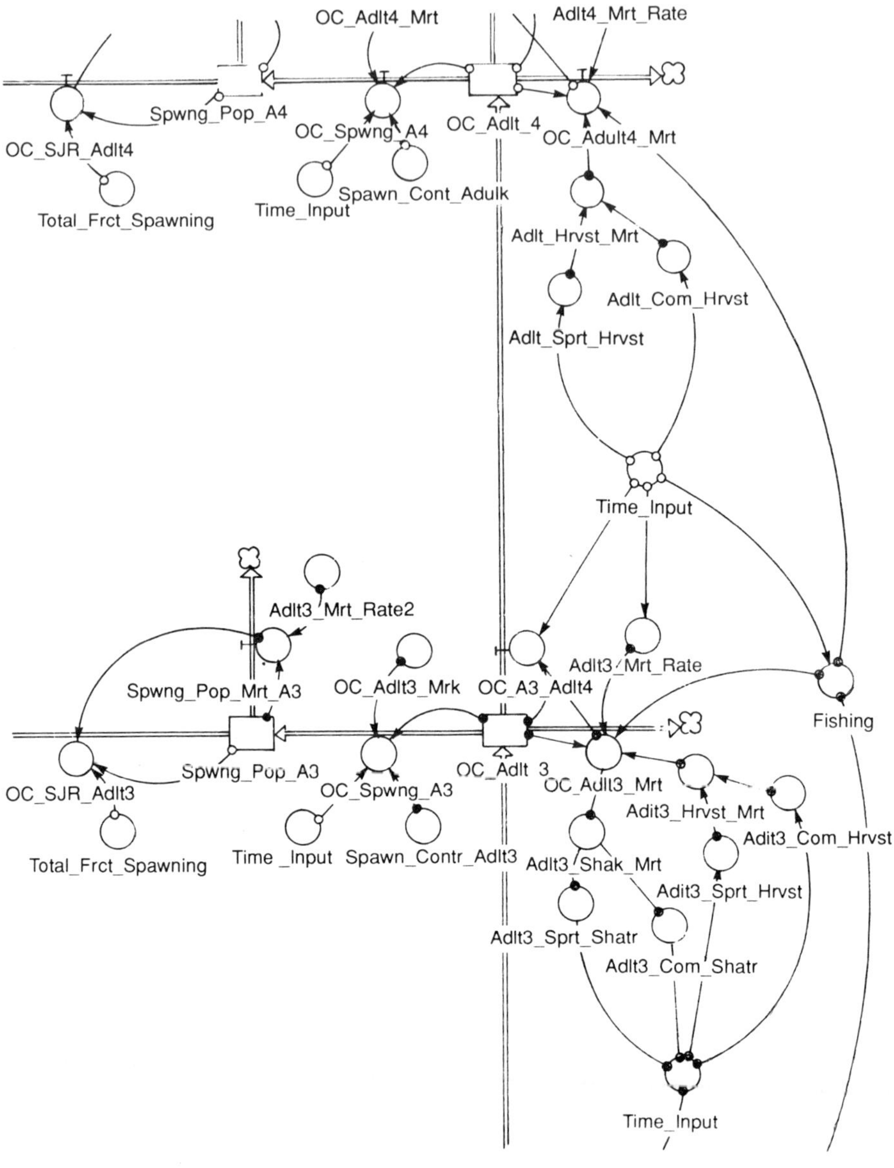

Figure 13.4 Ocean part of the EACH model

time according to a simple population model, and that there is a simple relationship between these recruit numbers and what we observe, which is the escapement (E_t) and an estimate (C_t) of the number of chinook salmon caught which originate from the San Joaquin river system. In order for the model to be first-order Markovian, we also need to include the number S_{t-1} of spawners in year $t-1$ as an unobserved state, this being the escapement in that year, apart from measurement error. Recruits are defined to be fish that survive to 2 years of age, these being the only ones we attempt to track. In the absence of further data it is simply impossible to estimate egg survival or fry mortality; all our data refer to mature fish, and so we take the numbers of such fish as our target. This model is depicted in Figure 13.5.

The model is defined by giving joint distributions for the states $x_t = (R_t^2, R_t^3, R_t^4, S_{t-1})$ and for the observables $y_t = (E_t, C_t)$ for $t = 1, \ldots, T$. These distributions are specified by giving an initial distribution for x_0, state-update distributions $p(x_{t+1}|x_t, \theta)$ for x_{t+1} given x_t, and observation distributions $p(y_t|x_t, \theta)$ for y_t given x_t, and then decreeing that the whole system is Markovian, in the sense that

$$p(x_{t+1}, y_{t+1}|x_t, y_t, \ldots, x_1, y_1, x_0, \theta) = p(x_{t+1}|x_t, \theta)p(y_{t+1}|x_{t+1}, \theta), \quad t = 1, \ldots, T.$$

The model has a set θ of unknown parameters and only a small number of these will be estimated from the data; the rest will be assigned values based upon studies in the literature.

The idea of the model is extremely simple. State updates are (as with the deterministic simulation model) simply book-keeping: 3- and 4-year-old recruits come from 2- and 3-year-old fish, respectively, after natural mortality, commercial and sport catch and escapement have depleted their numbers, while catch and escapement are obtained from the recruits of different ages in a fixed way using parameters specific to each age-class. Indeed let μ denote the fraction of an age-class lost to ocean mortality, where we note that the use of a single figure for all ages and years is already quite an oversimplification. If we had reliable estimates from other data of age-specific and year-specific ocean mortalities, we would use them, but we do not. Similarly, let ω denote the fraction of 2-year-olds lost due to fishing (even through such fish are rarely big enough to be part of the legal catch) and let γ denote the fraction of 3- and 4-year-olds caught by fisherman. Finally, let ρ_2 and ρ_3 denote the fractions of 2- and 3-year-olds which return in the fall to spawn; we will be supposing that *all* remaining 4-year-olds return to spawn. On the basis of these simple considerations, we write

$$\begin{aligned} R_{t+1}^3 &= (1-\mu)(1-\omega)(1-\rho_2)R_t^2, \\ R_{t+1}^4 &= (1-\mu)(1-\gamma)(1-\rho_3)R_t^3, \\ S_t &= (1-\mu)(1-\omega)\rho_2 R_t^2 + (1-\mu)(1-\gamma)(\rho_3 R_t^3 + R_t^4) \end{aligned}$$

for the (deterministic) state-updates of the unobservables R_{t+1}^3, R_{t+1}^4 and S_t, with

$$\begin{aligned} E_t &= S_t + \delta_{1t}, \\ C_t &= (1-\mu)\gamma(R_t^3 + R_t^4) + \delta_{2t} \end{aligned} \qquad (13.1)$$

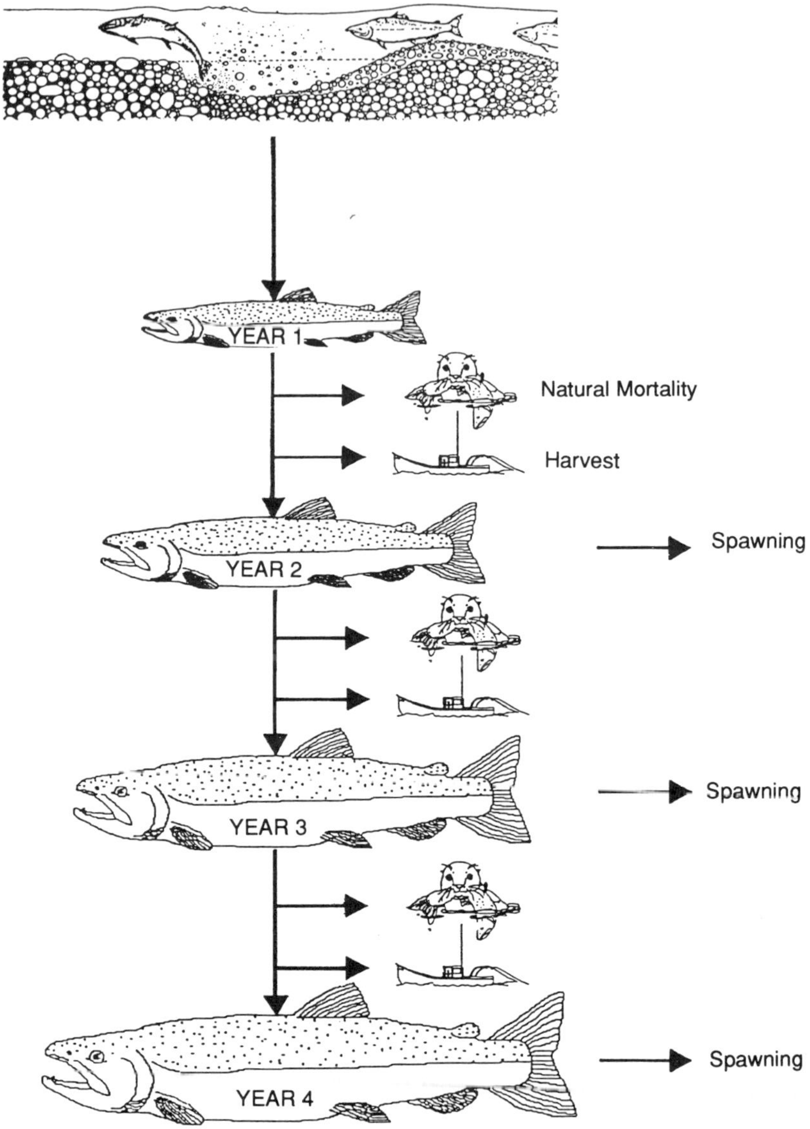

Figure 13.5 The distribution of fish from a single cohort assumed under model (schematic)

for the observation equations for escapement and catch, respectively. Here (δ_{1t}) and (δ_{2t}) are independent sequences of independently distributed errors, about which we will make various assumptions below. All that is necessary before the model is complete is a state-update equation for R^2_{t+1}, and as we shall see, this step is the heart of the model. It is here that we can incorporate both density-independent and density-dependent sources of mortality, and it is by considering different possibilities at this point that we can attempt to discriminate between alternative hypotheses concerning the growth and decline of the population. In Speed and Ligon (1993) we gave reasons for using the following

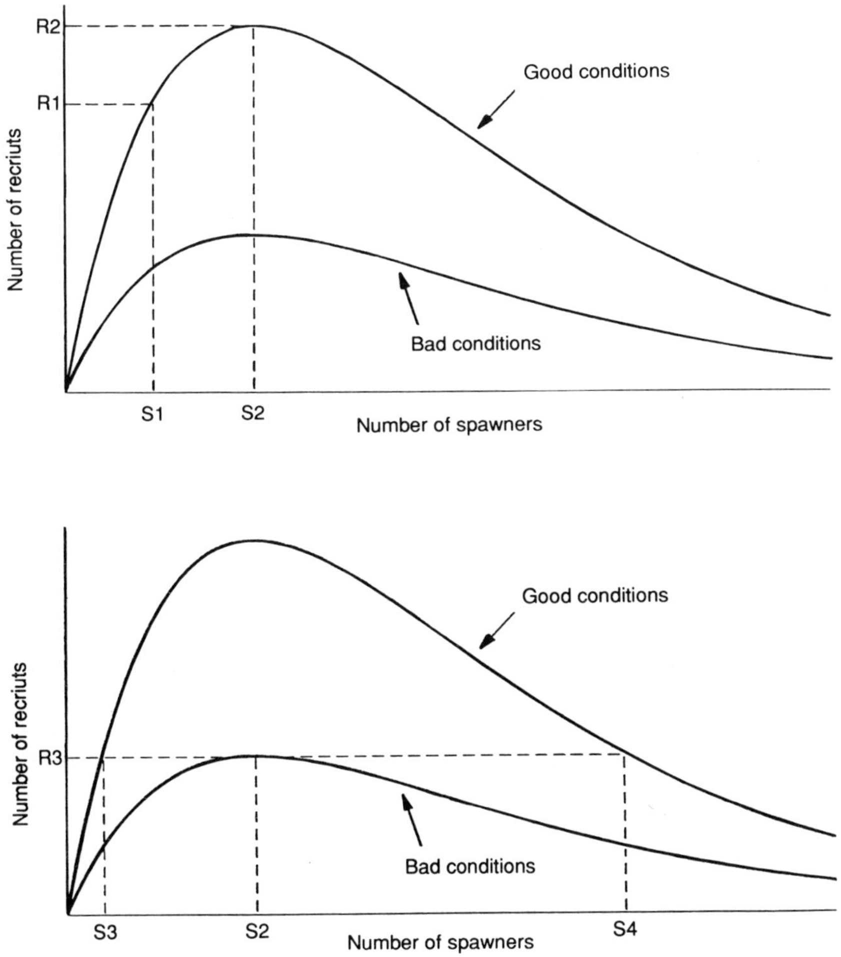

Figure 13.6 Some Ricker-type spawner–recruit curves modified by the environment

equation:

$$R_{t+1}^2 = \alpha Q_t S_{t-1} \exp(-\beta S_{t-1}) + \varepsilon_{t+1}, \quad t = 1, \ldots, T-1. \tag{13.2}$$

where Q_t is the average spring flow of the San Joaquin at a particular site (Vernalis) downstream of the three tributaries, and we are making use of the so-called Ricker form of spawner–recruit curve: $R = aS\exp(-bS)$ connecting the number of spawners S with the number of recruits R two years later. A number of such curves are plotted in Figure 13.6. Other forms of spawner–recruit curves can be found in the fisheries literature, see e.g. Ricker (1975) and Schnute (1985) for a general discussion, and one of particular interest to us is the so-called Beverton–Holt curve: $R = bS/(a+S)$. This would be incorporated into our model as

$$R_{t+1}^2 = \frac{bQ_t S_{t-1}}{a + S_{t-1}} + \varepsilon_{t+1} \tag{13.3}$$

and we note in passing that there is a simple 3-parameter family of curves which includes the 2-parameter Ricker form and the 2-parameter Beverton–Holt form as special cases, see Deriso (1980). Some Ricker, Beverton–Holt and intermediate curves are plotted in Figure 13.7 and it is apparent that although they differ considerably for large S, they can exhibit quite similar behaviour for small-to-moderate S. These curves have the equation $R = \alpha S(1 - \beta\gamma S)^{1/\gamma}$ for $\alpha = 6$, $\beta = 0.05$ and $\gamma = 0$, -0.5 and -1 respectively. Returning to our state-update equations, observe that there is no special reason why the Ricker curve was scaled by spring flow Q_t rather than by some function of flow such as $a + bQ_t$, some power Q_t^{ϕ} or something even more complicated. We chose to scale by Q_t for simplicity, after noting that the alternative $a + bQ_t$ did not perform noticeably better, although this does not rule out the possibility of other functions of Q_t doing better still. The errors (ε_t) in the state-update are to be regarded as mutually independent across years, and independent of the erors (δ_{1t}) and (δ_{2t}) and (δ_{2t}), and although we have written them additively in (13.2) and (13.3) and (13.1), multiplicative errors are considered as well. It would seem to me that the following assumptions concerning errors in these models are listed in order of increasing realism: additive normal errors with zero means and constant variances; additive normal errors with zero means and standard deviations proportional to the quantity they add to; and multiplicative errors whose logs are normal with zero mean and constant variance.

As soon as the form of error and spawner–recruit curve is fully specified, we can consider fitting one of these models. Unfortunately, there are too many parameters to estimate from two time series of 40 observations; indeed $\theta = (\alpha, \beta, \mu, \omega, \gamma, \rho_2, \rho_3, \sigma_\varepsilon^2, \sigma_\delta^2)$ is 9-dimensional, supposing that (δ_{1t}) and (δ_{2t}) have a common variance (σ_δ^2). Our solution to this problem was to get values for μ, ω, γ, ρ_2 and ρ_3 from the relevant fisheries data, and to estimate the key

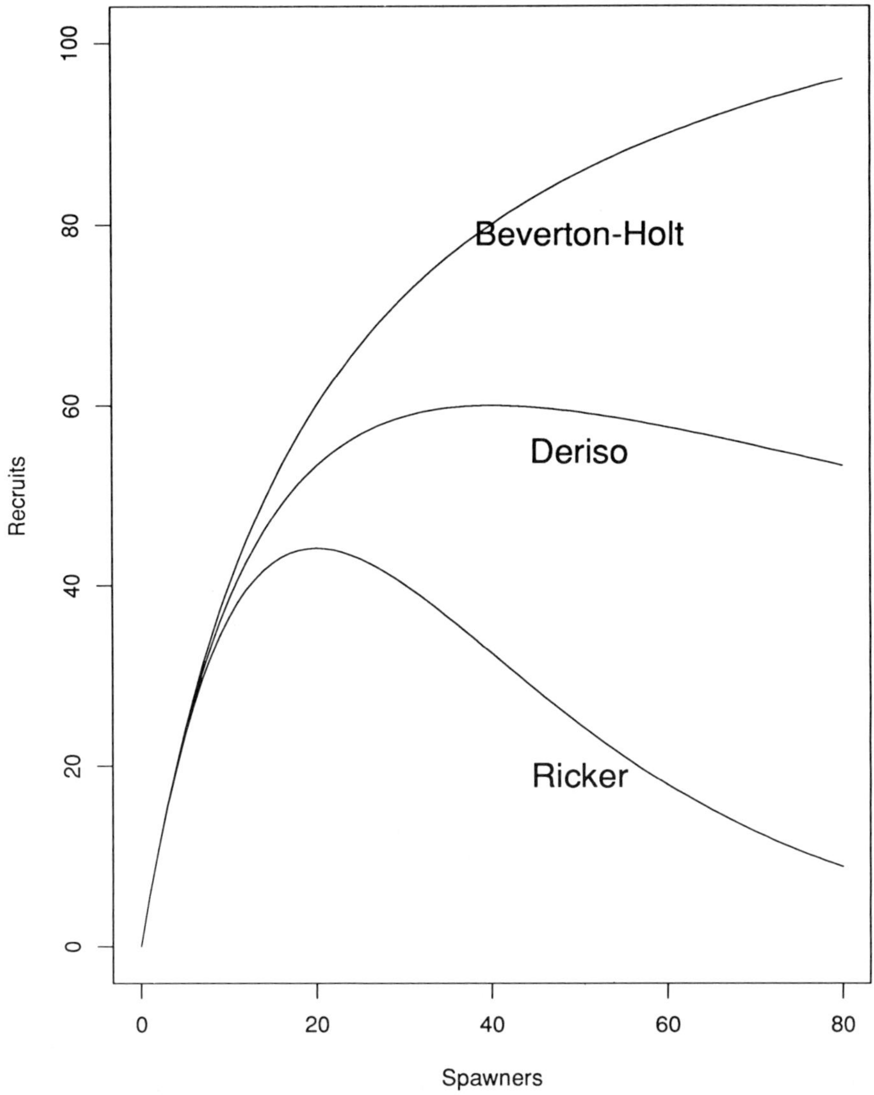

Figure 13.7 A Ricker, a Deriso and a Beverton–Holt curve

parameters α and β of the Ricker curve, and the error variances σ_ε^2 and σ_δ^2 by maximum likelihood.

We now explain briefly how an approximate maximum likelihood estimation procedure was carried out, and then describe some sample results. Forgetting the fact that we will not be estimating *all* of the components of θ, let us see how to calculate the likelihood $L(\theta|y^T)$ of θ based upon $y^T = (y_1, \ldots, y_T)$, i.e. how to calculate $p(y^T|\theta, x_0)$. We begin by noting that $p(y^T|\theta, x_0) = \Pi p(y_t|y^{t-1}, \theta, x_0)$, where $y^{t-1} \equiv (y_1, \dots, y_{t-1})$ and then we observe that each of the components in the product on the right-hand side can be calculated recursively via

$$p(y_t|y^{t-1}) = \int p(y_t|x_t)p(x_t|y^{t-1})\mathrm{d}x_t$$

as long as we have expressions for $p(x_t|y^{t-1})$. (Here we have dropped the θ and x_0, as they are present on the right side of the conditioning in all expressions which follow.) Now a recursion for $p(x_t|y^{t-1})$ can be developed using the equations

$$p(x_t|y^t) = p(y_t|x_t)p(x_t|y^{t-1})/p(y_t|y^{t-1})$$

and

$$p(x_{t+1}|y^t) = \int p(x_{t+1}|x_t)p(x_t|y^t)\mathrm{d}x_t, \quad t \geqslant 1,$$

see Kitigawa (1987). Indeed this development is a general form of the famous Kalman filter (Harvey, 1989), and reduces to simple mean and covariance matrix updates when all the state-updates and observation equations are linear, with additive normal errors. Note that in this linear case, we effectively obtain the likelihood as a by-product of the recursive calculation giving best predictors of the unobserved states.

The equation (13.2) updating R_t^2 is not linear, however, and so some generalisation of the Kalman filter is required. The simplest is the so-called extended Kalman filter (Young, 1984; Harvey 1989) which uses the linear Kalman filter recursion applied to a linearised form of the nonlinear state-update, this being essentially a one-term Taylor expansion of R in terms of S, about the best predictor of S available at that time. In our case we are effectively replacing the update (13.2) by $R_{t+1}^2 = \alpha Q_t s_{t-1} \exp(-\beta S_{t-1}) + \alpha Q_t(1 - \beta s_{t-1})(S_{t-1} - s_{t-1}) \cdot \exp(-\beta s_{t-1}) + \varepsilon_{t+1}$, where s_{t-1} is the mean of the distribution of S_{t-1} given y^{t-1}.

How does this model fit the data? With the following preassigned values: $\mu = 0.20$, $\omega = 0.18$, $\gamma = 0.60$, $\rho_2 = 0.17$ and $\rho_3 = 0.65$, additive normal errors in both the state-update for R_t^2 and (with equal variances) in the two observation equations, we find $\hat{\alpha} = 2.50$, $\hat{\beta} = 0.069$, $\sigma_\varepsilon = 55$ and $\sigma_\delta = 10$. In Figures 13.8 and 13.9 we plot the likelihood surface and log-likelihood contours, respectively, for this model. Furthermore, in Figure 13.10(a) we plot three confidence regions and in Figure 13.10(b) three spawner–recruit curves corresponding to the three dots in Figure 13.10(a). Note that these confidence regions for the parameters α and β do not reflect the uncertainty in σ_ε^2 and σ_δ^2, nor, of course, in the other parameters which were assigned values. As an indication of the fit of the model

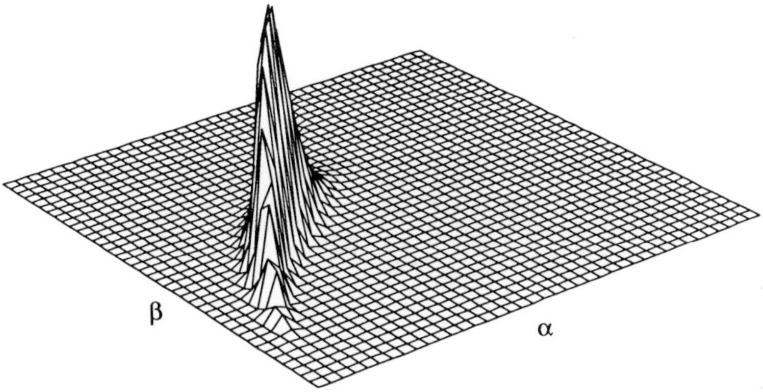

Figure 13.8 Profile likelihood surface for α and β (full model, both errors additive)

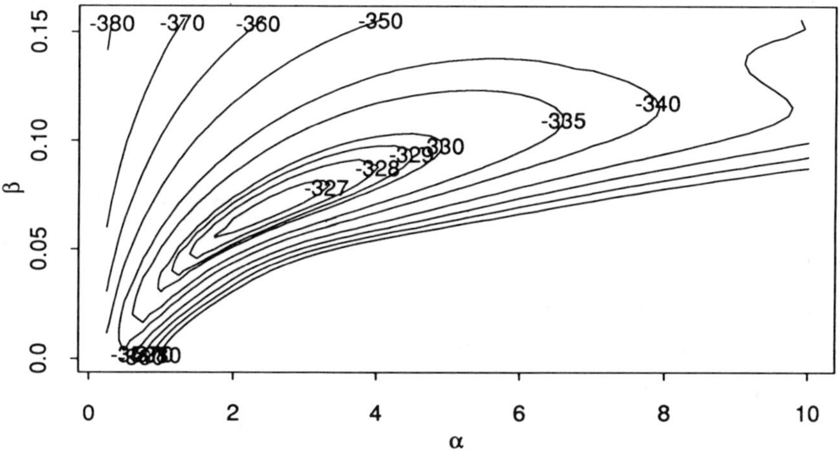

Figure 13.9 Profile log-likelihood contours for α and β (full model, both errors additive)

to the escapement and catch data, and to give an idea of the way the extended filter works, we plot in Figure 13.11(a) and 13.11(b) each observed series along with its *prediction* based upon the model, but not the data from the corresponding year, and also the so-called *filtered* value, which is the value expected under the model after the given year's observation has been incorporated into the estimate of the unobserved states. The last set of plots shows that the model captures the dynamics of recruitment fairly well.

As mentioned previously, the assumption of additive errors in both the state-update equation (13.2) for R_t^2 and in the observation equations (13.1) is

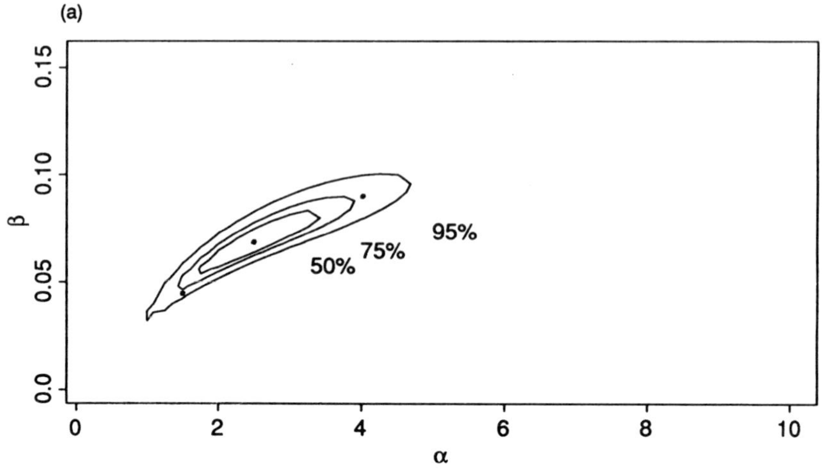

Figure 13.10(a) Three confidence regions for α and β based on the profile likelihood for the full model with both errors additive

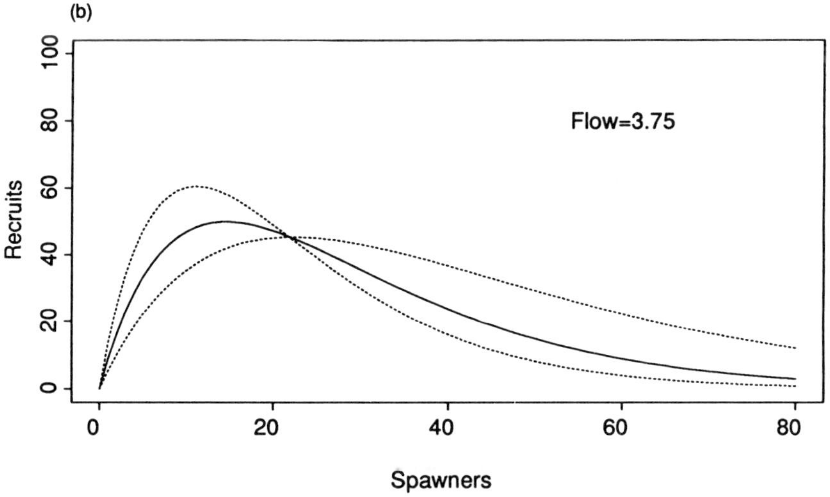

Figure 13.10(b) Spawner–recruit curves corresponding to the three points in Figure 13.10(a)

probably not the most reasonable one. The range of simple alernatives, such as multiplicative errors models, is somewhat restricted if we wish to keep using the extended Kalman filter. It turns out to be straightforward to incorporate multiplicative log-normal errors into (13.2), but harder to do so in the observation equations (13.1). Ultimately a more exact version of the nonlinear

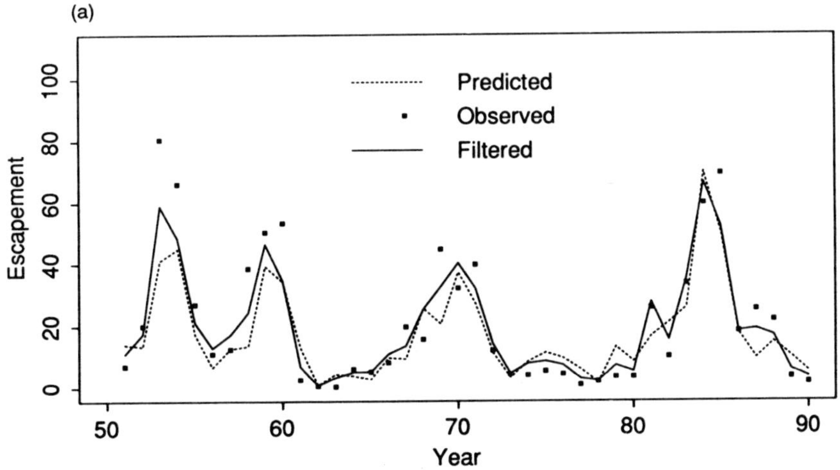

Figure 13.11(a) Predicted (one step ahead), observed and filtered series for escapement: full model; both errors additive

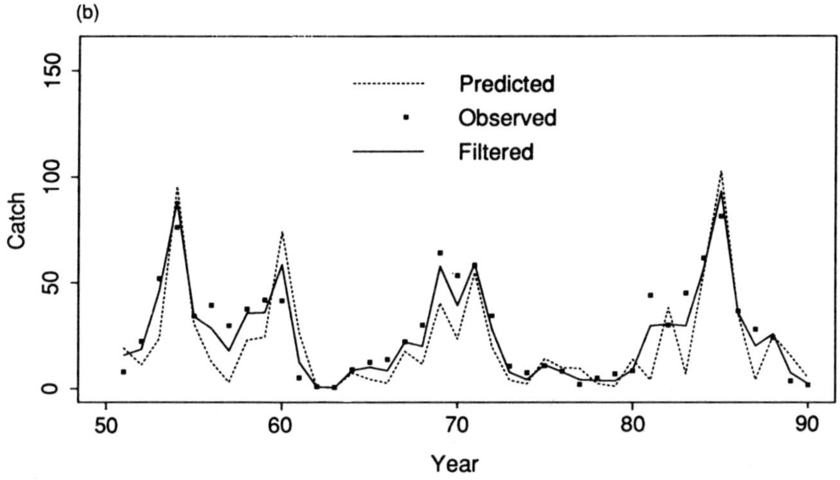

Figure 13.11(b) Predicted (one step ahead), observed and filtered series for catch: full model; both errors additive

Kalman filter will be used to incorporate any desired error structure for (13.2) and the observation equations (13.1), but until then we will be restricted to error structures which permit the likelihood to be calculated using the extended Kalman filter. For more on this topic, see the discussion and the rejoinder to Kitagawa (1987) and also Kitagawa (1991). Thus we can easily incorporate

Table 13.1 Model I; Additive normal errors in (13.2) and in (13.1)

	Full	No flow	No update
α	2.50	15.95	1.89
β	0.069	0.050	0.059
σ_ε	55	74	73
σ_δ	10	10	10
$\log L$	−326.4	−334.7	−333.3
SSPE			
esc	5742	12 076	7815
cat	11 589	12 333	15 058

Table 13.2 Model II; Multiplicative log-normal errors in (13.2), additive normal errors in (13.1)

	Full	No flow	No update
α	2.51	12.66	2.38
β	0.050	0.045	0.059
σ_ε	0.805	0.721	0.443
σ_δ	11	11	14
$\log L$	−332.7	−335.1	−334.8
SSPE			
esc	6578	12 479	6875
cat	19 483	13 102	17 646

Table 13.3 Model III; Multiplicative log-normal errors in (13.2), multiplicative normal errors in (13.1)

	Full	No flow	No update
α	3.54	15.45	2.27
β	0.069	0.055	0.043
σ_ε	1.353	0.741	2.523
σ_δ	0.774	0.568	0.385
$\log L$	−341.4	−334.1	−345.5
SSPE			
esc	5872	12 639	9739
cat	18 149	11 581	31 627

multiplicative errors into (13.2), whose *logarithms* are mean-zero normal deviates with variance σ_{ε}^2, and multiplicative normal errors can also be incorporated (13.1), whose means are 1 and variances are $\sigma_{\delta}^2 S_t^2$ and $\sigma_{\delta}^2 c_t^2$, respectively, where $E(C_t) = c_t$. Tables 13.1 13.2 and 13.3 display the parameter estimates, maximised likelihoods and sums of squared one-step prediction errors (SSPE) for the escapement (esc) and catch (cat) series, for the (full) state-space model as previously described, and for variants which (a) change the error

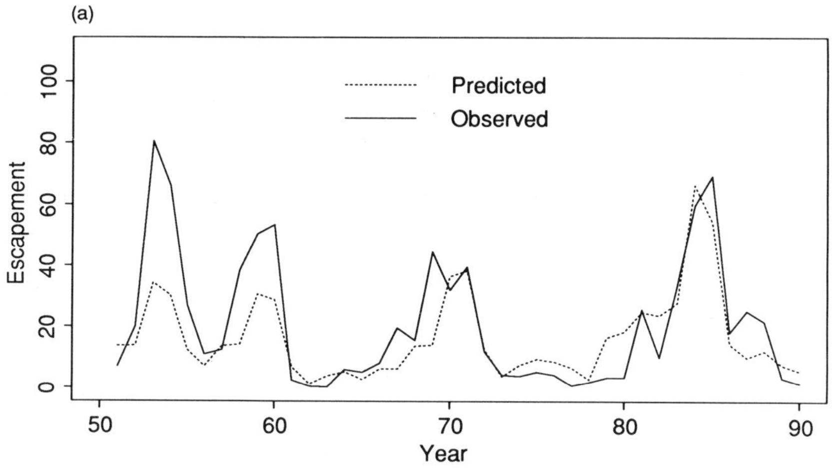

Figure 13.12(a) Predicted and observed series for escapement: full model; both errors additive

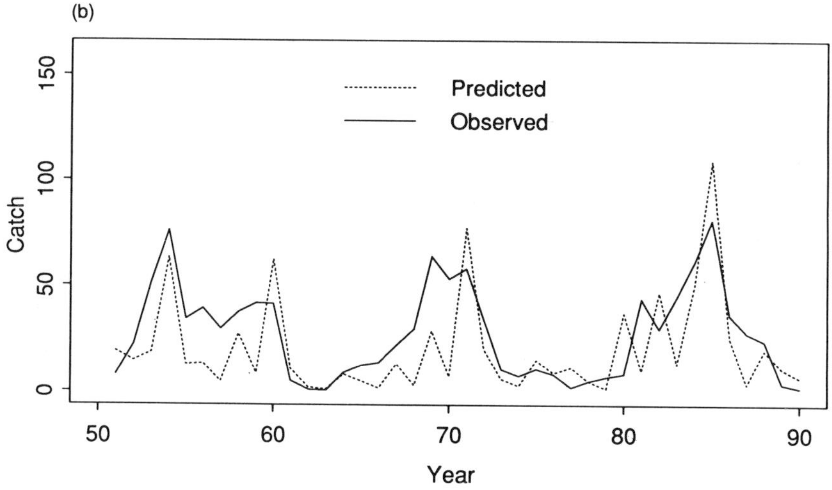

Figure 13.12(b) Predicted and observed series for catch: full model; both errors additive

structures to those indicated, and/or (b) omit the flow term Q_t from (13.2), thereby changing the magnitude of the parameter α, and/or (c) omit the update involving escapement.

The third column in each of Tables 13.1, 13.2 and 13.3 headed 'No update' gives the results of fitting a version of the model in which the escapement and catch series are not used in updating the state distributions. The flow series is used, as well as the unobserved state S_t, in (13.2), but escapement and catch only enter for the purpose of calculating the SSPEs. These results show that flow alone explains quite a lot of the variation in escapement and catch.

What are the general conclusions of this analysis? It seems that the different error models do not make a great difference to the overall fit of the model, and strangely, the one with both sets of errors additive seems to fit best by the likelihood and by the SSPE criterion. However, there are two problems here. Firstly, likelihoods computed under different error models are not directly comparable, although the SSPEs are, but, more importantly, in both series there are a small number of 'wild' or 'discrepant' observations which exert considerable leverage on the analysis. This is perhaps just a polite way of saying that the model does not fit, at least that it does not fit at these points, and it is perhaps no surprise that models with different error structures respond quite differently to the presence of such observations. We could have displayed plots such as those in Figures 13.11(a) and 13.11(b) for the different error models, but apart from showing that the different models do indeed treat these 'discrepant' observations differently, there seemed little else to be learned from them.

The differences between the various models with and without the flow term in (13.2) seem clearer and more consistent: the likelihood drops, and (what is

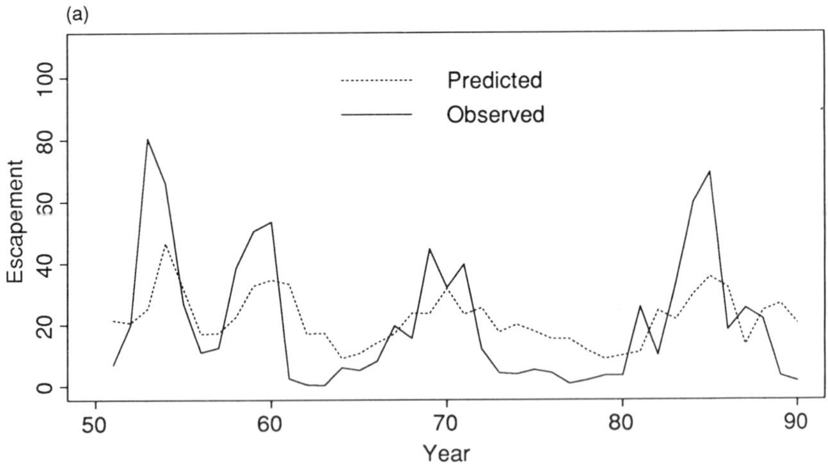

Figure 13.13(a) Predicted and observed series for escapement: model with flow omitted; both errors additive

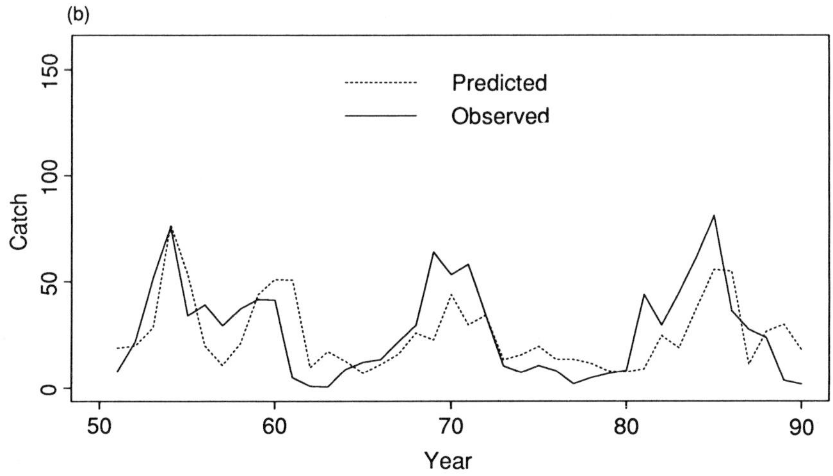

Figure 13.13(b) Predicted and observed series for catch: model with flow omitted; both errors additive

almost the same thing) the SSPE increases quite noticeably in all but one case. In Model III (Table 13.2) the likelihood increases, but it is mainly a trade-off, fitting the catch series more and the spawner series less closely. This is hardly a situation in which significance testing is appropriate, but a comparison of the peaks and (especially) the troughs of the series of predicted escapements and catches in Figures 13.12(a) and 13.12(b) with those of Figures 13.13(a) and 13.13(b) shows that the inclusion of the flow term *does* make a difference. What it *means* is discussed briefly in the next section.

13.4 DISCUSSION

Let us turn now from the technology of model fitting and assessment to the questions which motivated this study in the first place. Can we give answers to our initial questions on the basis of the analysis we have given?

Question 1: Is spring flow important in determining recruit numbers?

More fully, can we demonstrate convincingly that the level of spring flow has an important impact upon the number of recruits resulting from the juveniles maturing in that season? In a sense our analysis is no better placed to answer this question than the simple linear regression analysis or the simulation model EACH: the former makes use of observational data, and all statisticians know that we cannot deduce causal relationships from such data, while the latter essentially incorporates a causal assumption into its formulation, and also uses the same data. It is clear that spring flow may be acting as a surrogate for other important flow-related variables (e.g. temperature) which are causally connected to the survival of juvenile salmon, and no analysis along the lines of the one we have given can distinguish surrogacy from real causality. The answer to Question 1 will never come from a statistical analysis of this kind of data alone. Such analyses must be combined with other scientific knowledge, or the results of experimental interventions controlling alternative causes of the effect of interest. What our analysis has done is to include the size of the parent stock, which is itself related to spring-flow levels, in the model, and show that there is a discernible effect of spring flow (or related variables) on the number of recruits over and above that resulting from the size of the parent stock. The effect is not great, but it seems to be there. Such a conclusion seems rather lame, but it is probably the best one could hope for: the data have large 'measurement errors' associated with them, to put it mildly, and they are observational data, as has already been noted. There is a subtle point here, because parent stock (= spawners) and flow seem to be inextricably intertwined. We have seen that we can predict the observed escapement and catch series fairly well with these series alone, *without* flow, but we can also predict them fairly well from a model with flow, but *without* the lagged observations on escapement and catch. In thus seems difficult to say just what the effect of flow means, although we can assess the *extra* impact of including flow in the model, over and above that it exerts on parent stock. Any use of flow–population-size relationships in evaluating alternative policy measures designed to protect the salmon will need to be based on other, preferably experimental, data, as well as studies like ours.

Question 2: Is there evidence of density-dependent mortality of eggs or juveniles?

With this question we really can be more definite: there is clear evidence of density-dependence, but the data are unable to discriminate firmly between the Ricker and the Beverton–Holt form of spawner–recruit curve, when these curves are embedded in the model as in (13.2) or (13.3). A further analysis suggests that when the Deriso family of curves including both forms is fitted, a choice is made which is neither of these 'pure' forms but one 'midway' between the two. Both the Ricker and the Beverton–Holt curves incorporate density-dependence, albeit of different types. The possibility that recruitment actually *declines* when the number of spawners gets sufficiently large, as it does with the Ricker form, rather than simply levelling out as it does with the Beverton–Holt curve, is an matter of concern to policy-makers. Unfortunately our analysis does not throw any light on this issue.

As with Question 1, other information can sometimes help where a direct statistical analysis does not, and in the present situation it is possible to exhibit a simple and plausible simulation model which leads directly to the Ricker form of spawner–recruit curve (F. K. Ligon, pers. comm.) In this model the available spawning gravels are divided into a grid of small squares and the process of spawners arriving at the gravels, selecting an area on which to construct a redd, and 'protecting' the redd for a period after the eggs are laid, fertilised and covered over, is simulated in daily time-steps for a time corresponding to the fall spawning season. This model incorporates the process of superimposition of redds in a form which has been observed in the field, and when the offspring of salmon spawning at different times in the season are aggregated, with a lower survival assumed for the young of later spawners, a Ricker-type spawner–recruit curve does indeed result. Neither this mathematical analysis nor the statistical analysis done prove that Ricker-type density-dependent mortality is occurring, but when they are combined with other observations from the field, we see quite a strong case that such mortality is taking place.

Because they do offer insights into the population dynamics of salmon, it is inevitable that models such as ours will be used to assist in management or in the determination of policy questions. This will happen despite the protestations of statisticians who at times are forced to explain that the available data and scientific understanding of the issues under discussion might not be an adequate basis for answering such questions. A far more satisfactory situation is when the model is used in conjunction with a programme of field research into the mechanisms underlying the empirical relationships in the model. Frank Ligon has gone on to use the model in conjunction with research into suspected sources of density-dependent and density-independent mortality, to predict possible gains in recruit numbers following various actions. Those considered include: installing barriers, cleaning the gravels and implementing pulse flows. Barriers could be used to prevent later-arriving salmon from superimposing their redds on existing redds, effectively guiding them to vacant areas in the gravels. (If such barriers were used, the timing of their closure would need to be considered.) Cleaning the gravels would enhance the survival of the maturing eggs. Pulse flows, being high flows of short duration, resulting from specially planned and timed releases from dams of suitable quantities of water, would facilitate the downstream migration of smolts.

Even with additional scientific support, we cannot expect to take numerical predictions from models such as ours at their face value, but they may give us reasons to believe that one management option may have a more beneficial effect on the salmon population than another. While not satisfactory for most policy-makers, this seems a better approach to making decisions than is often used, and a desirable next step would be the experimental testing of management decisions. Statisticians and others will continue to be confronted with problems like the ones discussed in this chapter for many years to come, and it is hoped that by outlining one such experience, we have helped widen the appreciation of the difficulties.

ACKNOWLEDGMENTS

My primary acknowledgement is to Frank Ligon, who got me involved in this problem four years ago, and who, while not always agreeing or even being comfortable with my point of view, has remained a collaborator and friend. Special thanks are also due to Steve Rein, who has done many of the computations reported in this chapter, and to Mike Stewart and Takashi Ishikida, who have assisted in the past. I am also very grateful to Vic Barnett and Feridun Turkman for the invitation to present this work at the SPRUCE Conference. And last, but not least, many thanks are due to Chris Bush for her fast and efficient typing and willingness to work to a very tight deadline.

14

Ambiguous Inferences from Fisheries Data

Jon T. Schnute
Pacific Biological Station, Nanaimo, BC, Canada

14.1 INTRODUCTION

Marine fish populations live hidden in an environment where they experience a complex range of biological dynamics. They give birth, grow, and die; they migrate and are carried by currents; they act as both predator and prey. They also respond to environmental changes, due to both human and natural causes.

Data to investigate these populations usually come from a few scattered sources, such as commercial fisheries, sampling programmes, hydroacoustic surveys, mark–recapture studies, and environmental monitors. Furthermore, data collection often takes place in contexts where quality control is difficult; high levels of measurement error may be unavoidable. Thus, an analyst must try to reconstruct an invisible population from a few noisy clues.

Theoretically, the data should reflect various underlying biological processes. For example, births must somehow be related to the available mature population. Within one year, a fish inevitably becomes one year older; the ageing process may leave evidence in the form of annular marks on fish scales or other bony structures. Given a reasonable food supply, fish also grow as they age, although they commonly reach a limiting adult size. Relationhips usually exist between age, size and fecundity. If fish are abundant, they may be easier to catch; consequently, the catch per hour of fishing may serve as an abundance index.

Mathematical models have been proposed to describe all these processes analytically. Hilborn and Walters (1992) review much of the existing literature. An analyst typically begins with a deterministic model that theoretically relates

Statistics for the Environment. Edited by Vic Barnett and K. Feridun Turkman

the available data to a set of underlying biological parameters. The model is then rendered statistical and used for parametric inference. Often the method of introducing statistical error is rather *ad hoc*; because the deterministic model itself is rather uncertain, the precise role of statistical error seems at first relatively unimportant.

In fact, the method of introducing statistics into a deterministic model can substantially influence the model's conclusions. In this chapter, we demonstrate how seemingly innocent statistical assumptions can conceal important ambiguities inherent in a data set. For illustration, we consider examples taken or adapted from recent publications (Schnute, 1987, 1989, 1991; Schnute *et al.*, 1989, 1990). Thus, the chapter gives perspective to a problem that occurs frequently in the analysis of fishery data and perhaps is latent in many biological and environmental models.

A simple, artificial example illustrates the essential problem (Section 14.1), and data from a salmon stock demonstrate its practical consequences (Section 14.2). Similar features characterise a more complex example in which the data are serially correlated (Section 14.3). The concepts developed here pertain not only to parameter estimates, but also to forecast and control policies needed to manage fisheries optimally (Section 14.4). In summary (Section 14.5), failure to consider all sources of error can lead to conclusions that badly misrepresent biological reality.

14.2 A SIMPLE EXAMPLE

Ordinary regression illustrates some of the ambiguities possible in formulating statistical models. In this context, analysis begins with the deterministic linear relationship

$$y = a + bx \tag{14.1}$$

between variables x and y. For data (x_i, y_i), $i = 1, \ldots, n$, two possible statistical counterparts of (14.1) are

$$y_i = a + bx_i + \sigma\varepsilon_i, \tag{14.2}$$

$$x_i = \frac{1}{b}(y_i - a) + \tau\eta_i, \tag{14.3}$$

where the i.i.d. variates ε_i and η_i are presumed normal with mean 0 and variance 1. Model (14.2) ascribes error to y_i and assumes that x_i is known, while (14.3) reverses these assumptions. Both models include (14.1) as a deterministic special case ($\sigma = 0$ or $\tau = 0$).

Figure 14.1 illustrates a simple data set for model (14.2) or (14.3); the four

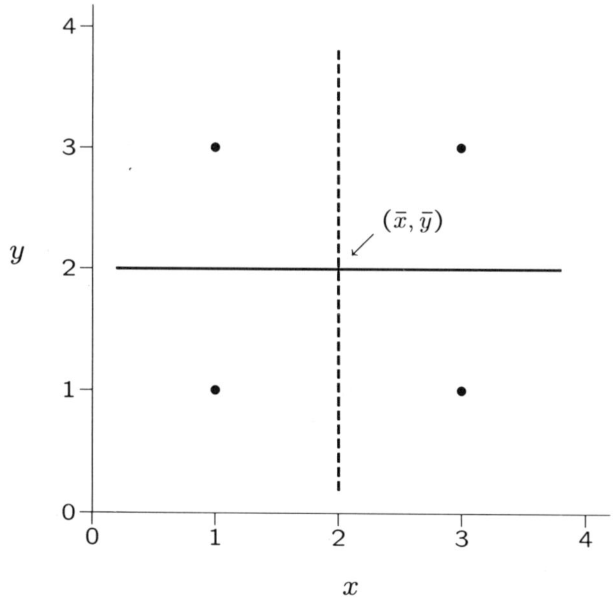

Figure 14.1 Data points (14.4) and regression lines obtained from models (14.2) (solid line) and (14.3) (broken line). The lines intersect at the data mean $(\bar{x}, \bar{y})$. *Adapted from Schnute, 1991, Figure 1*

points (x_i, y_i),

$$(1, 1), (1, 3), (3, 3), (3, 1), \tag{14.4}$$

define the corners of a square. If these data are interpreted using (14.2), then x is fixed at two values (1 or 3). When $x = 1, y$ takes two values (1 and 3), with mean 2. Similarly, when $x = 3$, the mean of y is again 2; consequently, y appears to be independent of x. Thus, model (14.2) leads to the estimates $\hat{a} = 2$ and

$$\hat{b} = 0. \tag{14.5}$$

An analogous argument based on (14.3) shows that x is independent of y and $\hat{b}^{-1} = 0$, i.e.

$$\hat{b} = \infty. \tag{14.6}$$

The estimates (14.5) and (14.6) constitute extremes for the relationhip (14.1). A line can only be horizontal, vertical, or somewhere in between. Conceptually, the line (14.1) represents the signal to be inferred from the data. Models (14.2) and (14.3) show that, in this case, the estimated signal depends *entirely* on the definition of noise. In general, the transition from a deterministic to a statistical model can dramatically influence the conclusions drawn from a data set.

The analysis of Figure 14.1 can easily be extended to produce confidence intervals associated with the estimates (14.5)–(14.6). For example, model (14.2) leads to the usual sum of squares,

$$S(b) = \sum_{i=1}^{n} [y_i - \bar{y} - b(x_i - \bar{x})]^2, \tag{14.7}$$

and a corresponding variance ratio,

$$R(b) = \frac{[S(b) - S(\hat{b})]/1}{S(\hat{b})/(4-2)}. \tag{14.8}$$

At the true value of b, $R(b)$ is known to be F-distributed. Thus, with probability α, the true value b lies inside the interval whose end-points are solutions b to the quadratic equation

$$R(b) = F_\alpha(1, 2), \tag{14.9}$$

where $F_\alpha(1, 2)$ denotes the upper α-point for the F-distribution with 1 and 2 degrees of freedom.

For the data in (14.4), a short calculation shows that $S(b) = 4 + 4b^2$ and $R(b) = 2b^2$. Also, the F-statistic in (14.9) can be expressed analytically: $F_\alpha(1, 2) = 2\alpha^2/(1 - \alpha^2)$. Consequently, (14.9) reduces to

$$b^2 = \frac{\alpha^2}{1 - \alpha^2}. \tag{14.10}$$

If θ is the angular inclination of the line (14.1), so that

$$b = \tan\theta, \tag{14.11}$$

then (14.10) reduces to the particularly simple result

$$\alpha = |\sin\theta| \tag{14.12}$$

associated with model (14.2). A calculation similar to (14.7)–(14.12) for model (14.3) begins with the sum of squares

$$S(b) = \sum_{i=1}^{n} \left[x_i - \bar{x} - \frac{1}{b}(y_i - \bar{y}) \right]^2$$

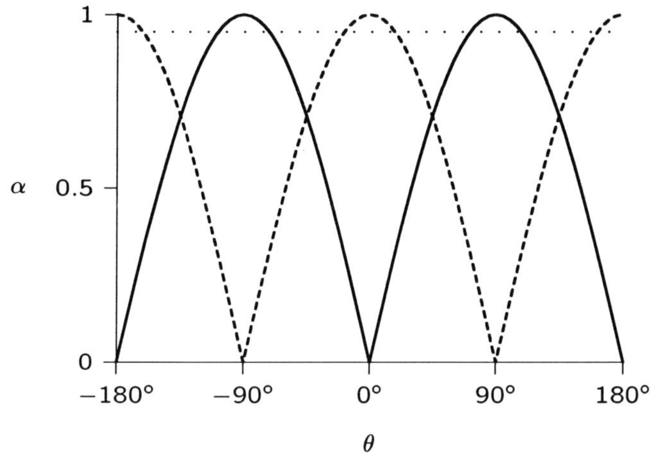

Figure 14.2 Probability profiles for the parameter $\theta = \tan^{-1} b$, based on the data (14.4) and models (14.2) (solid curve)) and (14.3) (broken curve). Analytically, the profiles are specified by (14.12) and (14.13)

and ends with the result

$$\alpha = |\cos \theta|. \tag{14.13}$$

Figure 14.2 shows the curves (14.12) and (14.13), which relate angles θ to rejection levels α. Each curve consists of a periodic set of arches; at a specified level α, the gap between arches defines an α-level confidence interval for θ. For example, models (14.2) and (14.3) attach 95% confidence to the conditions $\theta \in (-72°, 72°)$ and $\theta \in (18°, 162°)$, respectively. Each model excludes from possibility the estimate $\hat{\theta}$ obtained from the other; thus model (14.2) asserts that the estimate $\hat{\theta} = 90° = \arctan(\infty)$ from model (14.3) can be rejected with confidence $\alpha = 100\%$.

The parameter profiles in Figure 14.2 suggest a new interpretation for a ubiquitous icon of North American culture: McDonald's arches can serve as a reminder of statistical uncertainty. In the present context, two models' opposing views correspond analytically to the fact that $\sin(\theta)$ and $\cos(\theta)$ are 90° out of phase.

14.3 CHINOOK HATCHERY DATA

Schnute *et al.* (1990) analyse Pacific salmon hatchery data for which the ambiguities inherent in (14.1)–(14.3) have practical significance. Typically, young salmon leave a hatchery and swim downstream to the ocean, where they spend most of their adult lives. Salmon from a single hatchery can disperse

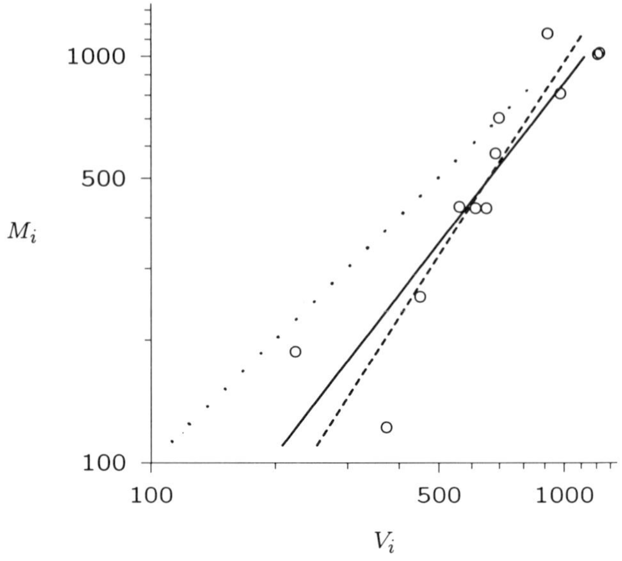

Figure 14.3 Marked estimate M_i v. visual estimate V_i of chinook salmon returning to Puntledge hatchery, for year $i = 1975–86$. Data points are represented as circles; solid and broken lines correspond to regressions from models (14.2) and (14.3), respectively. Dots indicate the 45° line, where $M_i = V_i$. *Adapted from Schnute et al., 1990, Figure 2*

to a large area of the ocean; however, on reaching maturity, they swim with remarkable fidelity back to their natal stream for spawning. Hatchery managers have considerable interest in the number of returning salmon. Two techniques are commonly used for estimating this number. First, direct visual counts are obtained at a suitable stream site. Second, a proportion of fish are marked as juveniles before they leave the hatchery, and the number of marked fish among adult returns is later used to estimate the total return.

For various reasons, including possible straying of hatchery and wild fish, these two estimates generally do not agree (Schnute *et al.*, 1990, p. 1465). Figure 14.3 portrays data on chinook salmon returning to Puntledge hatchery on Vanocouver Island in British Columbia, Canada. Estimates for 12 years i are plotted on a log-log scale, where V_i and M_i denote the visual and marked estimates, respectively. The data show a distinct linear trend, as in (14.1), where

$$x_i = \log_{10} V_i, y_i = \log_{10} M_i. \tag{14.14}$$

Most points (V_i, M_i) in Figure 14.3 lie below the 45° line, indicating a downward bias in marked estimates relative to visual counts. It is reasonable to ask if this relative bias is constant, that is, if

$$b = 1 \tag{14.15}$$

Table 14.1 Parameter estimates from models (14.2) and (14.3), based on the data in Figure 14.3. The 95% b-intervals are computed from (14.27) with $\lambda = \infty$ and $\lambda = 0$, respectively

Model	$\hat{a}$	$\hat{b}$	95% b-interval
(1.2)	−0.95	1.30	(0.89, 1.72)
(1.3)	−1.73	1.57	(1.19, 2.31)

in (14.1). If so, then (14.1) reduces to $y_i = a + x_i$, or

$$M_i = 10^a V_i. \tag{14.16}$$

Figure 14.3 shows the lines that result from conventional y-on-x and x-on-y regression (i.e. models (14.2) and (14.3)) applied to the data (14.14). Parameter estimates and 95% confidence intervals for b are listed in Table 14.1. As with the simple data set (14.4), model (14.3) suggests a steeper slope b than model (14.2). Furthermore, the two models produce different conclusions about the hypothesis (14.15); (14.3) suggests that (14.15) can be rejected with 95% confidence, but (14.2) does not.

In reality, neither model (14.2) nor (14.3) correctly represents the data in Figure 14.3. To accommodate error in both measurements x_i and y_i, Schnute *et al.* (1990) employ the functional errors-in-variables model (Kendall and Stuart, 1979, Section 29) defined by

$$Y_i = a + bX_i, \tag{14.17}$$

$$x_i = X_i + \sigma\delta_i, \tag{14.18}$$

$$y_i = Y_i + \tau\varepsilon_i, \tag{14.19}$$

$$\lambda = \tau^2/\sigma^2. \tag{14.20}$$

For $i = 1, \ldots, n$, the observed points (x_i, y_i) correspond to true, but unobserved, points (X_i, Y_i) that lie exactly on the line (14.17). The i.i.d. variates δ_i and ε_i in (14.18)–(14.19) are presumed normal with mean 0 and variance 1, so that x_i and y_i have variances σ^2 and τ^2, respectively. For parameter estimates to be well-defined, the variance ratio λ in (14.20) must be known *a priori*. The EV (errors in variables) model (14.17)–(14.20) reduces to (14.2) or (14.3) when $\lambda = \infty$ or $\lambda = 0$, respectively. In particular, example (14.1)–(14.4) illustrates the need for prescribing λ prior to analysis.

The EV model includes $3n + 1$ free parameters $(a, b, \sigma, X_1, \ldots, X_n)$, where these determine the remaining $n + 1$ parameters $(\tau, Y_i, \ldots, Y_n)$. The model's likelihood function is

$$L = \left(\frac{1}{2\pi\sigma\tau}\right)^n \prod_{i=1}^{n} \exp\left[-\frac{1}{2}\left(\frac{X_i - x_i}{\sigma}\right)^2\right] \exp\left[-\frac{1}{2}\left(\frac{Y_i - y_i}{\tau}\right)^2\right]. \tag{14.21}$$

Schnute *et al.* 1990 define

$$l(b) = \min_{a,\sigma,X_1,\cdots,X_n} [-\log L] \tag{14.22}$$

and prove that

$$l(b) - l(\hat{b}) = n\log\left[\frac{(\lambda + \hat{b}^2)S(b)}{(\lambda + b^2)S(\hat{b})}\right], \tag{14.23}$$

where

$$S(b) = \sum_{i=1}^{n} [y_i - \bar{y} - b(x_i - \bar{x})]^2. \tag{14.24}$$

They also show that, at the true value b, the log-likelihood ratio (14.23) conforms to the distribution

$$l(b) - l(\hat{b}) \sim n\log\left[1 + \frac{1}{n-2}F(1, n-2)\right] \tag{14.25}$$

exactly when $\lambda = 0$ or $\lambda = \infty$ and approximately otherwise. The accuracy of this approximation improves when σ^2 is small compared to the variance of X_i $(i = 1, \ldots, n)$ and τ^2 is small compared to the variance of Y_i $(i = 1, \ldots, n)$. Incidentally, the result (14.25) corrects an error in the number of degrees of freedom cited by Schnute *et al.* (1990).

The EV model is parametrically nonlinear; (14.17) includes products bX_i. The result (14.25) extends classical linear regression theory, as employed in (14.7)–(14.9), to an analogous theory for the EV model. The function $l(b)$ in (14.22) has a counterpart in classical regression defined from *twice* the negative log-likelihood. It follows from (14.21)–(14.25) that the maximum likelihood estimate for b is

$$\hat{b} = \frac{m_{02} - \lambda m_{20} + \sqrt{(m_{02} - \lambda m_{20})^2 + 4\lambda m_{11}}}{2m_{11}} \tag{14.26}$$

and that an approximate α-level confidence interval for b is defined by solutions

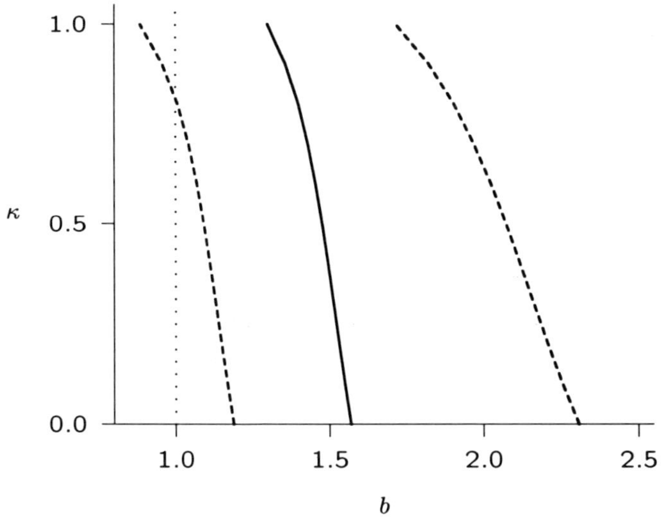

Figure 14.4 Parameter estimate $\hat{b}$ (solid curve) and 95% confidence interval for b (between broken curves) for values κ in (14.28). In particular, models (14.2) and (14.3) correspond to $\kappa = 1$ and $\kappa = 0$, respectively. *Adapted from Schnute et al., 1990, Figure 5*

to the quadratic equation

$$\frac{\lambda + \hat{b}^2 m_{02} - 2bm_{11} + b^2 m_{20}}{\lambda + b^2 m_{02} - 2\hat{b}m_{11} + \hat{b}^2 m_{20}} = 1 + \frac{1}{n-2} F_\alpha(1, n-2), \qquad (14.27)$$

where $m_{ij} = \sum_{k=1}^{n} (x_k - \bar{x})^i (y_k - \bar{y})^j$.

Figure 14.4 represents the estimate $\hat{b}$ and a corresponding 95% b-interval obtained from the data in Figure 14.3, based on the analytic results (14.26)–(14.27). The analysis depends on the value of

$$\kappa = \frac{2}{\pi} \tan^{-1} \lambda, \qquad (14.28)$$

which acts as a surrogate for λ, scaled to the range [0,1]. Thus, Figure 14.4 summarises results from a continuous range of models between (14.2) and (14.3), where $\kappa = 1$ and $\kappa = 0$, respectively. The hypothesis (14.15) can be rejected with 95% confidence if $\kappa > 0.82$ ($\lambda > 3.36$), but not otherwise.

14.4 SERIAL CORRELATION

Ambiguities similar to those described for the relationship (14.1) occur in the context of an autoregressive process

$$x_t = a + bx_{t-1} \tag{14.29}$$

for a univariate time series $x_t, t = 1, \ldots, n$. Stochastic counterparts of (14.29), analogous to (14.2) and (14.3), are the forward and reverse AR(1) processes

$$x_t = a + bx_{t-1} + \sigma\varepsilon_t, \tag{14.30}$$

$$x_t = \frac{1}{b}(x_{t+1} - a) + \sigma\varepsilon_{t+1}, \tag{14.31}$$

where thc variates ε_t are presumed standard normal i.i.d. The data

$$(x_1, x_2, x_3, x_4, x_5) = (1, 1, 3, 3, 1) \tag{14.32}$$

lead to points (x_{t-1}, x_t) identical to those listed in (14.4). Furthermore, for the series (14.32), models (14.30) and (14.31) yield the maximum likelihood estimates $(\hat{a}, \hat{b}) = (1.72, 0.0402)$ and $(-42.7, 24.9)$, respectively. Figure 14.5 shows that these define two lines intersecting near a point defined by the mean $\bar{x} = 1.8$ of the data (14.32).

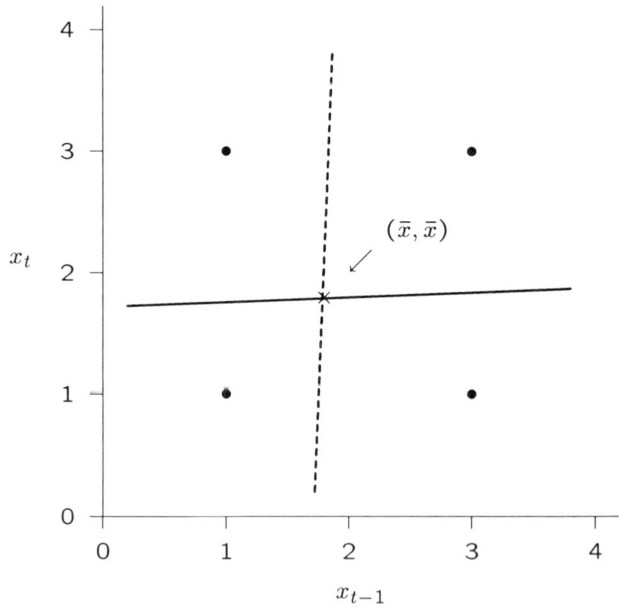

Figure 14.5 Data points (x_{t-1}, x_t) from the series (14.32) and autoregressive lines estimated from models (14.30) (solid lines) and (14.31) (broken lines). The lines intersect near the point $(\bar{x}, \bar{x}) = (1.8, 1.8)$, marked '×'. *Adapted from Schnute, 1991, Figure 1*

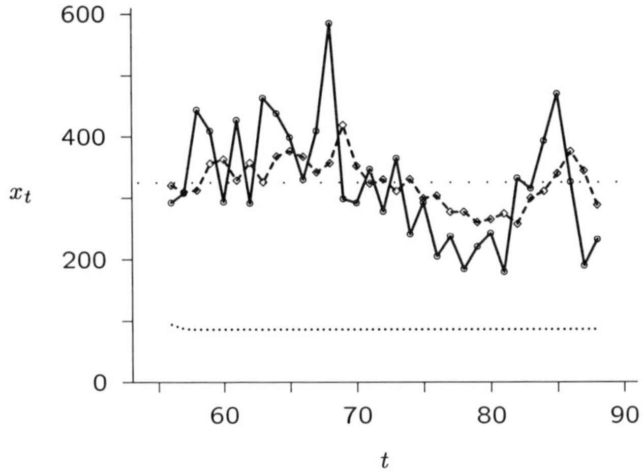

Figure 14.6 Lingcod CPUE data x_t (kg·hr^{-1}) v. year $t = 1956$–88. The figure shows original data (circles, solid line), along with maximum likelihood estimates $\hat{u}_t$ (diamonds; broken line) and $\sqrt{\hat{v}_t}$ (dotted line) computed from (14.37)–(14.38). A thinly dotted horizontal line indicates the series mean $\bar{x} = 324\,\text{kg}\cdot\text{r}^{-1}$. *Adapted from Schnute, 1991, Figure 2*

Figure 14.6 presents a series x_t corresponding to catch per unit effort (CPUE) in a lingcod fishery off Vancouver Island in British Columbia, Canada. Values x_t represent the annual catch (kg) per hour of fishing over the 33-year period 1956–88. Biologists often use CPUE as an index of abundance; thus, high catch rates theoretically indicate large populations. From this point of view, the parameter a in (14.29) indexes the annual recruitment to the population (i.e. the biomass of young fish entering the fishery) and b represents the fraction of biomass surviving from the previous year. More precisely, b includes the combined effect of survival and growth, because fish that survive usually grow. For further details, see Schnute *et al.* (1989) and Schnute (1991).

The AR(1) model (14.30) at first seems reasonable for the data in Figure 14.6. The population appears stable, and (14.30) describes the transition from one year to the next, allowing for environmental variability. However, because CPUE data invariably contain high measurement error, a possible alternative to (14.30) is the pair of equations

$$y_t = a + by_{t-1} + \sigma\varepsilon_t, \tag{14.33}$$

$$x_t = y_t + \tau\delta_t, \tag{14.34}$$

where both series ε_t and δ_t are standard normal i.i.d. Conceptually, (14.33) describes an AR(1) process for a series of true (but unknown) values y_t; by (14.34), these are measured with error to give the actual observations x_t. Process and measurement error variances are σ^2 and τ^2, respectively. We refer to (14.33)–(14.34) as a Kalman model, because of its similarity to the Kalman filter. It is equivalent to the classical ARMA(1, 1) model

$$x_t = a + bx_{t-1} + \kappa(\eta_t - \lambda\eta_{t-1}), \tag{14.35}$$

where, as usual, the series η_t is assumed standard normal i.i.d. The parameters (κ, λ) in (14.35) can be transformed to equivalent parameters (σ, τ) in (14.33)–(14.34) by the relations

$$\sigma^2 = \kappa^2(\lambda - b)(\lambda - b^{-1}), \quad \tau^2 = \lambda\kappa^2/b. \tag{14.36}$$

The equivalence (14.36) needs some qualification; as described by Schnute (1991), the parameter space for the Kalman model is somewhat more restricted than that for the ARMA(1, 1) model.

The likelihood for the Kalman model can be computed efficiently using concepts from Kalman filter theory. Define $u_t = E[x_t|x_1, \ldots, x_{t-1}]$ and $v_t = \text{var}[x_t|x_1, \ldots, x_{t-1}]$, where it is understood that u_1 and v_1 denote the unconditional moments $E[x_1]$ and $\text{var}[x_1]$, respectively. Then it can be shown recursively (Schnute, 1991) that

$$u_1 = \frac{a}{1-b}, \quad v_1 = \frac{\sigma^2}{1-b^2} + \tau^2, \tag{14.37}$$

$$u_t = a + bx_{t-1} + \frac{u_{t-1} - x_{t-1}}{v_{t-1}} b\tau^2, \; v_t = \sigma^2 + \tau^2 + \frac{v_t - \tau^2}{v_t} b^2\tau^2, \tag{14.38}$$

where (14.38) applies when $t > 1$. Based on these recursions, the likelihood for (14.33)–(14.34) is

$$L(x_1, \ldots, x_n|a, b, \sigma, \tau) = \prod_{t=1}^{n} \left[\frac{1}{\sqrt{2\pi v_t}} \exp\left[-\frac{(x_t - u_t)^2}{2v_t} \right] \right]. \tag{14.39}$$

The Kalman model reduces to the AR(1) model when $\tau = 0$; in this case the calculation (14.37)–(14.39) reduces to the classical AR(1) likelihood.

Table 14.2 lists maximum likelihood parameter estimates for the data in Figure 14.6, according to the Kalman and AR(1) models. By contrast with the analysis in Section 14.2, no prior knowledge of the variance ratio τ^2/σ^2 is needed here; serial correlation renders all parameters estimable. The Kalman model suggests greater measurement than process error ($\hat{\tau} = 64$ kg/h v. $\hat{\sigma} = 49$ kg/h).

Figure 14.7 portrays two estimated lines (14.29) in relation to the data (x_t, x_{t-1}). The AR(1) model suggests a relatively shallow line, corresponding biologically to a population driven by high recruitment a and low survival b.

Table 14.2 Parameter estimates from the AR(1) model (14.30) and the Kalman model (14.33)–(14.34), based on the data in Figure 14.6. The 95% b-intervals are computed from (14.42), i.e. from the profiles in Figure 14.8. Equivalent parameters from the ARMA(1, 1) model (14.35) are also shown, based on (14.36)

Model	$\hat{a}$	$\hat{b}$	95% b-interval	$\hat{\sigma}$	$\hat{\tau}$	$\hat{\kappa}$	$\hat{\lambda}$
AR(1)	197	0.39	(0.06, 0.70)	87	0	87	0
Kalman	94	0.71	(0.10, 0.98)	49	64	86	0.39

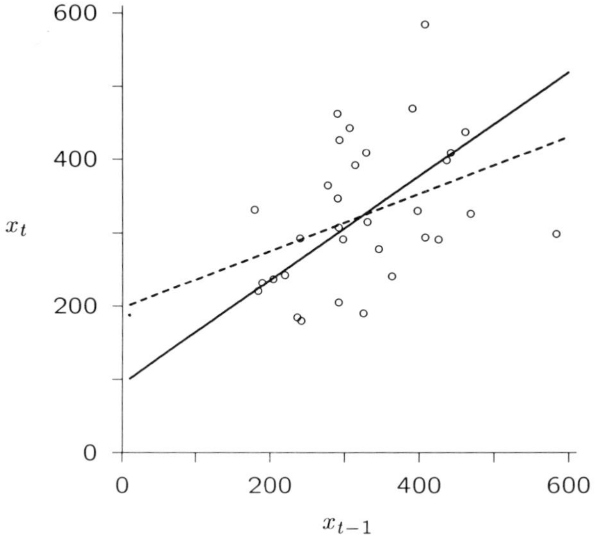

Figure 14.7 Data points (x_{t-1}, x_t) from the series in Figure 14.6 and autoregressive lines (14.29) estimated from models (14.33)–(14.34) (solid line) and (14.30) (broken line). *Adapted from Schnute 1991, Figure 3*

The Kalman model with measurement error ($\tau > 0$) indicates a distinctly different biological strategy, based on lower recruitment and higher survival.

The estimates cited here are computed by nonlinear maximisation of the likelihood function (14.37)–(14.39). This calculation also yields maximum likelihood estimates of u_t and v_t, which are illustrated in Figure 14.6. The quantity u_t can be interpreted as a prediction of x_t, based on past data and parameter estimates for the entire series; similarly, v_t is a predictive variance. Kalman filtering is more commonly used to estimate the unobserved value y_t based on data up to the present time, i.e. $E[y_t|x_1, \ldots, x_t]$. Schnute (1991) provides further details for the example here.

To assess uncertainty in the estimate $\hat{b}$, consider the functions derived from (14.39),

$$l_0(b) = \min_{a,\sigma} [-2 \log L(a, b, \sigma, \tau = 0)], \qquad (14.40)$$

$$l_1(b) = \min_{a,\sigma,\tau \geqslant 0} [-2 \log L(a, b, \sigma, \tau)], \qquad (14.41)$$

where (14.40) and (14.41) apply to the AR(1) and Kalman models, respectively. Asymptotically, at the true value b,

$$l_i(b) - \min_b l_i(b) \sim \chi^2(1), \qquad (14.42)$$

where $i = 0$ or 1 depending on the assumed model. Figure 14.8 shows likelihood profiles for b, based on (14.42). Additional measurement error in the Kalman model greatly increases uncertainty in the estimate $\hat{b}$; nevertheless, the 95% b-interval from the AR(1) model excludes the Kalman estimate (Table 14.2). Again, the assumed error structure substantially influences the outcome of the analysis. Schnute *et al.* (1989) and Schnute (1991) present further empirical evidence supporting the higher estimate $\hat{b}$ obtained from the Kalman model.

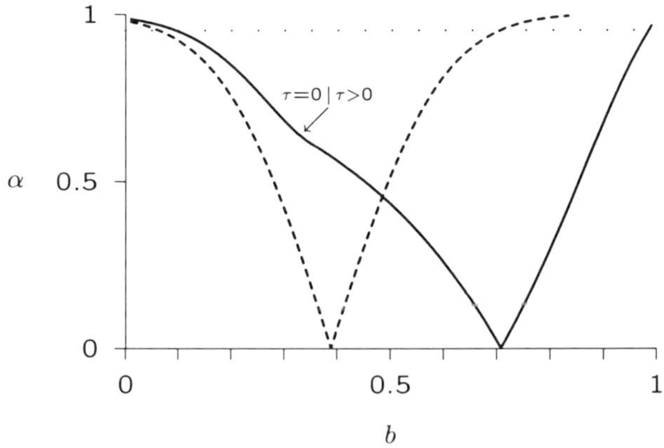

Figure 14.8 Probability profiles for the parameter b, based on the data in Figure 14.6 and either the Kalman model (14.33)–(14.34) (solid line) or the AR(1) model (14.30) (broken line). For each b, a confidence level α is computed from the asymptotic distribution (14.42). A thinly dotted horizontal line indicates the level $\alpha = 95\%$. A bend marked on the left limb of the Kalman profile corresponds to the point where the constraint $\tau \geqslant 0$ in (14.41) becomes operative; thus, $\tau = 0$ for $b < 0.34$. *Adapted from Schnute, 1991, Figure 4*

Incidentally, the Kalman profile in Figure 14.8 includes a sudden bend resulting from the constraint $\tau \geqslant 0$ in (14.41); for $b \leqslant 0.34$, the minimisation (14.39) requires $\tau = 0$. Theoretically, the asymptotic result (14.42) is derived without such constraints. Schnute (1991) uses simulations to explore the validity of (14.42) for the example here.

14.5 STOCK AND RECRUITMENT

Perhaps the most fundamental issue in fish-stock assessment concerns harvest strategy. How many fish can safely be caught, and how many must be left to perpetuate the stock? Clark (1972) provides an analysis of this question, based on the concept of an underlying relationship between the residual stock in one generation and the recruitment in the next. To be specific, consider the dynamic model

$$R_t = \alpha S_{t-1}(1 - \beta\gamma S_{t-1})^{1/\gamma} e^{\sigma\varepsilon_t}, \tag{14.43}$$

$$S_t = R_t - C_t, \tag{14.44}$$

that relates stock S_{t-1} in year $t-1$ to recruitment R_t in year t. In (14.44) the subsequent stock S_t consists of the recruitment R_t minus the catch C_t. The relationship (14.43) (proposed in various forms by Ricker (1954), Ricker (1975),

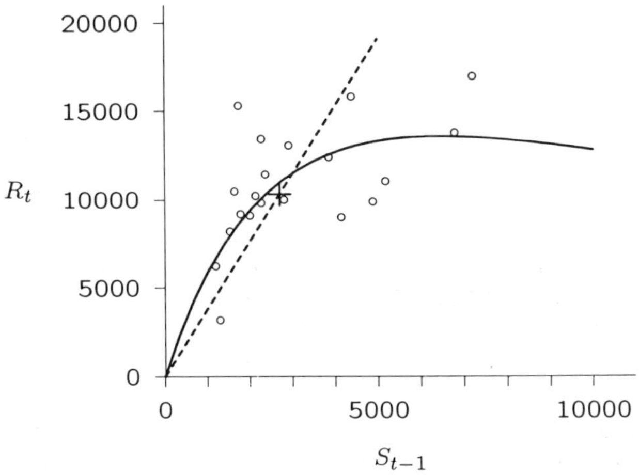

Figure 14.9 Simulated recruitment R_t in year t as a function of the previous stock S_{t-1}. The solid curve corresponds to a maximum likelihood estimate from model (14.43). In fact, simulation is based on the model (14.45)–(14.50). The symbol '+' marks the point $(\tilde{R}, \tilde{S})$ defined by geometric means of the data, and the broken line determined by this point and the origin corresponds to the slope estimate (14.51)

Deriso (1980) and Schnute (1985)) includes three parameters (α, β, γ), as well as the variance σ^2 associated with log-normal error. As usual, the variates ε_t are assumed standard normal i.i.d.

Figure 14.9 shows a data set apparently consistent with (14.43). The corresponding curve is computed from maximum likelihood parameter estimates: $\hat{\alpha} = 7.45$, $\hat{\beta}^{-1} = 6467$, $\hat{\gamma} = -0.392$. (Incidentally, $\hat{\sigma} = 0.274$.) A residual analysis indicates reasonable model fit. Both data and model suggest that the population can be harvested down to about 3000 fish without severe penalty. Stocks as high as 7000 or 8000 fish seem to produce recruitments similar to those obtained from stocks of only 2000 or 3000 fish.

In fact, the data in Figure 14.9 were generated by a model quite different from (14.43)–(14.44). Assume that the observed stock S_t is actually measured with error, where S_t^* denotes the true, but unobserved, stock. Similarly, let R_t and R_t^* denote the observed and true recruitment. Suppose that the fishery is conducted with the intent to maintain a fixed target stock T. Then, in this notation, the model used to generate the data in Figure 14.9 is

$$S_0^* = T, \tag{14.45}$$

$$R_t^* = aS_{t-1}^* e^{\sigma_1 \delta_t}, \tag{14.46}$$

$$C_t = R_t^* e^{\sigma_2 \varepsilon_t} - T, \tag{14.47}$$

$$S_t^* = R_t^* - C_t, \tag{14.48}$$

$$S_t = S_t^* e^{\sigma_3 \eta_t}, \tag{14.49}$$

$$R_t = S_t + C_t, \tag{14.50}$$

where δ_t, ε_t and η_t are standard normal i.i.d. By (14.45), the true stock begins at the target level; thereafter, simulation follows recursively via (14.46)–14.50). True recruitment in (14.46) is simply a linear multiple of the previous stock, with log-normal error. The catch is adjusted in (14.47) to achieve the target T, with error; the true remaining stock is then calculated in (14.48) to account for the lost catch. Thus, $S_t^* = T$ when $\sigma_2 = 0$. Assumption (14.49) introduces log-normal measurement error into the observation S_t, and (14.50) determines the observed recruitment from the observed stock.

Data in Figure 14.9 came from (14.45)–(14.50) with $T = 2500$, $a = 4$, $\sigma_1 = 0.07$, $\sigma_2 = 0.07$ and $\sigma_3 = 0.40$. Thus, the stock is held essentially at the level 2500. This produces about 10 000 recruits of whom roughly 7500 are harvested. Measurement error accounts for most data dispersion about the line $R_t = 4S_t$; because the error is log-normal, it has greatest influence on large stocks S_t, thus giving the appearance of an asymptotic relationship where none exists. The real message in the data can be captured from the geometric means $\tilde{R}$ and $\tilde{S}$ of the observed series R_t and S_t, respectively. The ratio

$$\hat{a} = \tilde{R}/\tilde{S} = 3.81 \tag{14.51}$$

provides a reasonable estimate of a in (14.46) and defines the line shown in Figure 14.9.

In practical terms, failure to detect measurement error in Figure (14.9) leads to the false impression that a larger fish stock would yield no significant advantage in the long run. For the example here, if catch rates were reduced to allow stock rebuilding, much higher yields would be possible in the long term.

14.6 CONCLUSIONS

Examples in this chapter have been kept relatively simple to highlight ambiguities inherent in fishery data. Although fishery models are typically more complex than those presented here, similar problems often are latent within them. For example, ageing error can have a profound effect on conclusions from an age-structured model. Measurement error in time-series data for bio-environment models can mask relationships that may exist. Similar remarks apply to multi-species fishery models. Predator–prey relationships between species may be disguised by processes analogous to the stock–recruitment example in Section 14.5. We believe that the statistical problems described here are typical, not exceptional, for fish population data. Quite possibly these problems extend generally to the field of biological and environmental modelling.

The results are also independent of the inference method and model diagnostics. We have stressed likelihood methods here, but the same ambiguities are inherent in other methods, such as Bayes models and bootstrap techniques. The example in Section 14.2 illustrates that no formal structure can alter the need for making a fundamental statistical choice.

Models may be constructed for estimation, forecasting, or control. Although our examples pertain largely to estimation, Section 14.5 illustrates how ambiguous error structure can influence both forecasting and control. A statistician investigating bio-environment data clearly does well to maintain a sceptical eye for unadvertised sources of error.

15

Recapture Data on Tagged Fish: some Questions of Analysis and Interpretation

R. M. Cormack
University of St Andrews, Scotland

15.1 INTRODUCTION

Marine fish populations are one of the more difficult types of population to study. Out of sight, with extremely high birth rates and early natural mortality, some populations disperse or migrate over large distances. Yet, quite apart from their intrinsic biological interest, fish form such a large part of the diet of their human predators that it is vital that we learn quickly enough about these populations to be able to maintain and rebuild their stocks. Our main source of data currently is the commercial catch, indirectly through analysis of its composition, directly through the recovery of tagged fish, at least one aspect of whose history is known.

The aim of this chapter is to discuss some of the statistical problems which arise in the analysis of such data, often now pertaining to artificially reared fish, tagged in large numbers in hatcheries. In marine studies data are patchy and non-sampling errors likely to be dominant. In such circumstances, it will be argued, most information will be gained from a robust exploratory approach to the analysis of the data, set within a statistical framework whose principal properties are understood. Two examples are used to show the benefits of this

Statistics for the Environment. Edited by Vic Barnett and K. Feridun Turkman

approach, using generalised linear models. Although in many cases the same main conclusions would be reached by a formal multinomial analysis, different questions come into focus on the way. The many options which face the scientist during data analysis are discussed. Although in most cases no definitive answer is provided, preferences will be apparent. The key feature is a flexible approach, being guided by the data, but aware of the dangers of overinterpretation.

15.2 APPROACHES TO RECAPTURE DATA

Formal statistical inference is based initially on the concept of a fixed population from which a sample of independent observations are obtained by a well-defined random mechanism. This bears little relation to the situation faced in many cases by the fishery scientist. Even when we have a fixed population of known, tagged, fish they do not behave independently, the random mechanisms affecting survival and movement are not well understood and, except in pure research studies, the process of random sampling is not as well controlled as would be desired. Moreover, inference to the population already released and recaptured is of very limited value. From the practical point of view the narrowest population to which an inference is wanted is a different batch of fish of the same genetic stock, handled in identical fashion, and released on a corresponding day in another year.

One statistical choice in this situation is somewhat akin to the division between reductionist and holistic approaches to biology. At the one extreme, the statistician tries to provide a detailed probabilistic description of one component of the variability—for example, the subsampling of the commercial catch for coded-wire tags, or some aspect of the process in space and time which leads to the presence of a fish in a particular area in a given week. The opposite extreme aims at a robust general model of the process transforming releases into recaptures. Any statistical analysis is a balance between believable assumptions and precise inferences. In attempting to produce a general methodology the statistician should accept that the decision on the point of balance has to be left to the scientist or policy-maker. The statistical methodology should draw attention to both principal and subsidiary features of the data, while having the flexibility to develop their analysis and interpretation in different ways.

One consequence of this approach is that data should be analysed initially at the most disaggregated level available. As soon as catches from two days or two boats are amalgamated, some information is lost: no model can then be considered in which the two boats are different, while discovery that they are different may suggest a different variance structure as appropriate for the aggregated data. This is particularly important when only subsamples of individual catches have been examined for tags, and the numbers of each type of tag are scaled up by an expansion factor, the inverse of the sampling fraction, to give an estimated number of tags in the whole catch. Application of an average expansion factor to an aggregated number of observed tags distorts either the mean or the variance of the expanded number of tags. This does not mean that samples cannot be composited for subsampling, if deemed necessary: only that the aggregation needs to be done before, not after, the subsampling, and that the proportions of different initial units in the composite need to be known (see e.g. Boswell *et al.*, 1988). Counts at the disaggregated level can be very small, causing difficulties with the application of asymptotically valid procedures of statistical inference. What we would like is a method of analysing disaggregated data which gives the same results as an analysis of aggregated data when the aggregation is over catches which are found to be have the same survival parameters under an acceptable model.

The release of a single batch of R fish, many of which will be recaptured dead in one of several catches, with others never being recaptured, suggests the use of a hypergeometric or multinomial distribution as statistical model. With several releases these become multihypergeometric (Seber, 1962) and product multinomials, presented by Seber (1970) and developed at length by Brownie *et al.* (1978, 1985) and, specifically in the fisheries context, by Burnham *et al.* (1987). Although hypergeometric distributions are still used to develop tests of models, conditional on the observed number of recaptures, most of the development of models and estimation has been under the multinomial umbrella, since in practice sample sizes are not fixed a priori. Models are developed for the probabilities that an individual from the ith release will be captured in the jth catch. These parameters can in turn be modelled as functions of quantitative covariates, such as water flow, temperature, release date, or distance from source. The multinomial likelihood is then maximised, numerically if explicit estimators cannot be found for the particular model.

A variety of specialised software packages has been developed specifically for the analysis of tag-recovery data—SURVIV (White, 1983), SURGE (Clobert *et al.*, 1987), and, more specifically for testing between treatments, RELEASE (Burnham *et al.*, 1987) and SURPH (Smith and Skalski, 1990).

15.3 MULTINOMIAL MODELS

The most extensive development and cohesive discussion of multinomial models is given by Burnham *et al.* (1987). Their methods are much more widely applicable than the recapture of young tagged salmon during their downstream migration with which their book is chiefly concerned. Recaptures at different locations are modelled, together with those fish never recovered, as jointly multinomial random variables given the number released with that tag. Probabilities are modelled as products of survival probabilities ϕ_k through a number of stages k and probabilities p_k of capture or q_k of its avoidance. A wide variety of protocols is presented, covering different levels of available data on multiple capture and release. In our context we are concerned with their 'first capture history protocol' with all captured fish being removed from the population. Their primary concern is with S, the ratio of survival between groups treated differently at or before release. Models are set up constraining various sets of survival and capture probabilities to be equal over treatment groups. Minimal sufficient statistics (MSS) are found for these models and maximum likelihood estimates obtained from the MSS. The fit of the models is tested by χ^2 goodness-of-fit tests on contingency tables formed by conditioning on the observed MSS, pooled as necessary to avoid small expected numbers. This approach was pioneered by Robson and Youngs (1971) and Brownie and Robson (1976).

For example, with a single release of R_t fish in a treatment group, and R_c control fish, caught downstream at three successive locations, the data are in the form of Table 15.1. The most general model is that:

$$(n_{t1}, n_{t2}, n_{t3}) \text{ are multinomial } (R_t; \phi_{t1}p_{t1}, \phi_{t1}q_{t1}\phi_{t2}p_{t2}, \phi_{t1}q_{t1}\phi_{t2}q_{t2}\phi_{t3}p_{t3})$$

$$(n_{c1}, n_{c2}, n_{c3}) \text{ are multinomial } (R_c; \phi_{c1}p_{c1}, \phi_{c1}q_{c1}\phi_{c2}p_{c2}, \phi_{c1}q_{c1}\phi_{c2}q_{c2}\phi_{c3}p_{c3})$$

where the $\phi_{\cdot j}$ are survival probabilities between successive sampling points and $p_{\cdot j} (q = 1 - p)$ are probabilities of capture at each.

With the assumption that $p_{tj} = p_{cj}$ for each j, the primary hypothesis of interest is that $\phi_{t1} = S\phi_{c1}, \phi_{t2} = \phi_{c2}, \phi_{t3} = \phi_{c3}$. The maximum likelihood estimator of S under this hypothesis is $(r_t/R_t)/(r_c/R_c)$ and the hypothesis that $S = 1$ (i.e. treatment does not affect survival) is tested by the 2×2 contingency table in

Table 15.1 Typical form of data for small experiment to compare 2 treatments

		Numbers captured			
	Number released	Location 1	Location 2	Location 3	Total
Treatment	R_t	n_{t1}	n_{t2}	n_{t3}	r_t
Control	R_c	n_{c1}	n_{c2}	n_{c3}	r_c

Table 15.2 Contingency tables for testing treatments and assessing goodness-of-fit

(a)

r_t	$R_t - r_t$
r_c	$R_c - r_c$

(b)

n_{t1}	n_{t2}	n_{t3}
n_{c1}	n_{c2}	n_{c3}

Table 15.2(a). The goodness-of-fit of the underlying model, essentially testing the equality between groups of the other parameters, is tested by the 2×3 contingency table in Table 15.2(b).

If the assumptions are rejected then a sequence of models can be examined, allowing the treatment to have a continuing effect on survival after the first recapture; for details see Burnham *et al.* (1987). Although they give first a confidence interval for S based on an assumption of asymptotic normality for $\hat{S}$, a later discussion recommends that confidence intervals be based on a normal assumption for $\log(\hat{S})$ with variance estimated by

$$\left(\frac{1}{r_t} - \frac{1}{R_t} + \frac{1}{r_c} - \frac{1}{R_c}\right).$$

Replication of treatment and control groups is strongly recommended, although true replicates in the sense of independent batches of fish which have been treated identically are often difficult to achieve. When available, a separate estimate of S can be obtained for each replicate, and the empirical sampling variance between the estimates used instead of the theoretical variance above in making inferences about S.

However, variation between batches before release is only one aspect of variation over and above that generated by random, independent movement of fish after release. Fish shoal, and groups share a common response to disturbing features in their environment. Overdispersion, variance greater than theoretically predicted by the multinomial distribution, is very commonly observed, whether due to clumping of individuals or variation between individuals due to some factor not explicitly recognised in the model. The two responses to this made by statisticians have taken one or other of the forms mentioned in the previous section—to find a specific model for the overdispersion distribution, or to develop a new type of analysis which requires only a general description of the form of the overdispersion. The latter, which we advocate strongly here, is justified by the principle of quasi-likelihood (Wedderburn, 1974). In its simplest form it permits the scaling up of all variances by a variance inflation factor (VIF) as discussed by Burnham *et al.* in their Section 4.1.3. Since it arises more naturally within the framework of generalized linear models, further discussion is deferred until they have been introduced. An alternative approach, with somewhat different philosophy has been intensively developed by Carroll and co-workers: see, for example, Davidian and Carroll (1988).

15.4 GENERALIZED LINEAR MODELS

A notable recent trend in the statistical analysis of counts has been towards the use of the Poisson, rather than the multinomial distribution, putting such analyses firmly within the framework of generalised linear models (McCullagh and Nelder, 1989). The close parallels between these and analysis of variance encourages the development of flexible analyses for counts including critical examination of patterns of residuals. Following Cormack's (1981) extended adaptation to open populations of Fienberg's (1972) log-linear formulation of multiple recapture models, Green and Macdonald (1987) showed how effective such an approach could be to the analysis of commercial recoveries of coded-wire tags.

We formulate a model for θ_{ij}, the average probability that an individual from release i is caught in the jth fishery stratum. If only a proportion f_j of the catch is inspected for tags, then the expected number of tags found from release i in that inspection is

$$\mu_{ij} = E(n_{ij}) = R_i \theta_{ij} f_j.$$

It is the varying sample fraction f_j which makes model formulation difficult for counts aggregated over fishery strata (Cormack and Skalski, 1992). The aim of the analysis is to find a parsimonious representation of the θ_{ij} in terms of a structure on the release suffix i (different hatcheries, days of release, treatments, etc.) and the recovery suffix j (different areas, times, types of fishery, etc.).

When appropriate to the study, for example in a research study with multiple recapture, the θ_{ij} can incorporate intermediate survival rates and capture probabilities, and the observations n are then numbers having a particular capture history (Cormack, 1989). If the structure of the θ_{ij} can be expressed as a product of parameters, say $\alpha_{ih}\beta_k\gamma_j$, then

$$\log \mu_{ij} = \log R_i + \log \alpha_{ih} + \log \beta_\mathrm{k} + \log \gamma_j + \log f_j.$$

With the further assumption that the n_{ij} are independent Poisson counts, this is a standard log-linear model for which maximum likelihood estimators, with asymptotic variances, can be obtained readily by use of many general statistical computer packages. As a fully interactive package specifically designed for generalised linear models, GLIM (Payne, 1987) has many advantages and will be used in the examples in this chapter. Covariates for any of these parameters are easily incorporated so long as the log parameter is a linear function of covariate parameters.

The Poisson assumption requires that the variance V of a count equals its mean μ, just as the multinomial assumption requires $V = \mu(1 - \mu/R)$. However, the problems posed by the presence of overdispersion ($V > \mu$) have been the subject of much study (see e.g. Breslow, 1990) and several approaches are now possible. The simplest is that of assuming $V = \phi\mu$ with $\phi > 1$, a variance inflation factor, where ϕ is estimated by dividing either the residual deviance or the Pearson goodness-of-fit χ^2 by the residual degrees-of-freedom. Although the former estimate is not consistent, Nelder and Lee (1992) point to other advantages. Alternatively it might be postulated that the variance is of negative binomial form $V = \mu + \alpha\mu^2$: a GLIM macro to fit this variance function was given by Breslow (1984). Analysis of data with these variance structures is justified by quasi-likelihood theory (Wedderburn, 1974). No distributional form is assumed for the random variable, only the postulated functional relationship between the variance of an observation and its mean. Cox (1983) avers that 'the method of quasilikelihood is likely to have high efficiency for modest amounts of overdispersion', but we then have to decide what is modest—the VIF of 2 for the model selected in Example 15.2, or the 11 found by Green and Macdonald (1987) for a hatchery return model? There are many further developments of this and other modern ideas for the analysis of counts which remain to be tried with fisheries data—joint modelling of variance and mean (Nelder and Pregibon, 1987), combining theoretical and empirical variances (Breslow, 1990), incorporating nested strata of variation (Morton, 1988), modelling the correlation structure of nonindependent clusters of observations by generalised estimating equations (Zeger *et al.*, 1988).

Whatever variance structure is adopted, questions arise in the course of the analysis, many of which are the same as arise in analysis of variance. With a variance parameter to be estimated, the correspondence between the two types of analysis is even closer than for the analysis of cross-classified categorical data (see e.g. Fienberg, 1980) assumed to have an absolute standard of residual variation—the residual deviance having a χ^2 distribution. Other questions are specific to the analysis of counts, or to studies in which each observation has a location in space and/or time.

(i) Population of interest

To what further situations do we want inferences to apply? Does this imply that factors have to be formally defined as fixed or random, with consequences for the choice of tests and errors appropriate to different comparisons?

(ii) Replicates

What is meant by 'replicates' in fisheries releases? Different tag codes randomly allocated to groups of fish otherwise treated identically are usually so regarded, but what does the phrase 'treated identically' really mean? If fish with duplicate codes are mixed in the one tank and treatments are necessarily applied to whole tanks, then the error appropriate to treatment comparisons involves between-tank variability whereas replicate differences are within tanks and so inappropriate. On the other hand if different codes are held in different tanks, they are unlikely to have been treated identically. In other situations one release will inevitably have a known relation to another in time or space.

(iii) Measures of goodness-of-fit

Which of the measures of overall goodness-of-fit of a model to the data should be used: Pearson's χ^2 goodness-of-fit or residual deviance (likelihood-ratio of the model relative to a full model)? Or should another of the family of power-divergence statistics (Read and Cressie, 1988) be considered? Should the method of estimation be matched to the measure of fit—maximum likelihood to deviance, minimum χ^2 to Pearson's χ^2—as Read and Cressie recommend? Or is it sensible to use maximum likelihood to estimate and Pearson's χ^2 to test, as practised by Burnham *et al.*, when the χ^2 value from one model can be greater than from a model within which it is nested? When overdispersion is allowed for by choosing the additional variance parameter to optimise the overall fit, how then should goodness-of-fit be measured?

(iv) Model selection

How do we select a good model from all subsets of interactions of factors and available covariates in a study: forward selection, backward elimination, or some other search? How do we balance precision against parsimony: Akaike's or some related criterion, or use of significance tests between alternatives?

(v) Residual analysis

Which of the residuals suggested for discrete data should we examine—Pearson, Anscombe, deviance, or some other? Will probability or added variable plots help to identify outliers? If they do, what then? Does knowledge of the temporal or spatial location of either release or recapture suggest specific correlations which should be sought among the residuals?

(vi) Interval estimates

Do we rely on asymptotic normality of the basic parameters of survival or capture, or of some transformation of them—say logit of capture probabilities, and log or logit of survival rates? Can we readily find profile-likelihood intervals, or do we use resampling techniques such as jack-knife or bootstrap?

(vii) Small-number problems

With sparse data, exacerbated by analysis at a disaggregated level, how wrong are the assumptions of χ^2 distributions for log likelihood-ratios or goodness-of-fit tests? For example, can an apparent Poisson variability arise from observations of small numbers from a clumped distribution? If so, are inferences about the parameters based on the fallacious assumption that there is no overdispersion seriously in error? Will the analysis of Monte Carlo samples, generated under the fitted model, provide more reliable inferences?

Questions (i) and (ii) concern fundamental principles of all scientific research. Cox (1958) discusses the principles of planned experimentation, while the elements of a controlled observational study, which are inseparable from much fisheries research, are challengingly discussed by Kish (1987) in terms of concepts of representation, randomisation and realism, and of controlled, disturbing and randomised variables modulating the effect of predictor variables on the predictand, the object of the study.

Questions (iii), (iv), (v) and (vii) impinge on one another in that a satisfactory model requires not only a balance between precision and parsimony, variance and robustness, in the model as an average representation of the whole data set, but also in its localised components. Residuals and their distribution can indicate that a model is misspecified even when the analysis of deviance suggests it is plausible. In the likelihood framework, deviance residuals are evaluated, for observations n with fitted value $\hat{\mu}$, as the square root of $2[n \log(n/\hat{\mu}) - n + \hat{\mu}]$, taking the same sign as $(n - \hat{\mu})$. They are very close numerically to those proposed by Anscombe (1953) for Poisson variables, have the justification of being individual contributions to the adopted measure of relative fit between models, and behave in a less extreme way than Pearson residuals with sparse data. Their use is advocated by Pierce and Schafer (1986). Particularly when the data are sparse, residual inspection should be combined with Monte Carlo simulations of the provisional model to obtain an empirical, not merely theoretical, assessment of terms included and excluded from the model and of the distribution of the residual deviance and of parameter estimates. The transformations $\log[(1-p)/p]$ and $\log(\phi)$ are the natural parameters for capture probability and survival within the generalised linear model framework—although $\log(\phi)$ retains an asymmetry between survival and death; in some circumstances $\log[(1-\phi)/\phi]$ might be more natural—and their estimates have a much less skew distribution than those of either p or ϕ. These asymptotic intervals can be checked against the Monte Carlo intervals. All these questions would repay more investigation, both theoretical and by simulation.

Two examples will now illustrate the pragmatic way in which these principles can be applied. They are selected to be different in kind, the first a controlled experiment with reasonably large numbers of tags recovered locally, the second an attempt to gain information from commercial catches over a wide area. Readers should be able to repeat the analyses and explore the data further.

15.5 EXAMPLE 1: EFFECT OF EXPERIMENTAL DAM MANIPULATION ON LOCAL SURVIVAL

Burnham *et al.* (1987, pp. 246–52) use a subset of data from Long *et al.* (1975) to illustrate their analyses. The aim was to examine the effect on survival of young coho salmon *Oncorhynchus kisutch* of the presence of perforated bulkheads at a dam turbine unit. On each of three days, branded fish from each of four folding tanks (one of which had equipment problems and is not reported) were released, some (treated) above and some (control) below a test structure.

Recaptures were made at each of two downstream dams. The data, and complete GLIM program for their analysis, are given in Appendix 15.1.

The most general loglinear model represents μ_{ijkl}, the number of the R_{ijk} fish released into treatment i from tank j on day k which are expected to be recaptured at dam l, as

$$\log(\mu_{ijkl}) = \log(R_{ijk}) + \xi_{ijkl}$$

where $\xi_{ijkl} = \log(\theta_{ijkl})$ are $2 \times 3 \times 3 \times 2 = 36$ additive parameters which we hope can be reduced to a smaller set of main effects $\tau_i, \rho_j, \beta_k, \delta_l$ and low-order interactions. The particular symbols chosen are designed to aid understanding, the holding tanks having been intended as replicates ρ_j of different batches (days) β_k. One way of searching for a parsimonious acceptable model is to employ backward elimination. The fully saturated model $T*R*B*D$ is fitted first (which of course fits perfectly), the four-factor interaction $T.R.B.D$ then removed, the choice of terms omitted subsequently being determined by comparison of the estimates with their theoretical standard errors (under the Poisson assumption), both given by the $dis e$ instruction after each model is fitted. The analysis of deviance formed by sequential removal of terms from the model is shown in Table 15.3

Table 15.3 Deviances from models fitted to Coho data by GLIM program in Appendix 15.1

Term removed from model	Degrees of freedom	Increase in deviance	Degrees of freedom	Overall deviance
T.R.B.D	4	6.74		
T.R.D	2	2.66		
T.B.D	2	5.89		
T.D	1	0.35		
T.R.B	4	6.11		
T.B	2	3.59	15	25.34
R.B.D	4	13.67		
R.D	2	0.46		
R.B.	4	9.72	25	49.20
Difference between Bath 2 and Batch 3	2	6.13	27	55.33
Total deviance from model:				
$T + R + BB + D + T.R. + BB.D$	27	55.33		
Removing *T.R.*	2	77.99	29	133.31
Removing *BB.D*	1	42.92		

Variation greater than Poisson, due to nonindependence of individual fish movement, is evident immediately in the four-factor and then the three-factor interactions, though none of the first six terms removed (all of the interactions of treatments with other factors, except for $T.R$) are separately significant at the 5% level as χ^2's. Unless one is going to retain the Poisson distribution with an irreducible model all of these six terms have to be omitted, giving an overall deviance of 25.44 with 15 degrees of freedom (d.f.), a VIF of 1.70. Continuing with this philosophy, the other three-factor interaction $R.B.D$, although highly significant with respect to a Poisson variance ($\chi^2_4 = 13.67$), is not significant with this inflated variance ($F_{4,15} = 2.02$). The next two two-factor interactions can also be omitted but, even relative to the VIF now of 1.97, the remaining two interaction terms, Days × Dams ($B.D$) and Treatments × Tanks ($T.R$), are very highly significant. Further reduction in model complexity can be achieved only by noticing that Days 2 and 3 are very similar, both in their main effects and in their interaction with dams, so that the factor B at three levels can be replaced by a factor BB at two levels contrasting Day 1 with Days 2 and 3.

This comprises the final model:

$$\log(\mu_{ijkl}) = \log(R_{ijkl}) + \tau_i + \rho_j + \alpha_{ij} + \beta_{k*} + \delta_l + \gamma_{k*l}$$

where $k^* = 1$ on Day 1 ($k = 1$) and $k^* = 2$ otherwise. The GLIM scale factor can be set equal to the VIF, mean deviance, of 2.049 before parameter estimates and their variance–covariance matrix, given in Table 15.5, are obtained by $dis e v $. Approximate confidence limits can be constructed from these, and exponentiated to give confidence limits for interesting survival rates and their ratios.

Because of the significant interactions, interpretation is not straightforward. It will doubtless be easy for the experimenter to produce possible explanations of

Table 15.4 Estimates and variances (including VIF) for selected model (relative to Dam1, Rep1, Trt1, Day1)

Effect	Estimate	Covariances $\times 10^4$								
1	−3.31	7.0								
DAM2	−0.31	−4.4	10.3							
DD2	0.027	−4.4	4.4	6.5						
DAM.DD	0.18	4.4	−10.3	−6.5	14.8					
TRT2	0.10	−3.4	0	0	0	9.7				
REP2	−0.20	−3.5	0	0	0	3.4	7.8			
REP3	−0.061	−3.4	0	0	0	3.4	3.4	7.1		
TRT2.REP2	0.26	3.4	0	0	0	−9.7	−7.8	−3.4	19.9	
TRT2.REP3	0.054	3.4	0	0	0	−9.7	−3.4	−7.1	9.7	19.7

why the ratio of captures at Dam 2 to those at Dam 1 is 0.88 for batches released on Days 2 and 3 but only 0.74 on Day 1, though it may prove of biological interest to give them critical consideration. This, however, is an issue aside from the main experimental aim of comparing treatments, to which the highly significant *T.R.* interaction between treatments and holding tanks is central. Why is the effect of treatment on survival so different: 0.904 for fish from Tank 1, 0.857 for those from Tank 4 ($j = 3$), but only 0.695 from Tank 2? Of course, these three values can be obtained from the original data as the ratios of the overall return rates of test and control fish from each of the three tanks separately, if the analyst chooses to look at these figures. The advantage of the form of analysis illustrated here is that it forces such questions on the experimenter's attention.

Until that question is resolved we cannot give a definite estimate for the effect of the treatment on the survival of young coho moving downstream through a turbine. Which of the three batches is more typical of the conditions which might apply if the treatment were implemented in future? If the experimenter discovers that there were some abnormality in the releases from Tank 2, then these data can easily be removed from the analysis by defining a suitable weight function $cal wt = %ne(r,2)$wei wt$ and repeating the analysis in the hope that the remaining interaction between tanks and treatments will no longer need to be included in the model. This is indeed the case, and a model

$$\log(\mu_{ijkl}) = \log(R_{ijk}) + \tau_i + \rho_{j^*} + \beta_{k^*} + \delta_l + \gamma_{k^*l}$$

holds for the data from Tanks 1 and 3 without any overdispersion parameter, giving an estimate of relative survival of 0.881 with 95% confidence interval (0.854, 0.908). However, throwing away half one's data (one tank was already omitted from the data in Appendix 15.1) requires very good cause. What conclusions are reached if Tank 2 is retained?

With the scale factor reset to the new mean deviance from a model including no interactions of treatment with any other factor, the main effect of the treatment together with its standard error (which of course incorporates the VIF as a measure of overdispersion) could be quoted as a summary of the evidence about the effect of the treatment on the logarithm of survival. Indeed, this may be the best practical strategy to adopt for a model which includes inexplicable interactions between survival and factors, such as day or batch, which are if not formally randomly selected levels of such factors, going to be different the next time fish are released. Explore different models to gain understanding of the system: then, guided by this understanding, fit a model in which treatments occur only as a main effect, their interactions being allowed to swell the standard errors of the estimated treatment differences by increasing the variance inflation factor (4.60 in the example, with 29 d.f.). Here this gives 0.817 with standard error 0.0224 ignoring the slight asymmetry in the exponentiation. The point estimate is just the ratio of the two overall capture rates

$$\left[\left(\frac{16\,269}{258\,094}\right)\Big/\left(\frac{10\,042}{130\,159}\right)\right],$$

the pooled estimate quoted by Burnham *et al.* Our standard error is larger than the asymptotic value of 0.0103 given by Burnham *et al.*—essentially the unscaled standard error here—but less than their 'empirical' standard error of 0.0329 which is the sample standard deviation among the nine ratios separately calculated for each tank on each day.

The choice here is related to one of the classic discussions about analysis of variance—the interpretation of random factors and the set of future cases to which inference is to be made. Here we have two arguably random factors, each at 3 levels, days and tanks, which the approach of Burnham *et al.* essentially regards as one factor at 9 levels, the interaction between the fixed treatment factor and the random factor being taken as the appropriate error for treatment comparison, with no pooling being allowed with higher-order error terms. Our analysis here allows pooling of terms, governed by their significance.

It needs to be remembered that the ratio of treatment to control survival rates does not in itself determine the experimental procedure which gives the young fish the greatest survival rate. Understanding of some of the other significant factors in the experiment may do more to increase survival rates than consideration of this particular treatment.

Data analysis is incomplete without consideration of residuals from the provisional model. Deviance residuals are calculated from the two selected models, including or excluding the interaction between treatments and tanks. Unfortunately, from the first these models there are two outlying residuals: too many fish are recaptured at Dam 1 from treatment release on Day 1 and at Dam 2 from control release on Day 2, both involving fish from Tank 2. Only the second of these remains an outlier from the second model.

The residuals provide information on whether or not the removals at Dam 1 significantly reduce the number available for recapture at Dam 2, other than in the average way incorporated in the Dam terms in the model. If so, one would expect a negative correlation between the residuals from the same group of fish at the two dams. In fact it is slightly positive, with a value of 0.08.

15.6 EXAMPLE 2: COMMERCIAL RECAPTURES OF CHINOOK IN SOUTH-EASTERN ALASKA WATERS

The second example illustrates the sparseness and lack of balance inherent in many studies of commercial recaptures. We use it both to suggest a possible approach to such data, and to investigate the performance of the methodology.

Table 15.5 Number of tags recovered from those released, $R_1 = 94\,723$ with code 1, $R_2 = 37\,737$ with code 2.

	Troll				Gillnet			G	F	S	Seine				
Week	NW(7)	SW(8)	NE(2)	SE(3)	6	11	5	1	1	1	4	9	10	12	13
1–16		1 0	1 1	5 0											
23	10 2	6 3	11 6	11 5											
24	30 10	18 8	15 6	26 9											
25					1 0	1 0		1 0							
26					5 1		0 1	4 3							
27	9 3	3 2	10 2	48 17	0 1	2 0								2 1	
28	10 3	6 1	13 4	18 5	4 1			1 0	4 0		1 0				
29	8 2	4 0	13 4	12 2	2 0			1 1	1 0	3 1				2 2	
30	6 1		5 0	6 4					1 0	2 1			1 0	1 0	1 0
31								2 0	0 1	8 4	1 0	0 1	1 0		
32									1 1	7 1	0 1				
33							1 0								
35	3 1		2 0	2 0				1 0							
Code	1 2	1 2	1 2	1 2	1 2	1 2	1 2	1 2	1 2	1 2	1 2	1 2	1 2	1 2	1 2

The data, taken from Appendix E, Tables 1 and 2 of Clark and Bernard (1987), concern the recapture times and locations off south-eastern Alaska of chinook salmon (*Oncorhynchus tshawytscha*) released in 1983 with duplicate tag codes from a single hatchery. Their aim was to develop a probability model for the variation between duplicates, since 'an estimate of... the total number of fish of a given release represented by that tag code in fishery catches without a measure of the accuracy or precision of the estimate is of little value'. Four fisheries—troll TR, purse seine S, gillnet G, and fish traps F—contribute to the catch, an attempt being made to sample for tags at least 20% of the fish on those boats whose catch can be assigned to a single area–time stratum. For the last three fisheries catches are attributable to a specific week and statistical district. Troll catches often cover more than one district and one week, and are recorded in aggregated form by time and area.

Table 15.6 Expansion factors for different catches

Troll

	Region			
Weeks of landing	**NW**	**SW**	**NE**	**SE**
1–16		3	2	2
23	6.45	6.42	1.61	3.61
24	2.84	1.70	1.63	2.86
27	3.68	5	2.78	1.91
28	3.09	4.06	3	2.30
29	3.14	2.5	2.02	2.10
30	21.14		2	1
35	5		6.37	4

Other

	Area										
Week	**1**	**1**	**1**	**4**	**6**	**9**	**10**	**11**	**12**	**13**	**15**
25		G4			G5			G3			
26		G2.4			G2.6						G2
27					G3			G1.5	S1		
28		G1	F2.7	S3	G1.8						
29	S2.2	G1.5	F3		G2				S1.2		
30	S2.3		F2				S1		S2	S3	
31	S1.9	G2	F2	S5		S3	S1				
32	S1.6		F1	S4							
33											G8
35		G20.6									

Different fisheries are open at different times in different places, the pattern being shown by the data in Table 15.5 and the expansion factors e—the inverse of the fraction f of the catch sampled from each recovery stratum—in Table 15.6. Although e, together with n and R, forms the raw data, it is commonly not recorded directly in published data, being replaced by an estimated contribution of that batch of fish to the total catch. That applies in this example. Values of e were back-calculated, with some element of rounding error. The nonorthogonal pattern of data suggests that different fisheries could be analysed separately, with some comparison between S, G and F coming from an analysis of Area 1.

If a joint analysis of all data was attempted, regarding labels for weeks as having the same meaning for troll fisheries as for the others, but regarding troll areas as different from those for other fisheries, the general model is:

$$E(n_{ijkl}) = R_i \theta_{ijkl} f_{jkl}$$

where suffix i represents the two tag codes C; suffix j represents the four fishery types T; suffix k represents the thirteen areas A; and suffix l represents the thirteen weeks W.

The analysis of deviance in Table 15.7 offers little hope of simplification, irrespective of the order in which the parameter sets are added to the model. The high level of aliasing arising from the presence of only 63 of the 676 possible TAW combinations makes interpretation and model selection difficult, and backward elimination infeasible. Even one main effect is aliased. We note the significant difference between tag codes, presumably due to some systematic difference in how the fish were held at the hatchery. The codes do not provide true replicates. They do not contribute at the same rate to the total fishery, although their relative contribution to any one fishery is constant.

Table 15.7 Analysis of deviance of all catches

	Degrees of freedom	Deviance
Between fisheries T	3	403.70
Between weeks W	12	169.54
Between areas A	11	92.64
Interaction T*W	31	74.83
Other interactions	5	6.14
Codes	1	6.72
Residual	62	51.25
Total	125	804.80

Troll fishery

The troll data are reasonably balanced over weeks and regions, with an average number of 7 tags recovered, six of the 58 observed counts being zero.

The partition of deviance is shown in the left half of Table 15.8, suggesting no immediate simplification of the general model, with κ_i the effect of tag code i:

$$E(n_{ikl}) = R_i \kappa_i \phi_{kl} f_{kl}$$

However, following our data-analytic paradigm, we might try to represent the change in $\log \phi$ over time by a quadratic response function rather than by unrelated weeks, with the implication of an optimal time for each fishery and an optimal harvest to be obtained at that time. The resulting residual deviance, in the right half of Table 15.8, is significantly larger, but the description given in Table 15.9 of the number of fish contributed to each regional troll fishery in week t by 10 000 fish released from the hatchery may be of interest. Variances of the parameters on the log scale can readily be obtained by the usual linearisation and converted, including the mean deviance as a VIF, to the 95% confidence limits for optimal time τ and optimal harvest v given in the table. In the NW fishery the fitted quadratic coefficient of time is positive, though not

Table 15.8 Analysis of deviance of troll catches

	Degrees of freedom	Deviance		Degrees of freedom	Deviance
Between areas	3	43.50			
Between weeks	7	169.36	Quadratic	2	147.02
Interaction	18	61.34	Quadratic	6	46.04
Between codes	1	6.11			
Residual	28	18.46		45	56.15
Total	57	298.78			

Table 15.9 Estimated return in week $(21 + t)$ from 10 000 released fish

Region	Harvest	CI for τ	CI for v
NE	$3.8 \exp[-0.068(t-6.3)^2]$	5.7, 7.0	0.9, 16.4
SE	$9.8 \exp[-0.114(t-5.5)^2]$	5.2, 5.9	4.1, 23.1
SW	$4.0 \exp[-0.105(t-5.1)^2]$	4.3, 5.9	0.8, 20.5
NW	$17.5 \exp[-0.26t + 0.007t^2]$		

significantly nonzero, so that it cannot be expressed in terms of an optimal capture time. Fish held in the manner of Code 2 contribute roughly 25% fewer.

Here t has been taken as consecutive integers, obviously suboptimal since the troll recaptures were aggregated over different lengths of time, and it is easy to find sensible alternative values for t which give a better fit. We shall not pursue this further here. The point is that this approach to the analysis of recapture data encourages such model simplification, and may indicate that a more careful examination of a particular formulation would be worthwhile.

Gillnet fishery

The gillnet data are less balanced and more sparse, the 15 pairs of observations being recorded in 7 weeks over 4 districts. The average number captured is 1.1, the maximum count being 5 and eleven of the 30 counts being zero.

The partition of deviance is shown in the left half of Table 15.10, suggesting the model

$$E(n_{ikl}) = R_i \omega_k \alpha_l f_{kl}$$

as provisionally acceptable, with residual deviance, pooling the last three rows, of 21.69 with 20 degrees of freedom. Examination of the parameter estimates suggests that weeks 25, 27, 28, 29 and 31 could be represented by a single parameter, weeks 26 and 35 by a second, with week 33 separate. Similarly areas 11 and 15, which happen to be neighbouring districts far up the same fjord system, can be represented by the same parameter. Although this would give a very parsimonious description of the data, with analysis in the right half of Table 15.10, the power of the data to distinguish some of these groupings is ludicrously low—notably areas 11 and 5, each with captures in only two weeks, neither in common with the other. This raises a further question: if the fishery in area 11 was open in week 26 should the fact that no tags were sampled in the catch not be part of the data? Unfortunately sampling fraction is often not recorded when no tagged fish at all are caught. Even when it is, it may still

Table 15.10 Analysis of deviance of gillnet catches

	Degrees of freedom	Deviance	Degrees of freedom	Deviance
Between weeks	6	9.89	2	9.39
Between areas	3	6.94	2	6.90
Interaction	5	6.64		
Between codes	1	0.69		
Residual	14	14.36	25	22.23
Total	29	38.52		

not be clear whether or not such zeros should be included in the analysis. Some of the problems are discussed by Cormack and Skalski (1992).

One hundred simulations, treating the parsimonious model with its fitted parameters as the true model, allow us to comment on the performance of the methodology with this fairly sparse data set, and to draw more robust inferences about the data. They show that:

(i) sparseness has no effect on the mean of either the residual deviance or goodness-of-fit χ^2—respectively 25.0 and 24.0;
(ii) the residual deviance has too small a variance, 31.5, to be acceptable as a χ^2 with 25 d.f., whereas the variance of the goodness-of-fit χ^2 is somewhat too large at 65.1;
(iii) the selected model is an acceptable fit to these data—the observed deviance lies between the 30th and 31st smallest simulated values;
(iv) the theoretical variances of the estimated effects are reliable on average: the mean of the theoretical variances lay within 15% of the empirical variance of the estimates, in either direction;
(v) the distributions of the estimated effects (on the log scale) were acceptably normal;
(vi) correlations between estimated effects and their theoretical standard errors tended to be negative and less than 0.25 in magnitude.

Corresponding simulations of the two models for the troll fishery reinforced conclusions (i) and (ii) above, except that the diminution in the variance of the residual deviance occurs only for the model fitting 30 parameters to 58 observations, not for that fitting 13 parameters. They also show:

(vii) the deviance difference between two models when the more restricted one is true is very well fitted by the relevant χ^2, but the difference in goodness-of-fit statistics has too high a variance, though the mean is reasonable.

15.7 EXTENSIONS

In an important paper Hilborn (1990) proposes extending the use of the Poisson distribution beyond the confines of the generalised linear model. The approach relies on the power of modern computing equipment, even quite small microcomputers, to find the maximum or minimum of any expression specified algebraically as a function of a number of parameters. Software is readily available for numerous algorithms to minimise nonlinear functions, the functions here being the Poisson log-likelihood $\sum_j (n_j \log \mu_j - \mu_j)$ where μ_j is the number of fish expected in any recapture category. Hilborn points out that, since the minimisation is carried out numerically, it is no greater disadvantage to compute the μ_j numerically, rather than algebraically. Thus any model can be formulated for the survival and migration of tagged fish so long as the eventual expected number in the population in any category can be computed by a sequence of numerical operations using trial values of unknown parameters. A second model, describing the sampling procedure, then permits the number μ_i expected to be caught in any category to be computed from the numbers expected to be present in different categories in the population.

This could lead to the idea that any of the multitude of deterministic models for fisheries processes could be made amenable to statistical analysis by superimposing a Poisson distribution on all counts. This is appealing but seems dangerous. Not only should careful consideration be given to the variance structure, but maximum likelihood has its limitations. Its desirable properties are asymptotic, and extremely uncertain both when the data are sparse and when the assumed probability distribution is unreliable. Some answers can be given to the questions raised in this chapter, and more are being sought, within the developing statistical frameworks of generalised linear models, quasi-likelihood, and related, or competing, theories. Global statistical optimisation of a numerical model has statistical properties which are largely unknown. Nevertheless such approaches may generate some insight into problems which could not otherwise be tackled.

Appendix 15.1 COMPLETE GLIM INPUT FOR DATA AND ANALYSIS OF EXAMPLE 1

```
$units 36 $
$cal day=%gl(3,3):rep=%gl(3,1):trt=%gl(2,9):dam=%gl(2,18) $
$fac trt 2 dam 2 day 3 rep 3 $
$data n $read
1014 969 966 1006 686 932 1198 932 1162
 548 606 578  619 597 552  643 698  651
 766 618 756  908 650 812 1072 856  966
 422 460 428  506 632 508  518 554  522
$var 18 rel $data rel $read
  28739 28856 28558 26395 23710 27293 31929 31663 30951
  13724 14577 14590 14122 14121 13665 15100 15404 14856
$assign r=rel,rel $
$yvar n           $err p $
$cal lr=%log(r)          $soffset lr $
$fit dam*day*rep*trt     $dis e $
$fit -dam.day.rep.trt    $dis e $
$fit -dam.rep.trt        $dis e $
$fit -dam.day.trt        $dis e $
$fit -dam.trt            $dis e $
$fit -day.rep.trt        $dis e $
$fit -day.trt            $dis e $
$fit -day.rep.dam        $dis e $
```

```
$fit -rep.dam                $dis e  $
$fit -day.rep                $dis e  $
$cal bb=1+%ge(day,2) $fac bb 2  $
$fit -dam*day+dam*bb  $dis e  $
$cal %d=%dv/%df             $look %x2 %d  $
$scale %d       $fit .          $dis e v  $
$extract %pe  $
$cal e=%exp(%pe):e(2):e(2)*e(9)  $
$cal 1/e(5):1/(e(5)*e(7)):1/(e(5)*e(8))   $
$cal dr=2*%gt(%yv,%fv)-0.5)*%sqrt(2*(%yv*%log(%yv/%fv)-%yv+%fv))  $
$cal or=%nd((%gl(%nu,1)-0.5)/%nu):res=dr   $
$sort res                    $plot res or  $
$dis r  $
$cal wt=%ne(rep,2)       $wei wt           $scale 1  $
$fit.: -trt.rep $
$cal %a=%exp(-0.1269):%b=%exp(1.96*0.01551):%a:%a*%b:%a/%b  $
$wei                         $fit . $
$cal %d=%dv/%df  $look %x2 %d       $scale %d  $
$fit .                       $dis e  $
$cal %exp(-0.2027):%exp(-0.2027+1.96*0.02721)  $
$cal %exp(-0.2027-1.96*0.02721)  $
$cal 0.0438/1.96  $
$cal %exp(-0.2027+1.96*0.01269)  $
$cal 0.0206/1.96  $
$cal dr=2*(%gt(%yv,%fv)-0.5)*%sqrt(2*(%yv*%log(%yv/%fv)-%yv+%fv))  $
$cal res=dr  $sort res   $plot res or  $
$look 32 dr $
$stop
```

PART VI

Forestry: Supply and Conservation

16

Statistical Analysis of Climatological and Ecological Factors in Forestry; Spatial and Temporal Aspects

Wolfgang Urfer
University of Dortmund, Germany

16.1 INTRODUCTION

The major challenge for research on forest decline over the next decade is the interpretation of the effects of multiple stress factors on the functioning of forest ecosystems. This can be analysed by models that are process-based and developed primarily from field observation and experimentation. This chapter discusses statistical methods for the analysis of longitudinal categorical response variables and possible effects of covariates, relevant to the forest ecosystem problem.

Marginal models give representations for the marginal distribution of response at each time point and transitional models give representations for the transition probabilities between outcomes at successive time points. The two approaches are illustrated through analysis of the data from 16 permanent observation plots for Norway spruce (*Picea abies*) established by the Institute for Ecology, Landscape Development and Forest Planning in North Rhine-Westphalia (Germany).

Statistics for the Environment. Edited by Vic Barnett and K. Feridun Turkman

The influence of ecological and climatological factors on the response variable *crown transparency* is investigated by logistic regression models and weighted least squares analysis. Further aspects of forest ecosystem study, such as sample-size considerations, genetic variability of tree species and the need for integrated European research projects on forest ecosystems, are also discussed.

16.2 DATA ANALYSIS IN THE PROCESS OF SCIENTIFIC INVESTIGATION: LONGITUDINAL STUDIES

In his paper 'Statistics and the Environment' Box (1975) said, 'Scientific method is a process of controlled learning. The object of statistical method is to make that process as efficient as possible.'

In this process of learning, a hypothesis H (or theory or model or conjecture) leads by a process of deduction to certain consequences which may be compared with known facts deduced from analysis of the available data. Usually the consequences and the facts fail to agree, leading by a process called induction to revision of the hypothesis H to the modified hypothesis H'. Consequences of H' may now be in accord with the data analysis or may still be discordant. When it is not clear what further modification should be made to an unsatisfactory hypothesis or, alternatively, when confirmation of an apparently satisfactory hypothesis is needed, further data must be sought. Depending on the context, the further data may come from a designed experiment, a sample survey, or already existing results.

While the conjectured state of nature at a particular stage may be false or at least inexact, the data themselves are generated by the true state of nature. It is because of this that the comparison of successively conjectured states of nature with actual data can lead to convergence on the truth.

The statistician's tasks in this process are:

- deciding what would be appropriate data to seek at each stage of the investigation (design problem);
- deciding what information is contained in the data at each stage of the investigation (analysis problem).

In longitudinal studies the response of each individual is observed on two or more occasions. Longitudinal studies are suited to the study of change of individuals over time and space including the effects of factors that affect change. There are two different types of models: *marginal* and *transitional* models.

Marginal models describe the time-specific distributions of the outcome variable and the dependence on individual characteristics and other independent variables. Statistical objectives for analysis by marginal models include:

(i) the evaluation of group effects;

(ii) the evaluation of time effects;
(iii) the evaluation of group × time interactions.

These objectives deal with changes in marginal distributions. They could be studied with cross-sectional data sets, provided that cohort effects were small. When they are investigated in longitudinal data sets, we have to consider the correlation between observations induced by repeated observation of single subjects.

Transitional models describe the distribution of individual changes over time in the state of response variables and the effects of individual characteristics or risk factors on these changes. Questions addressed by transitional models include:

(a) With what frequency is state l observed at annual examinations of individuals that were in state l' at the previous time point (the year before)?
(b) How does this frequency depend upon individual characteristics or risk factors?

Because these questions concern changes in the response states of individuals over time, they can be investigated only with longitudinal data. Models for transitions can be represented as probability distributions for the future state given the individual's history. They are sometimes called conditional models.

This conceptual distinction between marginal and conditional models has produced a corresponding split in the statistical methodology. In discussing conditional models, we distinguish between the analyses of processes defined at discrete times and those defined in continuous time.

The discussion of Ware *et al.* (1988) makes it clear that marginal models do not fully utilise the information in longitudinal data. They do not fully model individual changes over time or the effects of covariates on individual changes.

16.3 ANALYSIS OF CATEGORICAL DATA IN LONGITUDINAL STUDIES BY A MARGINAL MODEL

Longitudinal studies in ecology are concerned with intra- and inter-individual change for response variables that are measured in terms of discrete categories. Examples of such categorical response variables include stage of disease severity, trunk visibility index and aspects of the general impression of trees. The subjects in longitudinal studies often come from two or more subpopulations. The corresponding groups may represent the cross-classification of two or more factors or explanatory variables. The subjects in each group are observed over time.

Let y_{ijk} represent the response of the kth subject in the ith group for the jth time point ($i = 1, 2, \ldots, s; j = 1, 2, \ldots, d; k = 1, 2, \ldots, n_i$). It is convenient to consider the data structure shown in Table 16.1. The possible outcomes of each

Table 16.1 Data structure for longitudinal studies

Group	Subjects within groups	Time points 1	2	...	d
1	1	y_{111}	y_{121}	...	y_{1d1}
1	2	y_{112}	y_{122}	...	y_{1d2}
⋮	⋮	⋮	⋮		⋮
1	n_1	y_{11n_1}	y_{12n_1}	...	y_{1dn_1}
2	1	y_{211}	y_{221}	...	y_{2d1}
2	2	y_{212}	y_{222}	...	y_{2d2}
⋮	⋮	⋮	⋮		⋮
2	n_2	y_{21n_2}	y_{22n_2}	...	y_{2dn_2}
⋮	⋮	⋮	⋮		⋮
s	1	y_{s11}	y_{s21}	...	y_{sd1}
s	2	y_{s12}	y_{s22}	...	y_{sd2}
⋮	⋮	⋮	⋮		⋮
s	n_s	y_{s1n_s}	y_{s2n_s}	...	y_{sdn_s}

y_{ijk} are indexed by $l = 0, 1, ,2, \ldots, L$ for the classification of the corresponding subject into one of $L + 1$ categories of a binary, nominal, or ordinal scale. Each subject is represented by a sequence (profile) of d outcomes and there are $r - (L + 1)^d$ possible response profiles $(a_1, a_2, \ldots, a_d)$ where $a_j = 0, 1, 2, \ldots, L$ for $j = 1, 2, \ldots, d$.

For the comparisons (i)–(iii) of Section 16.2 attention can be restricted to the marginal probabilities $\varphi_{ijl} = P(y_{ijk} = l)$ for $i = 1, 2, \ldots s$; $j = 1, 2, \ldots, d$ and $l = 0, 1, \ldots, L$.

In the case of an ordinally scaled, or binary, response variable, we are interested in the variation of the mean scores $\mu_{ij} = \sum_{l=0}^{L} m_l \varphi_{ijl}$ with respect to the values $m_0, m_1, \ldots, m_L$, which are assigned to the possible outcomes $0, 1, 2, \ldots, L$.

These variations across the groups and time points are characterised by the objectives (i)–(iii) of Section 16.2 and the hypotheses corresponding to these objectives can be expressed in terms of the $\{\mu_{ij}\}$ using the following null models (hypotheses):

$$\text{(i)}\ H_{01}: \mu_{1j} = \mu_{2j} = \cdots = \mu_{sj}, \quad j = 1, 2, \ldots, d;$$
$$\text{(ii)}\ H_{02}: \mu_{i1} = \mu_{i2} = \cdots = \mu_{id}, \quad j = 1, 2, \ldots, s;$$
$$\text{(iii)}\ H_{03}: \mu_{1j} - \mu_{1d} = \mu_{2j} = \mu_{2d} = \cdots = \mu_{sj} - \mu_{sd}, \quad j = 1, 2, \ldots, d - 1.$$

The hypothesis H_{03} means no interaction between groups and time points and can be expressed by the additive model

$$\mu_{ij} = \mu + \xi_i + \tau_j,$$

with the restrictions $\xi_1 = 0$ and $\tau_1 = 0$; μ denotes the reference mean for group 1 and time point 1, ξ_i represents the increment for the ith group and τ_j represents the increment for the jth time point.

The weighted least squares analysis for estimating the parameters μ, ξ_i and τ_j and tests for the hypotheses H_{01}—H_{03} are given by Koch *et al.* (1977).

16.4 ANALYSIS OF CATEGORICAL DATA IN LONGITUDINAL STUDIES BY A LOGISTIC REGRESSION MODEL

In the models of Bonney (1987) the dependence of the outcomes is described in terms of conditional probabilities. The joint probability of the outcomes is decomposed as a product of successive conditional probabilities. Since the probability of an observation is conditioned on all preceding observations, the models will be referred to as regressive logistic regression models. The parametrisation of logistic regression models is such that the jointly dependent observations can be fitted with existing computer programs for the standard logistic regression such as the LOGIST procedure of the computer package SAS.

Consider a set of d binary variables $y = (y_1, y_2, \ldots, y_d)$, where each y_j is coded 1 or 0, and has associated with it measurements on m explanatory variables $[x_{j1}, x_{j2}, \ldots, x_{jm}]$. Let $x = (x_1, x_2, \ldots, x_d)$ be the matrix with $x_j = (x_{i1}, x_{i2}, \ldots, x_{jm})$. The elementary decomposition of the joint probability of y given x is a product of d probabilities given by

$$P(y|x) = P(y_1|x)P(y_2|y_1, x)\cdots P(y_d|y_1, y_2, \ldots, y_{d-1}, x). \tag{16.1}$$

The regressive logistic model defines a logistic function for each factor in (16.1) by

$$\theta_j = \log \frac{P(y_j = 1|y_1, y_2, \ldots, y_{j-1}, x)}{P(y_j = 0|y_1, y_2, \ldots, y_{j-1}, x)} = \alpha_j + \sum_{q=1}^{j-1} \gamma_{jq} z_q + x_j \beta \tag{16.2}$$

where

$$z_q = \begin{cases} 2y_q - 1, & \text{if } q < j \\ 0, & \text{if } q \geqslant j \end{cases}$$

and α_j, β and the γ's are parameters that can take any values from $-\infty$ to $+\infty$.

If the dependence parameter γ_{jq} is homogeneous we have

$$\gamma_{jq} = \gamma(k) \text{ for } j - q = k$$

and the model (16.2) is simplified to

$$\theta_j = \alpha_j + \sum_{k=1}^{j-1} \gamma(k) z_{j-k} + x_j \beta \qquad \text{for } j = 1, 2, \ldots, d. \tag{16.3}$$

The model for all d dependent observations can be written in matrix form by setting

$$\theta = (\theta_1, \theta_2, \ldots, \theta_d)', \lambda = (\alpha_1, \alpha_2, \ldots, \alpha_d, \gamma(1), \gamma(2), \ldots, \gamma(d-1), \beta')'$$

$$A = \begin{pmatrix} 1 & 0 & 0 & \cdots & 0 & 0 & 0 & \cdots & 0 & x_{11} & x_{12} & \cdots & x_{1m} \\ 0 & 1 & 0 & \cdots & 0 & z_1 & 0 & & & x_{21} & x_{22} & \cdots & x_{2m} \\ 0 & 0 & 1 & \cdots & 0 & z_2 & z_1 & & 0 & x_{31} & x_{32} & \cdots & x_{3m} \\ & \vdots & & & & \vdots & \vdots & & \vdots & \vdots & \vdots & & \vdots \\ 0 & 0 & 0 & \cdots & 1 & z_{d-1} & z_{d-2} & \cdots & z_1 & x_{d1} & x_{d2} & \cdots & x_{dm} \end{pmatrix}.$$

Then (16.3) becomes

$$\theta = A\lambda. \tag{16.4}$$

If $\gamma(k) = 0$ for $k > r$ our model (16.3) describes Markov dependence of order r. Markov dependence of order 1 implies

$$P(y|x) = P(y_1|x_1) \prod_{i=2}^{d} P(y_i|y_{i-1}, x_i)$$

or

$$\theta_j = \alpha_j + \gamma z_{j-1} + x_j \beta.$$

The likelihood function for the d dependent observations is

$$L(\lambda|y, x) = \prod_{j=1}^{d} \frac{\exp[\theta_j(\lambda) y_j]}{1 + \exp[\theta_j(\lambda)]},$$

which is a product of univariate logistic functions.

The log-likelihood is

$$\log L(\lambda) = \sum_{j=1}^{d} [y_j \theta_j(\lambda) - \log(1 + e^{\theta_j(\lambda)})].$$

The score vector is

$$\frac{\partial \log L(\lambda)}{\partial \lambda} = U(\lambda) = \sum_{j=1}^{d} [y_j - (1 + e^{-\theta_j(\lambda)})^{-1}] v_j$$

where $v_j = \partial\theta_j(\lambda)/\partial\lambda$ is the transpose of the jth row of A. The likelihood equations are obtained by setting $U(\lambda)$ to the zero vector.

Hypotheses may be tested by the likelihood-ratio criterion. It may be desirable to compare the fit of various patterns of dependence. A comparison based on the likelihood theory is obtained by the use of Akaike's AIC criterion defined as

$$\text{AIC} = -2\log(\text{likelihood}) + 2(\text{number of parameters estimated}).$$

Akaike's criterion adjusts the log-likelihood for the number of parameters estimated. Thus the model with the minimum AIC gives the best fit to the data and is useful for comparison of parametric models with different numbers of parameters as we will see in Section (16.5).

16.5 APPLICATION OF WEIGHTED LEAST SQUARES AND LOGISTIC REGRESSION ANALYSIS ON DATA FROM PERMANENT OBSERVATION PLOTS

Since 1970 there has been a decline in the health of forests in large areas of the Federal Republic of Germany. A forest inventory carried out in 1991 gave an indication of the extend of forest decline: more than 60% of the forested area in FRG was affected.

Various species of forest trees (spruce, pine, beech, and oak) are affected in different ways and at different rates. A single survey of forest injury does not answer the question of variation with time which is important for defining countermeasures and for forest protection. The missing information can be obtained by observing permanent plots, which allows the measurement of risk factors. The Institute for Ecology, Landscape Development and Forest Planning in North Rhine-Westphalia established 16 observation plots. At each location 30 spruce trees were randomly selected from a population of 150 trees.

In the following study we give the results of the analysis of the response variable *crown transparency* with three possible outcomes: 0 = healthy, 1 =

Table 16.2 p-values ($\alpha = 0.05$) for weighted least squares analysis for 6 explanatory variables

	Age of trees	Soil character	Soil type	Type of landscape	Nutrition	Humidity
Group	0.0005	0.2545	0.0005	0.0255	0.0071	0.4045
Time	0.0001	0.0001	0.0001	0.0001	0.0001	0.0001
Interaction	0.0001	0.0001	0.0028	0.0001	0.0001	0.1338

warning state and 2 = ill. The following categorical explanatory variables are considered: age of trees, soil character, soil type, type of landscape, nutrition and humidity. More details are given by Urfer *et al.* (1990).

The result of using the weighted least squares analysis of Section 16.3 is shown in Table 16.2. Using $\alpha = 0.05$ as the level of significance, time has a significant effect on the response variable crown transparency for all explanatory variables. The interaction effects of the explanatory variables are also significant with the exception of humidity. The group effects of the explanatory variables: age of trees, soil type, type of landscape and nutrition are significant.

A possible explanation for the nonsignificant effect of *humidity* (which is characterised by nine categories from 1 = wet to 9 = dry) is that in the available data only 4 categories were represented, which describe similar ecological conditions. The variable *soil character* (clay, loam, sand, gravel) has no direct significant effect but influences the response variable by a time and interaction effect. The *soil-type* is characterised by the categories: brown soil, podzolic brown soil and pseudo-gley.

The logistic regression model allows the statistical analysis of time-dependent covariates. Using the same data, the following explanatory variables were analysed by Woltering (1992):

$$x_1 = \begin{cases} 1 & \text{for podzolic brown soil} \\ 0 & \text{otherwise} \end{cases} \qquad x_4 = \text{height above sea level (60–670 m)}$$

$$x_2 = \begin{cases} 1 & \text{for pseudo-gley} \\ 0 & \text{otherwise} \end{cases} \qquad x_5 = \text{nutrition of soil}$$

$$x_{j3} = \text{age of stand (time-dependent covariate, 54-102 years)} \qquad x_6 = \text{humidity of soil.}$$

The response variable is again crown transparency with two possible outcomes: 1 = ill and 0 = healthy.

With the help of model (16.3) from Section 16.4 the following equation was examined:

$$\theta_j = \alpha_j + \gamma(1)z_{j-1} + \gamma(2)z_{j-2} + x_1\beta_1 + x_2\beta_2 + x_{j3}\beta_3 + x_4\beta_4 + x_5\beta_5 + x_6\beta_6 \quad \text{for } j = 1, 2, 3.$$

Table 16.3 shows the comparison of different models using the data from the 480 spruce trees at the 16 permanent observation plots for three time points. The three best fitting models, by Akaike's criterion, are characterised by the following restrictions on the parameters:

$$\beta_1 = \beta_2 = \beta_3 = 0, \quad \beta_5 = \beta_6 = 0 \quad \text{and} \quad \beta_6 = 0.$$

Table 16.3 Comparison of some models

Restrictions on parameters	$-2 \log L$	Number of parameters	AIC
None	1098.23	11	1120.23
$\gamma(2)=0$	1105.97	10	1125.97
$\gamma(1)=\gamma(2)=0$	1275.20	9	1293.20
$\alpha_1=\alpha_2=\alpha_3=\alpha$	1264.22	9	1246.22
$\beta_1=\beta_2=0$	1100.37	9	1118.37
$\beta_4=0$	1101.05	10	1121.05
$\beta_6=0$	1098.24	10	1118.24
$\beta_3=\beta_5=0$	1100.39	9	1118.39
$\beta_1=\beta_2=\beta_6=0$	1100.39	8	1116.39
$\beta_5=\beta_6=0$	1099.53	9	1117.39
$\beta_3=\beta_4=\beta_5=\beta_6=0$	1105.82	7	1119.82
$\beta_1=\beta_2=\beta_3=\beta_5=\beta_6=0$	1107.28	6	1119.28
$\beta_1=\beta_2=\beta_5=\beta_6=0$	1106.25	7	1120.25
$\beta_1=\beta_2=\beta_3=\beta_4=\beta_5=0$	1113.92	6	1125.92
$\beta_1=\beta_2=\beta_3=\beta_4=\beta_5=\beta_6=0$	1114.50	6	1125.50
$\alpha_1=\alpha_2=\alpha_3=\alpha, \beta_6=0$	1264.23	8	1280.23

The common restriction on the parameters is $\beta_6=0$. Therefore Table 16.4 gives the results of the maximum likelihood approach for all the other parameters.

The effects of $y_1, y_2, \ldots, y_{j-1}$ and x_j on y_j are additive on the logit scale. For the level of significance $\alpha=0.05$, the explanatory variables $x_1, x_2, \ldots, x_5$ were not significant. The dependence parameters $\gamma(1)$ and $\gamma(2)$ are both significant and allow the following interpretation:

Table 16.4 Maximum likelihood estimates, standard errors, χ^2 and *p*-values

Estimates	Standard error	χ^2-value	*p*-values
$\hat{\alpha}_1=0.85$	0.93	—	—
$\hat{\alpha}_2=-0.28$	0.92	—	—
$\hat{\alpha}_3=-0.60$	0.90	—	—
$\hat{\gamma}(1)=-1.32$	0.12	123.83	<0.01
$\hat{\gamma}(2)=-0.67$	0.25	7.21	<0.01
$\hat{\beta}_1=0.15$	0.11	2.04	0.15
$\hat{\beta}_2=0.09$	0.14	0.45	0.50
$\hat{\beta}_3=-0.01$	0.008	1.86	0.17
$\hat{\beta}_4=0.0008$	0.0005	3.36	0.07
$\hat{\beta}_5=-0.15$	0.13	1.30	0.25

the odds of $y_j = 1$ (ill trees) for trees with $y_{j-1} = 0$ (healthy trees) is increased by the factor $e^{1.32}/e^{-1.32} \approx 14$ compared to trees with $y_{j-1} = 1$ (ill trees); the odds of $y_j = 1$ for trees with $y_{j-2} = 0$ is increased by the factor $e^{0.67}/e^{-0.67} \approx 4$ compared to trees with $y_{j-2} = 1$.

The least squares analysis can also be applied to data with time-varying covariates, but sample-size problems arise. Moreover, the method requires the reduction of continuous covariates to categorical variables.

Stram *et al.* (1988) described a method for comparing responses in two groups of subjects observed repeatedly at a common set of observation times when the response is an ordered categorical outcome. This method, which allows both time-dependent covariates and missing observations, consists of two steps.

(i) In the first step, the data are analysed separately at each time point using a generalised linear regression model. The distribution of the estimates of these time-specific regression coefficients is obtained without imposing any parameteric model of dependence on the repeated observations.
(ii) In the second step this distribution is used to characterise the overall difference between groups and the variation of group differences over time.

The proportional odds model for marginal probabilities used in the first step was applied by Urfer (1988) and Knabe, *et al.* (1989) to analyse data from a survey of pollution impact on forests in North Rhine-Westphalia. Here it was found that the silicon content of needle tissue can be explained as residue of transpiration and is therefore an indicator of the impact of climate, air pollution and soil acidification on trees.

16.6 FURTHER ASPECTS OF FOREST ECOSYSTEM MEASUREMENTS

For any type of study, the nature of the sampling process is important. This identifies the basis for inference from the observed data to the population for which conclusions are to be formulated. Usually the data are assumed to be representative of a target population through some statistical model of its probability distribution. A worthwhile advantage of model-based inferences is their broader scope, while a noteworthy limitation is the usually unknown validity of their underlying assumptions.

For each subgroup, the n_i subjects in Table 16.1 are assumed to be representative of a corresponding infinite subpopulation in the sense of simple random sampling. This is fulfilled only for the trees at each plot, for they are a simple random sample of 30 out of 150 trees.

The key to the weighted least squares approach is the initial stratification of subjects into s groups with identical covariate values. Planning of a monitoring system for forest ecosystem measurements therefore has to consider this sampling aspect.

An important aspect of the data structure for longitudinal studies is the number of time points d for which the subjects are observed. If d is relatively small, e.g. 12 or less, statistical analyses can be carried out to describe the pattern of variation across time and groups. Weighted least squares methods are useful for this purpose provided that good estimates for the covariance matrices of the estimates of $\{\mu_{ij}\}$ can be constructed. When the sample sizes $\{n_i\}$ for all groups in Table 16.1 are sufficiently large, $n_i \geqslant d + 25$, say, then this assumption is fulfilled.

Urfer and Pohlmann (1992) discuss the determination of sample sizes required in environmental monitoring to determine the mean of quantitative measurements with sufficient accuracy, when data are spatially and serially correlated.

For the analysis of forest ecosystem measurements the following genetic aspects are relevant:

- different genotypes of the same species exposed to the same pollution regime under identical environmental conditions differ in sensibility (genotypic variation);
- the sensitivity of a given genotype is influenced greatly by environmental factors other than the chosen pollutant (environmental variation):
- different genotypes vary in their reaction to different environments (genotype–environment interaction).

Thus the phenotypic variation of response of trees to air pollutants contains genetic, environmental and interaction components. At present, the net of permanent observation plots consists of different genotypes in each plot, so that the observed symptoms can not clearly be related to the concentration of the pollutants.

Knabe *et al.* (1990) evaluated the genetic variability of Norway spruce under different pollution levels. The development of dieback, height growth and needle retention was analysed. The 100 investigated spruce varieties planted at three locations showed marked differences in pollution tolerance.

The majority of field studies of forest ecosystems carried out in northern and central Europe involve one species, namely spruce. Conclusions reached about the role of air pollutants on physiological processes are largely based on this one species. The extent to which they can be applied to other major European forest species is unknown and will require detailed investigation of a much wider range of species and site conditions throughout Europe. Ashmore *et al.* (1990) provided a review of current scientific understanding of air-pollutant effects on forest ecosystems in Europe. The major challenge for research on forest decline over the next decade is to set up large integrated field programmes involving both observation and experiment at a number of sites representing the range of climates, pollution conditions, forest types and soil conditions found within the European Community. The objective of such programmes will be to develop models which explain the role of different pollutants in a wide range of forest ecosystems. Statistical methods will of course help to make these programmes as efficient as possible.

17

Uncertainty and the Conservation or Destruction of Natural Forests and Wilderness

William J. Reed
University of Victoria, Canada

17.1 INTRODUCTION

In the last few years much attention has been focused on the adverse effects of man's activities on the world he inhabits. In the eyes of many of major threat to the continuation of life on the planet is no longer that of thermonuclear war but rather one brought about by the by-and-large peaceful economic activities of man. As the century draws to a close the converging threats of global climate change, the destruction of the ozone layer, the unchecked growth in world population and the loss of biodiversity loom ominously as portents of an ecological crisis unprecedented in human history.

No single economic activity has drawn so much obloquy for its adverse environmental effects as that of the destruction of natural forests. This activity has been blamed for aggravating the greenhouse effect, for destroying multitudes of plant and animal species, for the desertification of formerly fertile lands, for soil erosion, for flooding, for the silting of rivers and estuaries and, no doubt, for many other things besides. This is not a new phenomenon. Plato in the fourth century BC, described the results of logging on the hills around Athens

Statistics for the Environment. Edited by Vic Barnett and K. Feridun Turkman

(Attenborough, 1987): 'What now remains compared with what existed is like the skeleton of a sick man, all the fat and soft earth wasted away and only the bare framework of the land being left.' The destruction Plato was describing was on a relatively local scale. By the time of the fall of the Roman Empire, the environmental damage attributable to logging was on a much larger scale, and was possibly one of the causes of that fall. For example, the Roman provinces of North Africa, which had formerly been the bread basket of the empire, producing half a million tons of grain a year and boasting some six hundred cities, were reduced to a hot barren desert, the harbours of their once bustling ports silted up and abandoned, victims of the Ancient World's attitude to Nature and its insatiable demand for wood. Wood was virtually the only known fuel: it was indispensable in the construction of buildings, furniture, chariots and ships. The Ancient World's dependence on wood was perhaps even greater than the modern world's dependence on fossil fuels. And to the Romans the natural world was inexhaustible; they saw no reason for not extracting from it as much as they wanted. Indeed the state gave legal title to undeveloped land to anyone who cleared it of forest. The essential role of the forest in maintaining the ecology of the region were not understood.

The destruction of the forests which once ringed the Mediterranean was complete by the time of the decline of the Venetian Empire. The demand for wood, especially for building ships, was as great as in Roman times. For example, it has been estimated that the fleets fighting in the Battle of Lepanto in 1571 had been constructed at the expense of the destruction of over a quarter of a million mature trees. By the seventeenth century, after the centres of power, especially martime power, had moved to Western Europe, much of that territory had been denuded of trees. Unlike the Mediterranean where soil erosion and the grazing of goats prevented the re-establishment of forests, in Western Europe some afforestation was possible. John Clare, a courtier in the reign of Charles II, published in 1664 his classic textbook *Sylva* exhorting landowners to create artificial forest plantations; today virtually all of Britain's woodland consists of replanted forests.

Today most of the natural forests of Europe and eastern North America have been cleared and converted into agricultural land, while the tropical rainforests of Central America, Amazonia and South-East Asia are rapidly being destroyed. As the scale of logging operations has increased, so have the environmental consequences. It has been suggested that the serious monsoon flooding along the Ganges in India and Bangladesh in recent years is a result of large-scale deforestation in the Himalayan watershed some hundreds of miles away. In Panama deforestation has been blamed for a continuing drop in rainfall and the consequent drop in the water level of the lakes feeding the Panama Canal. It has also been blamed for silting of the Canal itself. Perhaps most serious of all, the destruction of natural forests especially tropical rainforest, has been blamed for between 10 and 30% of the carbon released annually into the atmosphere. Not only does the destruction of forest cause a one-time increase in atmospheric carbon, it also removes an important carbon sink, which otherwise could be relied upon to absorb atmospheric carbon on an ongoing basis.

Another global effect of deforestation is the irreversible loss of genetic resources. Tropical forests are incredibly rich in their diversity of species. A single hectare of Malaysian jungle may contain more than 800 species of woody plants—about as many as in all of North America—while between a half and three-quarters of all animal species live in tropical forests, including 90% of all insects (*The Economist*, 2 September, 1989). The rate of loss of species is staggering. Writing about a decade ago Myers (1979, 1983) estimated that species were being destroyed at the rate of 'more than one a day' with the possibility that over the last quarter of this century it could average 'over 100 extinctions a day'. He estimates that by the year 2000 as many as a million species of the current 5–10 million species could have been annihilated.

What are the reasons behind this continued deforestation? In many ways, in the poorer, tropical countries at least, they are not that different to those prevalent in the Ancient World—an ethos based on the notion of Nature being infinite, and the imperatives for ever more wood and ever more land driven by a burgeoning population. For millions of poor villagers around the world firewood is the only source of fuel. It is estimated that worldwide 50% of trees cut down annually are for fuel use (Dasgupta, 1982). Usually in poor countries sources of wood are common property resources and so there is no incentive for the transfer of land to fuel-producing forest. With a growing population the forests are inexorably destroyed often with horrific consequences such as those which have occurred in Ethiopia and Haiti.

With respect to land, similar laws to those of Rome which gave title to the clearer of forest land continue to operate. Such laws helped to clear the forests of North America and are now operative in Brazil and many other countries. In addition there are often other indirect incentives and subsidies for clearing the forest. For example, Brazil's tax law provides credits for livestock projects, encouraging the landless poor to the new frontiers in Amazonia.

No doubt there has always been conflict between those who want to log or clear forest land, and those who want to preserve it, but in recent years this conflict has greatly intensified, resulting in a polarisation of the positions of the advocacy groups and an excessive politicisation of the issues. Part of the reason for this is that there appears to be little common ground for rational discourse between the opposite factions, each basing its claims on different beliefs and principles. The arguments put forward by the proponents of logging are usually expressed in economic terms—the wealth generated for society, the number of jobs created or preserved, the increase in the supply of land for food production and the consequent value of its product, etc. To the North American logging industry and its supporters, forest land is a natural resource like any other which can be converted into wealth for the benefit of the industry and indeed for all of society. To governments of tropical Third World countries forests represent not only a source of timber wealth, but also a potential source of agricultural land. The conversion of forest land to agricultural land is seen as a step on the road to economic development, one taken by industrialised countries early on in their development. Indeed in the 1960s and 1970s international development agencies such as the World Bank encouraged and financed many such deforestation projects.

Those who oppose the logging of natural forest usually eschew economic arguments, basing their claims instead on philosophical or moral grounds. To many environmentalists and recreational users virgin forest and wilderness are irreplaceable natural treasures, valuable for their own sake, and worthy of considerable expense and effort for their preservation. Many scientists, especially those in the biological sciences, have lent support to the conservationists, pointing out the important role forests play in preserving genetic diversity and in maintaining environmental and ecological balances.

It would appear at first sight that many of these aspects defy quantification and therefore render economic analysis as being of little help in decision-making with respect to conservation. However if, as Samuelson (1975) describes it, economics is the study of 'how men and society end up choosing, with or without the use of money, to employ scarce productive resources which could have alternative uses', then it would appear that the questions facing society and indeed all of mankind, with respect to conservation or otherwise of natural forests, fall squarely within the realm of economic inquiry, and indeed in recent years economists have begun to grapple with some of the difficulties involved in environmental decision-making.

In principle at least the normative theory of welfare economics (see Mäler, 1985, for a discussion of welfare economics and the environment) and the applied methodology of cost–benefit analysis provide a framework for addressing such decision problems. None the less, there are many difficulties in applying cost–benefit analysis to decision problems such as those concerning the fate of natural forests and wilderness. Not least of these is the evaluation of the nonmarketed services which standing forest provides. However, this issue has been addressed in other areas of environmental regulation with, in some jurisdictions, cost–benefit analysis now being a mandated requirement for policy decisions. For example, in the United States an Executive Order of 1981 requires (for regulations concerning pollution control) 'the measurement in monetary terms of all the beneficial and adverse effects of the proposed regulations and stipulates that, unless otherwise precluded by law, regulatory policies should be designed to maximize net economic benefits' (Freeman, 1985). It would indeed be a big step forward in forest and wilderness management if such criteria were applied to harvesting decisions. This is patently not the case at present. Environmental costs, although often enormous, are typically considered as externalities and are often ignored. Furthermore even many of the direct costs are not properly assessed or allocated. For example, in North America the costs of building access roads and preparing logging sites are usually paid out of the public purse. Recently the US Forest Service sold logging rights on the Tongass rainforest of Alaska for 2% of what it spent in preparations for logging

(*The Economist*, 22 June, 1991). It was claimed in 1989 that each year the American taxpayer was providing $100 million to pay for the harvesting of timber too cheap to cover its costs (*The Economist*, 2 September, 1989).

Also considered as externalities are the costs associated with the disappearance of a valuable natural asset. In addition to the 'user cost' in its technical sense there is the cost associated with the termination of the flow of amenity benefits from the forest on its destruction. The amenity benefits which tropical forest provide include the means to a livelihood for forest dwellers through hunting and gathering of fruit, nuts, etc., in addition to the value to mankind of the forest as a respository of wildlife and genetic diversity and as a carbon store. These benefits are also part of the amenity services of temperate-zone forests. In addition temperate-zone forests have a value as sites for recreational and touristic activities. It has been suggested that the benefits through tourism and recreation alone exceed those of logging in forests managed by the US Forest Service (*The Economist*, 22 June, 1991), and that the value of the output of fruit, latex etc. in tropical rainforest can exceed by a factor of three that of land used for cattle pasture (*The Economist*, 2 September 1989).

If these benefits and costs were correctly allocated, it is clear that an application of cost–benefit methodology would lead to a great reduction in the rate at which natural forests are being destroyed. None the less, there are serious problems with the application of cost–benefit analysis. The difficulty in assessing the value of nonmarketed amenity services has been touched upon above, but methods such as those of contingent choice have been developed for handling such question (see e.g. Freeman, 1985, for a discussion of such methods). Another difficulty with cost–benefit analysis is that it does not deal with distributional and equity issues. Decisions involving the fate of forests involve redistribution of benefits. But cost–benefit analysis which seeks to maximising overall utility is unable to deal with problems such as how to find a trade-off between the needs of a poor villager for firewood and those of mankind for preserving biodiversity and keeping carbon locked up; or between the benefits of preserving endangered species with those of maintaining jobs and viable communities in areas traditionally dependent on logging. Furthermore there is the problem of intergenerational equity. How does one strike a balance between the benefits that can be enjoyed by people living now with the rights or needs of generations yet unborn? These are undoubtedly serious issues, which impose limits on the applicability of cost–benefit analysis to problems of this type. As Kneese and Schulze (1985) suggest, public-policy questions involving environmental issues are probably best dealt with by using cost–benefit analysis, but constrained by libertarian and egalitarian considerations.

Apart from the difficulties discussed above with respect to cost–benefit methods, there is another technical problem on which we wish to focus in this chapter. It concerns the inadequacy of naive cost-benefit methods when actions taken are *irreversible* and are taken in the face of future *uncertainty*. As an example consider the problem of deciding whether a particular area of natural forest should be logged or not. A naive application of cost-benefit analysis would involve evaluating the expected present value of benefits through logging (sale of timber plus future rents from the land), and the expected present value of costs (logging costs, environmental costs and amenity benefits forgone), and would prescribe logging if the expected present value of benefits exceeded that of costs. It is easily shown that this procedure is not optimal in the sense of maximising the expected present value of benefits net of costs. The reason is that it ignores the essential *dynamic* or temporal nature of the problem. If a decision is made not to log this year, it leaves open the option of logging next year or at some other future date. However, future benefits and costs are uncertain, because timber prices, the demand for amenity services, etc. are uncertain. One way around this difficulty is to use *expected values* for future benefits or costs, and then choose the date now or in the future for logging which maximises the benefits net of costs. This is known as a *certainty-equivalence* procedure. If applied in a *feedback manner* (i.e. with the expected values of future benefits and costs being updated as new information arrives), it will provide a decision rule which, dependent upon current information on timber prices, amenity values etc., will prescribe either logging or not logging at each point in time. Even this rule, however, is not optimal, because it ignores the essential *stochastic* nature of the decision problem. The *optimal* decision rule will take into account *all possible future values of timber and amenity services etc., with their associated probabilities*, and not just the *expected values* of these quantities (as does certainty-equivalence).

How can the optimal decision rule be obtained? One answer is through the use of *stochastic dynamic programming* (SDP) (see e.g. Ross, 1983). In the example discussed above where there are only two actions possible at each time (log or do not log), the problem reduces to one of *optimal stopping* (see e.g. Brock *et al.*, 1988). In more general cases where logging or development of part of the forest area is possible, the problem does not so reduce, but SDP can still be used to obtain a solution.

The shortcomings of naive cost–benefit analysis for irreversible decisions made in the face of future uncertainty, were first identified by Arrow and Fisher (1974) and Henry (1974). They used simple two-period and three-period models, with linear net benefit structures, to demonstrate the phenomenon. They identified a value (termed *quasi-option value*) associated with not taking an irreversible action. By not taking the irreversible action, one still has the option of taking it at a later date, when more information is available. Conrad (1980) showed that quasi-option value is equivalent to the expected value of information conditional on the irreversible decision not being taken immediately.

In this chapter the suboptimality of the cost–benefit methodology in its naive form, and in its more refined certainty-equivalence form are discussed for two models of the decision problem concerning the conservation or destruction of natural forest and wilderness. In both models the risk-neutral policy which maximises the expected present value of benefits net of costs is obtained, and it is shown how the naive and certainty-equivalence versions of cost–benefit methodology lead to premature destruction of the forest. These results are entirely in accord with those of Arrow and Fisher (1974) and Henry (1974). However, the models used are more detailed and realistic than the simple models used by these and other authors. Both are in continuous time with an infinite time-horizon.

The first model (Section 17.2) considers in isolation a stand of old-growth forest from which amenity benefits flow, but which is vulnerable at all times to the possibility of random catastrophic destruction. It is assumed that current timber values and amenity values are known, but that future values are not known with certainty. It is assumed that for each the growth rate is made up of a fixed and a random component. Optimal stopping methods are used to determine the decision rule which maximises expected net present value. It is shown how optimally the expected benefits must exceed the expected costs by a factor exceeding unity before harvesting should take place. This factor can be quite large (in numerical calculations given, it is shown that it could quite easily be as high as 13), and increases with increased uncertainty. It follows that naive cost–benefit analysis could be seriously suboptimal. The same is true for the certainty-equivalence method. Numerical calculations are given to indicate the degree of suboptimality both in terms of the reduction in expected present value and in the expected survival time of the forest stand.

The second model (Section 17.3) allows for partial harvesting or clearing of the forest or wilderness land, and assumes that the flow of benefits from wilderness land and 'developed' land depend on the quantities of each. While the current values of these flows are assumed known, the future values are assumed to be uncertain. The possibility of unforeseen benefits from the wilderness land, such as those resulting from technological breakthroughs utilising the genetic resources of the wilderness, is also included in the model. The dynamic-programming equation is used to find an analytic form for the feedback rule which maximises expected net present value. It is shown that optimally development takes place down to a level at which the *marginal* expected present value of developed land exceeds that of wilderness land by a factor again bigger than one, and which increases with increasing future uncertainty. The suboptimality of the naive cost–benefit procedure and the certainty-equivalence procedure are again discussed.

The results of the analysis of the two models are entirely consistent, and mathematically remarkably similar. The two models complement each other. While the first model does not permit partial harvesting, it does allow for uncertainty in future timber prices and does include the possibility of random catastrophic destruction. It is perhaps more appropriate for a stand of old-growth forest in western North America. The second model ignores uncertainty in timber benefits and harvest costs but does include flows of benefits from both forest land and developed land which are size-dependent and which are subject to future uncertainty. This model is perhaps more appropriate for tropical forests, which can be cleared for agriculture.

Neither model is meant to be an operational model for use by forest or land managers. The purpose for which they are used is to investigate analytically how uncertainty coupled with irreversibility interact in conservation-development decisions concerning natural forests and wilderness, and in particular to demonstrate how uncertainty should lead to policies which are much more conservative than would be indicated by simple cost–benefit analysis.

17.2 A SINGLE-STAND MODEL FOR THE HARVEST-CONSERVATION DECISION PROBLEM

Consider an area of old-growth forest which can be either conserved *in toto* or clear-cut harvested. Clearly the decision to conserve is reversible, in that the forest can be harvested at a later date, while the decision to harvest is irreversible. Suppose that if the forest is harvested at time t a net revenue $V(t)$ is realised. This includes the revenue from the sale of timber net of harvest costs, plus the value of the bare land on which future rotations may possibly be grown. If there are external environmental costs they should be included in the net revenue, V, thereby reducing it. It is assumed that $V(t)$ is an observable stochastic process following *geometric Brownian motion*, governed by the stochastic differential equation

$$\frac{\mathrm{d}V}{V} = b\,\mathrm{d}t + \sigma_1\,\mathrm{d}w_1 \tag{17.1}$$

where b is a drift parameter, $\{w_1(t)\}$ is a standard Wiener process, and σ_1^2 is a variance parameter. Thus it is assumed that *the proportional change in net value over time dt is a constant b dt plus or minus a random component* $\sigma_1\,dw_1(t)$ (Reed, 1992). Large value of the parameter σ_1 will correspond to a large amount of uncertainty in future timber values. If harvest and replanting costs are neglected, $V(t)$ should be proportional to the current price of timber, since both the revenues from the sale of timber and the land value will then be proportional to timber price (see Clarke and Reed, 1989). The assumption that a commodity price follows geometric Brownian motion is very reasonable. A similar assumption for asset prices lies at the heart of much of the recent literature in financial economics.

The realisation of the immediate benefit $V(t)$ resulting from harvesting at time t is offset by the forgoing of the option to harvest at a later date when timber prices might conceivably be higher, as well as by the forgoing of the flow of amenity services from the old-growth forest at all times beyond t.

Suppose that the flow of amenity services at time t has social valuation $A(t)$. It should be emphasised that $A(t)$ is a flow and has units such as dollars per month. It represents the rent that society is willing to pay to preserve the old-growth forest. To reflect the fact that future valuations of amenity services are uncertain, it is assumed that $\{A(t)\}$ is an observable stochastic process with proportional changes comprising fixed and random components. Thus it is assumed to follow geometric Brownian motion,

$$\frac{\mathrm{d}A}{A} = a\,\mathrm{d}t + \sigma_2\,\mathrm{d}w_2. \tag{17.2}$$

As before, a and σ_2^2 are drift (mean growth rate) and variance (uncertainty) parameters while $\{w_2(t)\}$ is another standard Wiener process. To reflect the fact that future changes in amenity value and timber value may be correlated, it is assumed that the white-noise processes $\{\mathrm{d}w_1(t)\}$ and $\{\mathrm{d}w_2(t)\}$ have correlation ρ. Positive values of ρ will reflect a positive association between fluctuations in timber and amenity values, and negative values of ρ a negative association.

Note that the assumption that the stochastic process $\{A(t)\}$ is observable is something of a simplification since societal valuations of amenity services can be determined only through sampling methods and thus will contain an element of sampling error. Similarly, even though current timber prices may be known exactly, the net revenue $V(t)$ to be realised through an immediate clear-cut harvest of the old growth cannot be known exactly prior to the harvest. For example, harvesting costs can only be estimated and the volume of usable timber is usually known only via sampling methods. However, uncertainty in future values of $V(t)$ and $A(t)$ is much greater than uncertainty concerning current values. Abstracting from the former source of uncertainty captures the essence of the problem of when, if ever, a clear-cut harvest should take place.

The fact that a harvest does not take place does not guarantee the survival of the forest for ever. There is always the possibility that it may be destroyed by some natural catastrophe such as fire, tempest, or pest infestation. In assessing the total value of amenity services obtained from the forest before a scheduled clear-cut harvest, this fact must be taken into account.

If there were to be no harvest, amenity services would accrue up until the time at which catastrophic destruction occurred, if indeed it ever did. Thus the expected present value of amenity services, using an *instantaneous discount rate*, δ, would be

$$\bar{A}_{\mathrm{f}}(0) = E\left\{ \int_0^{\tau} \mathrm{e}^{-\delta t} A(t)\,\mathrm{d}t \right\} \tag{17.3}$$

where τ is a random variable denoting the time of destruction (τ would be infinity if destruction never occurs). The expectation in (17.3) is taken with respect to both τ and the stochastic process $\{A(t)\}$. The integral can be re-expressed as

$$\bar{A}_{\mathrm{f}}(0) = E\left\{ \int_0^{\infty} \mathrm{e}^{-\delta t} A(t) S(t\,|\,0)\,\mathrm{d}t \right\} \tag{17.4}$$

where $S(t|0)$ is the *survivor function* denoting the probability that the stand survives catastrophic destruction until time t, given that it is 'alive' at time 0.

The survivor function is determined by the *hazard rate*. Any functional form could be used here (see e.g. Thompson, 1988). For simplicity, and in the absence of any direct indication to the contrary, we assume a constant hazard rate λ, i.e. assume that the hazard of catastrophic destruction is time-independent. In this case the survivor function can be written as

$$S(t\,|0) = \mathrm{e}^{-\lambda t}$$

and the expectation in (17.4) can be written as

$$\bar{A}_{\mathrm{f}}(0) = E\left\{\int_0^\infty \mathrm{e}^{-(\delta+\lambda)} A(t)\,\mathrm{d}t\right\} \qquad (17.5)$$

which can be evaluated (after solving (17.2) using the Itô integral—see Reed, 1992 for details) to give the expected present value, at time zero, of amenity services as

$$\bar{A}_{\mathrm{f}}(0) = \frac{A(0)}{\delta + \lambda - a}. \qquad (17.6)$$

More generally we define $\bar{A}_{\mathrm{f}}(t)$ *as the expected present value (at time t) of future amenity services conditional on the stand not being destroyed at time t, and on the value A(t) of the amenity flow at that time.* Specifically

$$\bar{A}_{\mathrm{f}}(t) = E\left\{\int_t^\infty \mathrm{e}-{}^{\delta(z-t)} A(z) S(z\,|\,t)\,\mathrm{d}z\,|\,A(t)\right\} \qquad (17.7)$$

$$= \frac{A(t)}{\delta + \lambda - a.}$$

Note that the presence of a constant risk of destruction affects the evaluation of future amenity services in the same way as the addition of a premium λ to the discount rate δ.

17.2.1 The cost–benefit prescription

A naive cost–benefit analysis would compare the immediate net benefits through harvesting, $V(0)$, with the cost (17.6) of amenity benefits forgone and would prescribe a harvest if the benefits exceeded the cost, i.e. if

$$V(0) - \bar{A}_{\mathrm{f}}(0) > 0. \qquad (17.8)$$

If this condition were not met, cost–benefit analysis would prescribe no harvest. If this procedure were applied in an ongoing fashion one would arrive at a *harvest (stopping) rule* which prescribed a harvest at time t if and only if

$$V(t) > \bar{A}_f(t). \tag{17.9}$$

We shall see that *this cost–benefit rule is considerably suboptimal from the point of view of maximising the expected net benefit from timber harvest and amenity services, and that it unambiguously prescribes premature harvests.*

17.2.2 The optimal harvest rule

To determine the optimal harvest rule, consider the *expected present value* (EPV) of benefits if a harvest is scheduled to take place at time T. They include the EPV of timber benefit

$$e^{-\delta T} S(T|0)V(T) = e^{-(\delta+\lambda)T} V(T) \tag{17.10}$$

and the EPV of amenity benefits

$$E\left\{\int_0^T e^{-\delta t} A(t) S(t|0)\,dt\right\} = E\left\{\int_0^\infty e^{-(\delta+\lambda)} A(t)\,dt\right\} - \left\{\int_T^\infty e^{-(\delta+\lambda)t} A(t)\,dt\right\}$$

$$= \bar{A}_f(0) - e^{-(\lambda+\delta)T} \bar{A}_f(T). \tag{17.11}$$

Thus the net EPV of a harvest scheduled at time T is

$$e^{-(\lambda+\delta)T} V(T) - e^{-(\lambda+\delta)T} \bar{A}_f(T) + \bar{A}_f(0). \tag{17.12}$$

Note that to ensure convergence of the integrals in (17.11) we require the condition

$$\delta + \lambda - a > 0. \tag{17.13}$$

It is assumed that this condition, and in addition

$$\delta + \lambda - b > 0 \tag{17.14}$$

holds. If either of these conditions is not met it will be optimal to never harvest the forest. In the technical parlance of optimal stopping the problem would not be *stable* (Ross, 1983, p 53). Thus sufficient conditions for *conservation in perpetuity* to be optimal are that $\delta + \lambda < a$ or $\delta + \lambda < b$. If the former is met, amenity services are growing faster in expectation that the risk-adjusted discount rate and it therefore pays to conserve for ever. In this case the stopping problem resembles the famous St Petersburg Paradox. While at each point in time it pays to defer the harvest, the eventual return in timber benefits will be zero since if the hazard rate λ is positive, with probability one the stand will be destroyed.

If the latter condition is met, timber values are growing faster in expectation than the risk-adjusted discount rate, and so at all times it will pay to postpone a clear-cut harvest.

The problem of determining an optimal harvest rule can be expressed as that of finding a *stopping time* T to maximise the total expected present value:

$$e^{-(\delta+\lambda)T}[V(T) - \bar{A}_f(T)] + \bar{A}_f(0). \tag{17.15}$$

This can be formulated as a problem of *optimal stopping* (see e.g. Brock *et al.*, 1988) with *intrinsic value function*

$$R(V, A) = V(T) - \bar{A}_f(T) = V(T) - \frac{A(T)}{\Delta - a} \tag{17.16}$$

and the objective of maximising the expected present value

$$E\{e^{-\Delta T} R(V(T), A(T))\} \tag{17.17}$$

where $\Delta = \delta + \lambda$ is the *risk-adjusted discount rate.* (Note that (17.16) omits the last term on the right-hand side of (17.15) which is a constant independent of T, $A(T)$ and $R(T)$. Note also that in general a stopping time T may depend on past and current values of the bivariate stochastic process $\{V(t), A(t)\}$, and that the expectation in (17.17) depends on the initial values (V_0, A_0) of this process.

It is shown in Reed (1992) that the *optimal stopping rule* involves a harvest at time t if and only if

$$V(t) \geqslant \left(\frac{1+\theta}{\theta}\right)\bar{A}_f(t) \tag{17.18}$$

where θ is the *positive* root of the characteristic equation

$$\frac{1}{2}\sigma^2\theta^2 + \left(b - a + \frac{1}{2}\sigma^2\right)\theta + (b - \Delta) = 0 \tag{17.19}$$

and σ^2 is the variance of the $\left\{\ln\left(\frac{V(t)}{A(t)}\right)\right\}$ stochastic process, i.e.

$$\sigma^2 = \sigma_1^2 - 2\rho\sigma_1\sigma_2 + \sigma_2^2. \tag{17.20}$$

An alternative way of specifying the optimal harvest rule is: harvest if and only if the ratio of current value to amenity value, $V(t)/A(t)$ exceeds a critical value $c^* = (1+\theta)/[\theta(\Delta - a)]$.

Compare the optimal rule (17.18) with the cost–benefit rule (17.19). Since θ is positive it is clear that *the optimal rule is more conservative than the cost–benefit rule*, i.e. optimally one harvests if and only if the net timber revenues ('benefits') exceed the EPV of revenues forgone ('costs') by a factor

$$q^* = \left(1 + \frac{1}{\theta}\right) \tag{17.21}$$

which is bigger than one.

It is fairly straightforward (Reed, 1992) to show that this critical factor q^* *increases* with increases in

(a) the expected growth rate of timber values b;
(b) the uncertainty in the growth of timber values σ_1;
(c) the uncerainty in the growth of amenity values σ_2.
Conversely q^* *decreases* with increases in
(d) the expected growth rate of amenity values a;
(e) the discount rate δ;
(f) the hazard rate λ;
(g) the correlation ρ between amenity service values and timber values.

Table A.1 displays the critical q^* for various values of the parameters a, b, δ and the total variance parameter σ^2. In all cases the hazard rate was set at $\lambda = 0.005$ per year corresponding approximately to a 1 in 200 chance of destruction in any given year. The blank entries (where a or b exceeds $\Delta = \delta + \lambda$) correspond to cases where it is never optimal to harvest. Note how the critical ratio q^* is large when the difference $b - a$ in expected growth rates for timber values and amenity value is large, and how q^* increases with the variance σ^2. For example, with $a = 0$, $b = 0.05$, $\delta = 0.05$ and $\sigma_2 = 0.02$ the benefits (the current timber value) would have to exceed the costs (expected present value of amenity services forgone) by a factor of more than 13 before, optimally, a harvest would be undertaken.

With σ^2 set at 0.02 one could be fairly confident ($\simeq$95% confidence) that within one year the ratio of timber value to amenity flow $V(t)/A(t)$ would lie within between 0.75 and 1.33 times its initial value; with $\sigma^2 = 0.01$ the corresponding range would be from 0.82 to 1.22 times the initial value. Such fluctuation seems well within the range of possibility.

17.2.3 Suboptimality of the cost–benefit procedure

The degree of suboptimality of the cost–benefit procedure can be measured in a number of ways. The most natural (since the optimal rule maximises EPV) is to look at the reduction in EPV through using the cost–benefit rather than the optimal rule.

In cases where the critical ratio q^* of $V(t)/\bar{A}_f(t)$ is finite ($a, b < \Delta$), it is easy to compute the percentage loss in EPV by using the cost–benefit rule. It

Table 17.1 Values of the critical ratio, q^*, of timber value to expected present value of amenity benefits forgone. In all cases the hazard rate is $\lambda = 0.05$. Blank entries indicate that it is never optimal to harvest

		$\delta = 0.03$			$\delta = 0.05$			$\delta = 0.07$		
		$\sigma_z^2 = 0$	$\sigma_z^2 = 0.01$	$\sigma_z^2 = 0.02$	$\sigma_z^2 = 0$	$\sigma_z^2 = 0.01$	$\sigma_z^2 = 0.02$	$\sigma_z^2 = 0$	$\sigma_z^2 = 0.01$	$\sigma_z^2 = 0.02$
	$b = 0$	1.00	1.46	1.70	1.00	1.35	1.53	1.00	1.29	1.44
$a = 0$	$b = 0.025$	3.50	4.16	4.77	1.83	2.15	2.40	1.50	1.73	1.91
	$b = 0.05$	—	—	—	11.00	12.09	13.16	3.00	3.24	3.51
	$b = 0$	1.00	1.19	1.36	1.00	1.17	1.31	1.00	1.16	1.28
$a = 0.025$	$b = 0.025$	1.00	2.00	2.62	1.00	1.50	1.77	1.00	1.37	1.55
	$b = 0.05$	—	—	—	6.00	7.16	8.27	2.00	1.32	2.62
	$b = 0$	—	—	—	1.00	1.10	1.20	1.00	1.10	1.19
$a = 0.05$	$b = 0.025$	—	—	—	1.00	1.19	1.38	1.00	1.17	1.32
	$b = 0.05$	—	—	—	1.00	2.62	3.73	1.00	1.55	1.86

depends on the initial values V_0 and A_0, or, what amounts to the same thing, on the ratio $V_0/\bar{A}_f(0)$. If this is greater than or equal to q^* then both procedures prescribe an immediate harvest and there is no loss in EPV. If $1 < V_0/\bar{A}_f(0) < q^*$, then the cost–benefit procedure prescribes an immediate harvest while the optimal procedure does not. Finally, if $V_0/\bar{A}_f(0) \leqslant 1$ then neither procedure prescribes a harvest. Formulae for the percentage loss in EPV are straightforward to obtain in the latter two cases. It can be shown that the percentage loss is maximum when $V_0/\bar{A}_f(0) = 1$ and that this maximum is given by the simple formula

$$\frac{100\theta^\theta}{\theta^\theta + (1+\theta)^{1+\theta}} \tag{17.22}$$

which is always $\leqslant 50\%$. Table 17.2 displays the maximum percentage loss in EPV for various values of the parameters a, b, σ^2 and δ. It can be seen that it increases with increases in b and σ^2, and with decreases in a and δ. Thus qualitatively the maximum loss in EPV through using the cost–benefit procedure behaves like the critical ratio q^*. Note especially that when uncertainty is large either in future growth of amenity values or of timber values (σ_1 or σ_2) the suboptimality of the cost–benefit procedure can be considerable.

It has been noted above that the cost–benefit procedure will always lead to premature harvesting. Another way of characterising the suboptimality of this procedure is to examine for how much longer the old-growth forest would survive under the optimal harvesting rule. It can be shown from the results in Reed (1992) that the *expected* extra survival time using the optimal harvest rule is

$$E(\tau_{ex}) = \frac{1}{\lambda}[1 - (q^*)^{-\phi}] \tag{17.23}$$

where ϕ is the *positive* root of

$$\frac{1}{2}\sigma^2\phi^2 + \left(b - a + \frac{1}{2}\sigma^2\right) - \lambda = 0. \tag{17.24}$$

Table 17.3 presents values of this expected excess survival time when the hazard rate λ is set at 0.005. In the absence of any harvesting the expected survival time of the old-growth forest would be $1/\lambda = 200$ years. It can be seen that in all cases (when $a < \Delta$, $b < \Delta$), harvesting under the cost–benefit rule is considerably premature with the expected difference ranging from a low of 17 years to a high of 115 years. However, unlike the maximum proportional loss in EPV through using the cost–benefit procedure (Table 17.2), the expected loss in survival time increases with increases in a and δ, but decreases with increases

Table 17.2 Maximum percentage loss in EPV through using the cost–benefit harvest rule, rather than the optimal one. In all cases the hazard rate is $\lambda = 0.005$. Blank entries correspond to situations in which it is never optimal to harvest

		$\delta = 0.03$			$\delta = 0.05$			$\delta = 0.07$		
		$\sigma^2 = 0$	$\sigma^2 = 0.01$	$\sigma^2 = 0.02$	$\sigma^2 = 0$	$\sigma^2 = 0.01$	$\sigma^2 = 0.02$	$\sigma^2 = 0$	$\sigma^2 = 0.01$	$\sigma^2 = 0.02$
	$b = 0$	0	12.2	16.2	0	9.9	13.5	0	8.5	11.8
$a = 0$	$b = 0.025$	30.2	32.6	34.3	18.0	18.5	16.4	12.9	16.6	19.0
	$b = 0.05$	—	—	—	41.7	42.3	42.8	27.8	29.0	30.3
	$b = 0$	0	6.0	10.1	0	5.5	9.0	0	5.2	8.3
$a = 0.025$	$b = 0.025$	0	20.0	25.4	0	12.9	17.2	0	10.3	13.8
	$b = 0.05$	—	—	—	36.8	38.5	39.7	20.0	23.1	25.4
	$b = 0$	—	—	—	0	3.4	6.3	0	3.4	6.0
$a = 0.05$	$b = 0.025$	—	—	—	0	6.0	10.6	0	5.5	9.2
	$b = 0.05$	—	—	—	0	25.4	31.1	0	13.8	18.4

Table 17.3 Expected additional survival time (in years) through using the optimal harvest rule rather than the cost–benefit rule, when the hazard rate for catastrophic destruction is set at $\lambda = 0.005$. In the absence of any harvesting the expected lifetime of the forest would be $1/\lambda = 200$ years. Cases marked with an asterisk (*) are for zero variance (deterministic model) with $a \geqslant b$. In such cases the optimal harvest rule collapses to the cost–benefit rule. However, with zero variance and $a < b$ this is not the case (see Section 17.2.4). Dashes (—) occur in cases when both the cost–benefi rule and the optimal rule never prescribe a harvest ($a \geqslant \Delta$), while daggers (†) correspond to cases ($b \geqslant \Delta$, $a < \Delta$) where the cost–benefit rule prescribes a harvest while the optimal rule never does

		$\delta = 0.03$			$\delta = 0.05$			$\delta = 0.07$		
		$\sigma^2 = 0$	$\sigma^2 = 0.01$	$\sigma^2 = 0.02$	$\sigma^2 = 0$	$\sigma^2 = 0.01$	$\sigma^2 = 0.02$	$\sigma^2 = 0$	$\sigma^2 = 0.01$	$\sigma^2 = 0.02$
	$b = 0$	0*	39.1	35.2	0*	33.8	28.8	0*	29.4	24.8
$a = 0$	$b = 0.025$	78.8	41.2	38.6	103.4	23.4	22.8	111.2	17.2	17.2
	$b = 0.05$	†	†	†	42.6	40.2	38.2	86.6	20.4	19.8
	$b = 0$	0*	103.6	84.6	0*	97.2	76.4	0*	92.2	70.8
$a = 0.025$	$b = 0.025$	0*	69.7	59.4	0*	44.4	37.6	0*	35.4	30.0
	$b = 0.05$	†	†	†	60.4	54.6	50.2	100.0	25.8	25.0
	$b = 0$	—	—	—	0*	115.4	104.6	0*	113.0	100.8
$a = 0.05$	$b = 0.025$	—	—	—	0*	105.6	87.2	0*	98.6	78.2
	$b = 0.05$	—	—	—	0*	89.6	76.4	0*	48.0	40.8

in σ^2. The greater the variability in amenity values $A(t)$ or timber values $V(t)$, the shorter the *expected* time to reach the critical barrier. In cases with no variability in both $A(t)$ and $V(t)$ ($\sigma^2 = 0$—deterministic model) the optimal rule collapses to the cost–benefit rule when $a \geqslant b$ (hence the zero loss in survival time), but not when $a < b$.

17.2.4 The certainty-equivalence procedure

A more sophisticated application of cost–benefit methodology might recognise that the net difference between benefits (net timber revenues) and expected costs (amenity services forgone) could change over time and thus seek a harvest time to maximise the present value of the difference. The harvest rule that does this is of course the optimal rule discussed in Section 17.2.2. However, in practice, rather than using the methodology of stochastic dynamic programming and optimal stopping, such problems are frequently treated via the use of the '*certainty-equivalence*' (CE) procedure. This involves replacing random variables by their expected values and solving the resulting deterministic optimisation problem. For some stochastic optimisation problems this procedure provides a good approximation to the optimal solution. However, it is well known (Arrow and Fisher, 1974; Henry, 1974; Conrad, 1980) that for problems involving *irreversibility* and *uncertainty* that the CE procedure can be considerably suboptimal.

In the context of the model used here, the CE procedure can be shown (Reed, 1992) equivalent to:

(a) the cost–benefit rule of Section 17.2.1 if $b \leqslant a$, i.e. if amenity values are growing in expectation as fast as or faster than timber values;
(b) the rule which prescribes harvesting as soon as the ratio $V(t)/\bar{A}_f(t)$ of timber benefits to amenity services forgone reaches the barrier

$$q_w = \frac{\Delta - a}{\Delta - b}, \tag{17.25}$$

if $b > a$, i.e. if amenity values are growing in expectation slower than timber values.

In either case the CE procedure will lead to a premature harvest.

Table 17.4 displays the critical ratio q_{CE} of $V(t)/\bar{A}_f(t)$ for the certainty-equivalence rule and the degree to which q_{CE} is too small compared with the optimal q^*. Table 17.5 displays the expected additional survival time through using the certainty-equivalence rather than the optimal rule; this is determined from the formula

$$\frac{1}{\lambda}[1 - (q_{CE}/q^*)^{\phi}]. \tag{17.26}$$

Table 17.4 Values of the critical ratio, q_{CE}, of timber value to expected present value of amenity benefits forgone for the certainty-equivalence harvest rule and in brackets the degree, r, to which q_{CE} is too small compared with the optimal q^*; $r = 100(q^* - q_{CE})/q^*$. Dashes in brackets occur in cases when it is never optmal to harvest

		$\delta = 0.03$			$\delta = 0.05$			$\delta = 0.07$		
		$\sigma_z^2 = 0$	$\sigma_z^2 = 0.01$	$\sigma_z^2 = 0.02$	$\sigma_z^2 = 0$	$\sigma_z^2 = 0.01$	$\sigma_z^2 = 0.02$	$\sigma_z^2 = 0$	$\sigma_z^2 = 0.01$	$\sigma_z^2 = 0.02$
	$b = 0$	1.0	1.0	1.0	1.0	1.0	1.0	1.0	1.0	1.0
		(0%)	(31.5%)	(41.2%)	(0%)	(25.9%)	(34.6%)	(0%)	(22.5%)	(30.1%)
$a = 0$	$b = 0.025$	3.50	3.50	3.50	1.83	1.83	1.83	1.50	1.50	1.50
		(0%)	(15.9%)	(26.6%)	(0%)	(14.9%)	(31.1%)	(0%)	(13.3%)	(21.5%)
	$b = 0.05$	1.17	1.17	1.17	11.00	11.00	11.00	3.00	3.00	3.00
		(–)	(–)	(–)	(0%)	(9.0%)	(16.4%)	(0%)	(7.4%)	(14.5%)
	$b = 0$	1.0	1.0	1.0	1.0	1.0	1.0	1.0	1.0	1.0
		(0%)	(16.0%)	(26.5%)	(0%)	(14.5%)	(23.7%)	(0%)	(13.8%)	(21.9%)
$a = 0.025$	$b = 0.025$	1.0	1.0	1.0	1.0	1.0	1.0	1.0	1.0	1.0
		(0%)	(50.0%)	(61.8%)	(0%)	(33.3%)	(43.4%)	(0%)	(27.0%)	(35.5%)
	$b = 0.05$	1.08	1.08	1.08	6.00	6.00	6.00	2.00	2.00	2.00
		(–)	(–)	(–)	(0%)	(16.2%)	(27.4%)	(0%)	(13.8%)	(23.7%)
	$b = 0$	1.0	1.0	1.0	1.0	1.0	1.0	1.0	1.0	1.0
		(–)	(–)	(–)	(0%)	(9.1%)	(16.7%)	(0%)	(9.1%)	(16.0%)
$a = 0.05$	$b = 0.025$	1.0	1.0	1.0	1.0	1.0	1.0	1.0	1.0	1.0
		(–)	(–)	(–)	(0%)	(16.0%)	27.5%)	(0%)	(14.5%)	(24.2%)
	$b = 0.05$	1.0	1.0	1.0	1.0	1.0	1.0	1.0	1.0	1.0
		(–)	(–)	(–)	(0%)	(61.8%)	(73.2%)	(0%)	(35.5%)	(46.3%)

Table 17.5 Expected additional survival time through using the optimal harvest rule rather than the certainty-equivalence rule, when the hazard rate for catastrophic destruction is set at $\lambda = 0.005$. In cases where $a \geqslant b$ the entries coincide with those of Table 17.2, since in this case the certainty-equivalence rule coincides with the cost–benefit rule. Also in cases with zero variance the optimal rule reduces to the cetainty-equivalence rule and so the additional survival time is zero. Dashes occur in cases where neither the certainty equivalence rule nor the optimal rule prescribes a harvest ($b \geqslant \Delta$ or $a \geqslant \Delta$).

		$\delta = 0.03$			$\delta = 0.05$			$\delta = 0.07$		
		$\sigma^2 = 0$	$\sigma^2 = 0.01$	$\sigma^2 = 0.02$	$\sigma^2 = 0$	$\sigma^2 = 0.01$	$\sigma^2 = 0.02$	$\sigma^2 = 0$	$\sigma^2 = 0.01$	$\sigma^2 = 0.02$
	$b = 0$	0	39.1	35.2	0	33.8	28.8	0	29.4	24.8
$a = 0$	$b = 0.025$	0	5.5	8.3	0	5.0	7.3	0	4.7	6.6
	$b = 0.05$	—	—	—	0	1.7	2.9	0	1.6	2.8
	$b = 0$	0	103.6	84.6	0	97.2	76.4	0	92.2	70.8
$a = 0.025$	$b = 0.025$	0	69.7	59.4	0	44.4	37.6	0	35.4	30.0
	$b = 0.05$	—	—	—	0	5.7	8.6	0	5.1	7.5
	$b = 0$	—	—	—	0	115.4	104.6	0	113.0	100.8
$a = 0.05$	$b = 0.025$	—	—	—	0	105.6	87.2	0	98.6	78.2
	$b = 0.05$	—	—	—	0	89.6	76.4	0	48.0	40.8

It can be seen by comparing Tables 17.1 and 17.4 that in cases where $b > a$, the certainty-equivalence barrier q_w is closer to the optimal barrier q^*, than the unit barrier of the cost–benefit rule. Also, from Tables 17.3 and 17.5 one can see that the degree to which the certainty-equivalence rule prescribes premature harvesting in the case $b > a$ is less than that of the cost–benefit rule. Thus, *in the (fairly unlikely) case of timber values having a higher mean growth rate than amenity service values, the certainty-equivalence procedure would provide a reasonable approximation to the optimal policy.* However, *in the (much more likely) case of amenity service values having a higher mean growth rate than timber values the certainty-equivalence procedure would provide no improvement over the naive cost–benefit procedure.* It could prescribe considerably premature harvests and result in a large loss in expected present value.

17.3 A MODEL FOR CONSERVATION-DEVELOPMENT DECISIONS FOR A LARGE AREA OF WILDERNESS LAND ALLOWING FOR PARTIAL DEVELOPMENT

A major shortcoming of the model discussed in Section 17.2 is that it treats a stand of old-growth forest in isolation, with the value of amenity services determined exogenously. Because old-growth forest, and wilderness in general, is an exhaustible resource one would expect its unit value (as an amenity) to increase as it becomes scarcer. This aspect, which was ignored in Section 17.2, will be incorporated in the model developed in this section, which will be for a large tract of wilderness land which can be partially developed, rather than for a single stand. In addition to services provided by wilderness land, the model will include services from developed land (e.g. from agriculture). The social valuations of both types of services in the future will be considered uncertain. However, for mathematical simplicity the revenues from harvesting (and/or costs of clearing the land) will be considered fixed and known. Also the possibility of catastrophic destruction will be ignored; on the other hand the model will include the possibility of unpredictable benefits arising from scientific or technological 'breakthroughs' utilising the genetic resouces of the wilderness. The probability of such a benefit occurring will be assumed to depend on the quantity of wilderness remaining, reflecting the fact that the destruction of wilderness leads to reduction in biodiversity.

17.3.1 The model

The model and optimisation analysis in this Section is from Clarke and Reed (1990). Suppose that, at time t, a land development/conservation planner has to make development decisions on a partitionable homogeneous area of wilderness $X(t)$ and an area of developed land $1 - X(t)$. Assume that, apart from benefits associated with random breakthroughs, the instantaneous flow of utility accruing to society when land is developed to this extent is described by the additively separable benefit function:

$$W(t, X(t)) = \pi(t)P(1 - X(t)) + v(t)Q(X(t)) \tag{17.27}$$

where $\pi(t)$ and $v(t)$ are the *social valuations* placed on the services yielded by developed land and wilderness respectively; and $P(\cdot)$ and $Q(\cdot)$ measure the size of the *service flows* from the respective land-resource assets. Assume that P and Q are twice continuously differentiable, nondecreasing concave functions with $P(0) = Q(0) = 0$. If $P'' < 0$, $Q'' < 0$ for some X then the marginal values will be nonconstant and land-service flows will be *size-dependent*; otherwise they are *size-independent* (i.e $P'' \equiv Q'' \equiv 0$) and the welfare function is *linear* in $X(t)$.

To reflect future uncertainty in social valuations assume that $\{v(t)\}$ and $\{\pi(t)\}$ are observable stochastic processes which evolve through time as geometric-Brownian-motion processes governed by the stochastic differential equations

$$dv/v = r_v + \sigma_v dw_v, \ v(0) = v_0; \tag{17.28}$$

$$d\pi/\pi = r_\pi + \sigma_\pi dw_\pi, \ \pi(0) = \pi_0; \tag{17.29}$$

where r_v and r_π are drift parameters representing constant upward or downward trends in relative valuations of wilderness and developed land and σ_v^2, σ_π^2 denote the instantaneous variances in the relative valuation processes. The stochastic processes $\{w_v(t)\}$ and $\{w_\pi(t)\}$ are standard Wiener processes so that $\{dw_v(t)\}$ and $\{dw_\pi(t)\}$ are white-noise processes. Thus social valuations are assumed to grow at a rate made up of a fixed component (r_v or r_π) and a random component ($\sigma_v dw_v$ or $\sigma_\pi dw_\pi$).

Consider now the development dynamics and assume that, at time t, land is being developed at the rate $u(t)$, and that this development is *irreversible*. Thus

$$\frac{dX}{dt} = -u(t) \leqslant 0. \tag{17.30}$$

Suppose also $X(0) = 1$, so that initially the total area of land is in a wilderness state. Suppose also that the cost associated with developing at a rate $u(t)$ is

$$cu(t) \tag{17.31}$$

where c is a positive constant.

Finally assume that, apart from the socially valuable 'consumption' service stream yielded by wilderness, it may also have a value in that the genetic pool contained therein may lead to scientific 'breakthroughs' of social value. For example, a wilderness species may yield an improved corn variety or a new drug. Suppose that such breakthroughs occur in a point process with intensity.

$$\lambda(t) = \lim_{\Delta \to 0} \{P(\text{breakthrough in } (t, t+\Delta]/\Delta)\} \tag{17.32}$$

dependent on the area of wilderness remaining. Specifically assume that

$$\lambda(t) = \psi(X(t)) \tag{17.33}$$

where ψ is twice-differentiable, nondecreasing and concave with $\psi(0) = 0$. Thus the larger the area of wilderness land the higher the probability of a breakthrough at any time, but with diminishing returns operative.

If the expected present value of rewards from a breakthrough at any time t is $\bar{R}$ and $T_1, T_2, \ldots$ denote the random times at which breakthroughs occur then the *net expected present value of the services of land* when both development costs and breakthrough possibilities are accounted for is

$$J = E\left\{\int_0^\infty e^{-\delta t}[W(t, X(t)) - cu]\,dt + \sum_{i=1}^{\infty} \bar{R}e^{-\delta T_i}\right\} \tag{17.34}$$

where E is the expectation operator, δ is the instantaneous positive discount rate, and W is as given in (17.27).

Socially optimal development could be considered as that which maximises (17.34) subject to (17.31), and the constraint

$$u \geqslant 0 \tag{17.35}$$

to reflect the irreversibility of wilderness development.

17.3.2 The cost-benefit prescription

The cost–benefit procedure would prescribe, at any time, development up to the level which maximised the difference between the expected present value (EPV) of benefits from developed land and the cost of development including the expected present value of wilderness benefits forgone, i.e. developing down to the level at which marginal expected benefits equalled marginal expected cost.

The marginal EPV of benefits at level X at time zero is

$$P'(1 - X)E\left\{\int_0^\infty e^{-\delta t}\pi(t)\,dt\right\} \tag{17.36}$$

which (cf. evaluation of (17.5)) can be evaluated (by first solving (17.29) using the Itô calculus) as

$$P'(1-X)\frac{\pi_0}{\delta - r_\pi}. \tag{17.37}$$

The marginal EPV of wilderness benefits forgone at level X is similarly

$$Q'(X)\frac{v_0}{\delta - r_v} \tag{17.38}$$

If the potential benefits of scientific 'breakthroughs' are ignored (which is likely to be the case in a cost–benefit analysis since they are hard to quantify), the cost–benefit prescription would prescribe at time t development to the level at which X units of wilderness remain, where X satisfies

$$P'(1-X)\frac{\pi(t)}{\delta - r_\pi} - Q'(X)\frac{v(t)}{\delta - r_v} - c = 0. \tag{17.39}$$

To include the benefits from random scientific breakthroughs the expectation

$$E\left\{\sum_{i=1}^{\infty} \mathrm{e}^{-\delta T_i}\right\} \tag{17.40}$$

occurring in (17.34) must be evaluated under the assumption that X is held fixed (so that $\lambda(t) \equiv \psi(X) \equiv$ constant). This is easily performed (Reed, 1984); the result being simply

$$\psi(X)/\delta. \tag{17.41}$$

The inclusion of these costs (forgone benefits) in the cost–benefit analysis would require (17.39) to be modified to the form

$$P'(1-X)\frac{\pi(t)}{\delta - r_\pi} - Q'(X)\frac{v(t)}{\delta - r_v} - \frac{\bar{R}\psi'(X)}{\delta} - c = 0. \tag{17.42}$$

We shall show that *this procedure prescribes overdevelopment (compared to the optimal) and that the degree of overdevelopment increases with increased uncertainty in future values of the services from wilderness land or developed land.* As in the single-stand model of Section 17.2 it is the combination of uncertainty and irreversibility that leads to the suboptimality of the cost–benefit procedure. Of course, if the potential benefits from scientific breakthroughs are ignored in the cost–benefit procedure (equation (17.39)) then the degree of overdevelopment will be even greater.

17.3.3 The optimal development rule

The development rule which maximises the EPV of all benefits net of costs is found by solving the stochastic optimisation problem of maximising (17.34) subject to the state equations (17.28), (17.29) and (17.31) with the constraint (17.35). It has not been possible to solve this problem analytically in full generality. However, a slightly simplified version has been solved, which is quite adequate for the purposes of demonstrating the sub-optimality of the cost-benefit procedure. This simplification involves removing the uncertainty of future valuations of one or the other of services from wilderness land or developed land. Thus consider either

(A) *wilderness valuation uncertainty* in which $\sigma_v > 0$ with $\sigma_\pi = 0$, or

(B) *developed-land valuation uncertainty* in which $\sigma_v > 0$ with $\sigma_v = 0$.

The method of solution used for the stochastic optimal control problem is *stochastic dynamic programming* which provides a second-order partial differential equation for the value function (maximum of (17.34) regarded as a function of the initial values of X, π and v). This equation can be solved analytically in the region for which optimally there is not development (where the optimal $u^* \equiv 0$); upon invoking the appropriate boundary conditions an explicit equation for the boundary of the no-development region can be found (Clarke and Reed, 1990). In the case (A) of uncertainty only in future wilderness valuation the optimal policy involves development if and only if

$$\frac{\pi(t)P'(1-X)}{\delta - r_\pi} - \frac{\bar{R}\psi'(X)}{\delta} - c > \left(\frac{1+\alpha}{\alpha}\right)\frac{v(t)Q'(X)}{\delta - r_v} \qquad (17.43)$$

where α is the *positive* root of the quadratic equation

$$\frac{1}{2}\sigma_v^2 z^2 - \left(r_v - r_\pi - \frac{1}{2}\sigma_v^2\right)z - (\delta - r_\pi) = 0. \qquad (17.44)$$

If the above condition is met, the optimal policy prescribes developing to the level (of X) for which the left-hand side of (17.43) is equal to the right-hand side.

The corresponding condition for case (B) in which there is uncertainty only in developed-land values the condition (17.43) is replaced by

$$\frac{\pi(t)P'(1-X)}{\delta - r_\pi} > \left(\frac{\beta}{\beta - 1}\right)\left[\frac{v(t)Q(X)}{\delta - r_v} - c - \frac{\bar{R}\psi'(X)}{\delta}\right] \qquad (17.45)$$

where β is the *positive* root of

$$\frac{1}{2}\sigma_\pi^2 + \left(r_\pi - r_v - \frac{1}{2}\sigma^2\right)z - (\delta - r_v) = 0. \tag{17.46}$$

It can easily be shown from the assumption that $\delta > r_\pi$ that the positive root β of (17.46) greater than one and hence that the factor $\beta/(\beta - 1)$ in (17.45) is also greater than one.

Conditions (17.43) and (17.45) have a simple interpretation in terms of marginal costs and benefits. Since in (17.45) $\alpha > 0$, it follows that for development to take place the marginal EPV of services from developed land net of development costs and losses related to scientific breakthrough must exceed the marginal EPV of wilderness services forgone by a factor $(1 + \alpha)/\alpha > 1$. Condition (17.45) requires that the marginal EPV of developed land must exceed the marginal loss (made up of the EPV of wilderness services forgone, benefits from scientific breakthroughs forgone, and development costs) by a factor $\beta/(\beta - 1) > 1$. Thus in both cases the optimal development rule is seen to be more conservative than that which simple cost–benefit analysis would prescribe. The factors $(\alpha + 1)/\alpha$ and $\beta/(\beta - 1)$ indicate the degree of which the optimal policy is more conservative than that of cost–benefit analysis.

A third, special, case in which it is possible to obtain an analytic solution to the stochastic optimal control problem is:

(C) *zero development costs and no benefits from scientific breakthroughs:*

$$c = 0; \quad \bar{R} = 0.$$

In this case the optimal development policy involves development only when

$$\frac{\pi(t)P'(1 - x)}{\delta - r_\pi} > \left(\frac{1 + \gamma}{\gamma}\right)\frac{v(t)Q'(x)}{\delta - r_v} \tag{17.47}$$

where γ is the *positive* root of

$$\frac{1}{2}\bar{\sigma}^2 z^2 + \left(r_\pi - r_v - \frac{1}{2}\bar{\sigma}^2\right)z - (r_\pi - \delta) = 0 \tag{17.48}$$

where

$$\bar{\sigma}^2 = \sigma_\pi^2 - 2\rho\sigma_\pi\sigma_v + \sigma_v^2 \tag{17.49}$$

is the variance parameter for the stochastic process $\{\ln[\pi(t)/v(t)]\}$ (ρ is the correlation between the valuation processes (17.28) and (17.29). In this case development takes place only when the marginal EPV of a unit of developed land (the left-hand side of (17.47)) is bigger than the marginal EPV of a unit of wilderness land by a factor $(1+\gamma)/\gamma$ bigger than 1.

17.3.4 The sub-optimality of the cost-benefit procedure

The degree to which the cost–benefit procedure prescribes overdevelopment is determined (on comparing (17.42) or (17.39) with (17.43), (17.45) and (17.47) by the factors $(1+\alpha)/\alpha$, $\beta/(\beta-1)$ and $(1+\gamma)/\gamma$ in the three cases (A), (B) and (C). Through examination of the characteristic equations (17.44), (17.46) and (17.48) it can be shown that

(i) α increases with σ_v^2;
(ii) β increases with σ_π^2;
(iii) γ increases with $\bar{\sigma}^2$.

Thus *increases in the uncertainty of the future valuation of services from wilderness land or developed land have the effect of making the optimal development rule more conservative, thereby making the cost–benefit procedure more suboptimal.*

This result concurs with the results from the model of Section 17.2. Indeed the similarity of the difference between the optimal rule and the cost–benefit rule for the two models is striking. A comparison of (17.43), (17.45) and (17.47) with (17.18) shows that in each case the costs of development (in terms of future amenity benefits forgone etc.) must be inflated by a factor greater than one in order to make cost–benefit analysis yield the optimal harvest rule. In each case this factor depends on the root of a quadratic equation dependent upon, and increasing with, a variance term reflecting the degree of future uncertainty.

17.3.5 The certainty-equivalence procedure

An approach to the development problem which would recognise the dynamic irreversibility of the process but not the inherent uncertainty would be that of certainty equivalence in which random variables would be replaced by their expected values and the resulting deterministic optimisation problem solved by the methods of deterministic optimal control. The optimal control would then be applied in a feedback fashion.

The expected values of the $v(t)$ and $\pi(t)$, given initial values v_0 and π_0, are

$$E(v(t)|v_0) = v_0 e^{r_v t}, \quad E(\pi(t)|\pi_0) = \pi_0 e^{r_p t}. \tag{17.50}$$

The deterministic optimal control problem to be solved for the certainty-equivalence procedure is to maximise

$$\int_0^{\infty} \mathrm{e}^{-\delta t}[\pi_0 \mathrm{e}^{r_\pi t} P(1-X) + v_0 \mathrm{e}^{r_v t} Q(X) + \bar{R}\psi(X) - c]\,\mathrm{d}t \qquad (17.51)$$

subject to

$$\frac{\mathrm{d}X}{\mathrm{d}t} = -u; \quad X(0) = 1 \qquad (17.52)$$

and

$$u \geqslant 0. \qquad (17.53)$$

This is a linear (in u) control problem and so its solution can involve only components of bang-bang control and singular control. Here this means either control with $u = 0$, pulses with $u = \infty$, or control to stay on the singular path given by

$$\left(\frac{\delta - r_\pi}{\delta}\right)\left[\frac{\pi_0 \mathrm{e}^{bt}}{\delta - r_\pi} P'(1-X)\right] = \left(\frac{\delta - r_v}{\delta}\right)\left[\frac{v_0 \mathrm{e}^{r_v t}}{\delta - r_v} Q'(X)\right] + \frac{\bar{R}\psi'(X)}{\delta} + c. \qquad (17.54)$$

If the singular path (17.54) is feasible the certainty-equivalence procedure would involve initially developing down to the level x satisfying

$$M_1\left[\frac{\pi_0}{\delta - r_\pi} P'(1-X)\right] = M_2\left[\frac{v_0}{\delta - r_v} Q'(X)\right] + \frac{\bar{R}\psi'(X)}{\delta} + c \qquad (17.55)$$

where

$$M_1 = \frac{\delta - r_\pi}{\delta} < 1 \text{ and } M_2 = \frac{\delta - r_v}{\delta} < 1. \qquad (17.56)$$

Subsequently, as the values of $\pi(t)$ and $v(t)$ evolved, development would take place only when the left-hand side of (17.55) (with π_0 replaced by $(\pi(t))$ exceeded the left-hand side (with v_0 replace by $v(t)$) and the development would be down to the level at which equality occurred.

However, situations can arise in which it is not feasible for a trajectory of (17.52) to coincide with the singular path, because the latter is *increasing in t*. As an example consider the simplified case (C) in which $\bar{R} = c = 0$. In this case the singular path can be written

$$\frac{P'(1-X)}{Q'(X)} = \frac{v_0}{\pi_0} \mathrm{e}^{(r_v - r_\pi)t}. \qquad (17.57)$$

The left-hand side is increasing in X. If $r_v > r_\pi$ the right-hand side will also be increasing in t. This implies that the singular path has X increasing with time. Clearly this is not feasible. In such cases the optimal control will be of bang-bang type, i.e. it will involve pulses of instantaneous development ($u^* = \infty$) and intervals of no development ($u^* \equiv 0$). It can be shown that in fact the deterministic optimal solution involves initially developing down to the level x which solves

$$\frac{\pi_0}{\delta - r_\pi} P'(1 - X) = \frac{v_0}{\delta - r_v} Q'(X) + \frac{\bar{R}\psi'(X)}{\delta} + c \tag{17.58}$$

and subsequently staying there. If this solution is applied in a feedback fashion development will take place whenever the left-hand side (with π_0 replaced by $\pi(t)$) exceeded the right-hand side (with v_0 replaced by $v(t)$). But this rule is just the cost–benefit development rule discussed above (equation (17.42)). Thus certainty-equivalence can yield either the cost–benefit rule or the rule (17.55) in which the EPV of returns from developed and wilderness land are modified by the factors M_1 and M_2. This is analogous to the application of certainty-equivalence to the single-stand harvesting model of Section 17.2 in which certainty-equivalence leads to the cost–benefit rule, when amenity values vaues of old-growth forest are growing faster in expectation than timber values ($a > b$), and another rule in the other case. Here we have the situation that when amenity values of wilderness are growing faster in expectation then those of developed land ($r_v > r_\pi$) certainty-equivalence can lead to the cost–benefit rule, but can lead to a different rule in the opposite case.

17.3.6 Size-independent land-service flows

Consider the case in which land-service flows are size-independent, i.e. $P' \equiv Q' \equiv 0$. Without loss of generality assume $P(1 - X) \equiv 1 \equiv Q(X)$; and for simplicity assume $c = 0$ and $\psi(x) \equiv 0$, i.e. ignore benefits from scientific breakthrough and development costs. In this case (from (17.47)), it can be seen that the optimal policy involves no development until

$$\frac{\pi(t)}{\delta - r_\pi} > \left(\frac{1 + \gamma}{\gamma}\right) \frac{v(t)}{\delta - r_v} \tag{17.59}$$

at which time *all* of the land is developed.

This is essentially the same as the optimal stopping rule (17.18) of Section 2 (the only difference being that the immediate benefits from harvesting $V(t)$ are replaced by the EPV of the developed land; and that the hazard of catastrophic destruction is not present, $\Delta = \delta$), since the equation defining γ (17.48) is identical to that defining θ (equation 17.19)).

Furthermore the certainty-equivalence procedures given by equations (17.55) and (17.58) reduce to those of Section 17.2 (see 17.2.4) in the two cases $r_\pi \geqslant r_v$ and $r_\pi < r_v$.

17.4 FINAL REMARKS

The purpose of this chapter has been to demonstrate analytically that applying cost–benefit analysis methodology to decision problems concerning the conservation or destruction of natural forest or wilderness, can lead to premature destruction of the resource, if the inherent irreversibility and uncertainty of the system are ignored. While this observation is not new, it has to date apparently only been demonstrated using somewhat artificial two- or three-period models, which involve all-or-nothing harvest decisions. In contrast, here we have analysed two different models of greater realism which explicitly recognise uncertainty in either (a) future timber and amenity values along with the hazard of catastrophic destruction, or (b) size-dependent service flows from wilderness land and developed land, along with the possibility of unpredictable benefits resulting from scientific breakthroughs utilising the genetic resources of the wilderness.

For each model the optimal feedback policy, which maximises expected present value of benefits net of costs, has been determined. Also for each model a policy, based on cost–benefit ideas, which prescribes harvesting or development whenever the expected (marginal) present value of benefits exceeds that of costs, has been determined. This has been referred to as *the naive application of cost–benefit methodology*, and it has been shown that it always leads to premature harvesting or development. It is naive not only because it ignores the uncertainty inherent in the problem, but also because it ignores the temporal aspect, i.e. that if one does not harvest a unit of the forest now, or in any given period, there is always the option of doing so later on.

A more sophisticated procedure, which allows for temporal choice, but again ignores uncertainty is the *certainty-equivalence procedure*. Here uncertain future quantities are replaced by their expected values, and the resulting deterministic temporal decision problem is solved. However, it has been shown in the chapter that under certain circumstances (essentially if the benefits from conservation are growing faster in expectation than those of development) that this certainty-equivalence procedure reduces to the naive cost–benefit procedure. It seems likely that as societies develop, appreciation of environmental services will increase, and that amenity benefits from conservation will grow faster than the benefits from development. Thus application of certainty-equivalence will provide no improvement over naive cost–benefit methods.

To fully incorporate the temporal aspect (with the irreversibility of wilderness destruction) and the future uncertainty in the benefits from conservation and development, the problem must be treated as one of *stochastic optimal control*.

Such problems can be solved, at least in principle, by stochastic dynamic programming. This has been the methodology used here to find the optimal development or harvesting rules (in the single-stand model of Section 17.2 we have used a characterisation of the optimal stopping rule, which is derived from the dynamic programming equation). In all cases the optimal policy is very similar in form to that of the naive cost–benefit procedure, the difference being that the costs of development (or the benefits of conservation) are inflated by a factor that is greater than one. This factor is shown to increase with increasing uncertainty in future valuations. Thus the more uncertainty there is in the future benefits from conservation or development, the more conservative the optimal policy will be, and furthermore the more suboptimal the naive cost–benefit procedure will be. Since in reality there is considerable uncertainty as to the future values of the amenity services of both natural forest and developed land and also of timber products, it seems from the point of view of public policy that a high degree of conservation should be employed when making irreversible decisions concerning the destruction of natural forest and wilderness.

It should be stressed that these results do *not* depend on risk-aversion. In fact they have been derived assuming risk-neutrality with the conventional goal of maximising expected present value of benefits, net of costs, both expressed in monetary terms.

One argument often put forward by the proponents of logging of old-growth forest is that if the forest is not harvested it will eventually be lost through fire or some other natural catastrophe (the 'use it or lose it' argument). In the single-stand analysis of Section 17.2, this component is explicitly incorporated into the model. It is shown how the presence of risk does indeed make the optimal policy less conservative, acting like the addition of a premium to the discount rate. There are some cases (e.g. when the amenity services provided by the forest are growing in expectation faster than the discount rate) where the presence of the risk of destruction could change the optimal policy from one of never harvesting to one which involves harvesting whenever timber values are suitably high compared with amenity service values. However, by no means does the presence of risk always indicate that prompt harvesting should be carried out.

In the body of this chapter attention has been focused on one particular technical shortcoming of cost–benefit analysis, and it has been indicated how the correct application of stochastic optimal control methodology can overcome this difficulty. However, as indicated in the Introduction, there are many other difficulties involved with cost–benefit analysis. Problems of measuring amenity service benefits and costs of environmental and ecological damage have been assumed away in this chapter. This does not means that they are not important—they have simply been ignored in order to focus on the one issue under consideration. Indeed, as has been indicated in the Introduction, including the true costs of logging or clearing natural forest or wilderness would likely invalidate most of the destruction currently taking place, even if cost–benefit analysis were applied in a naive way. Using stochastic optimal control methods to handle future uncertainty would make for still more conservative policies.

Also, distributional and equity issues have been ignored. Again, this is not because they are not important, but simply because they are not amenable to analysis of this sort. To handle these issues other methods are required. At present there seems to be no consensus as to how this should be done (see Kneese and Schulze, 1985) for a critique of cost–benefit analysis and a discussion of alternative environmental ethics and principles). Economists and social and moral philosophers are currently wrestling with these issues. Perhaps one day new principles will be established. In a sense one could view the current uncertainty over environmental criteria as being like the uncertainty discussed in this chapter. It seems reasonable, because we don't know yet exactly what we want, to be yet more conservative with respect to irreversible decisions. It would indeed be a tragic irony if a consensus on environmental goals were reached after the environment had been so irreversibly degraded as to make these goals unachievable.

18

Quantitative Assessment of Growth in the Tropical Rainforests

Sagary Nokoe

International Centre of Insect Physiology and Ecology, Nairobi, Kenya

18.1 INTRODUCTION

18.1.1 The tropical forest

The tropical rainforest has been described as a complex biological community with an unparalleled diversity in plants and animals (Richards, 1966), and extensive variation in the phenological characteristics of the species at various times of the year. The tropical forests make up only 6% of the land surface, but are inhabited by an estimated 90% of all species on earth. The forest is usually multi-layered with species exhibiting either shade-tolerant or light-demanding properties. Such complexities, as they relate to species differences in growth requirements, structure and mixture, complicate attempts aimed at studying the growth of the forest—on either an individual-tree or a stand (group of trees) basis.

The tropical rainforest is represented by a continuous belt of vegetation around the earth between latituded 24 °N and 24 °S though these boundaries are not very rigid especially in parts of Indo-China and Australia. Longman and Jenik in their book *Tropical Forest and its Environment* (1074) have grouped the tropical rainforests into four regions:

(i) the African tropical rainforest covering the southern part of West Africa,

Statistics for the Environment. Edited by Vic Barnett and K. Feridun Turkman

the Congo basin, parts of East Africa (mainly Uganda) and a few islands;
(ii) the American tropical forest region (mainly in South and Central America);
(iii) the Indo-Malayan region covering India, Burma, the Malay Peninsula and some South-East Asian islands; and
(iv) The Australian region (including eastern Australia, New Zealand and the Pacific islands).

Studies on the growth or the dynamics of the tropical forests are rare. However, relevant studies in this direction have been undertaken in Nigeria and Papua New Guinea, and to some extent in the Amazonian region of the American tropical forests. Such notable studies were carried out and reported by MacGregor (1934), Jones (1956), Keay (1961), Mervart (1972), Enright (1978), Kio (1978) and Bada *et al.* (1989). In all these studies, a major issue that was raised and which could not be handled effectively, was that of inadequate knowledge of the ages of the trees as 'annual' growth rings were either unreliable or not discernible. The issue of age indeterminacy is largely due to lack of growth data and ineffective monitoring, coupled with the large diversity in species and economic classification of the species at the initiation of monitoring. Usually, only the species which at the initiation of monitoring are considered to be of economic value have been studied. Thus, when markets change and other previously 'uneconomic' species become 'economic', no growth information would have been obtained.

Despite the lack of adequate information on the ages of the trees, it has been shown conclusively that the tropical rainforest takes a considerable time to attain an appreciable size of commercial utility, usually as defined by the

Table 18.1 MacGregor's estimates of species' ages at selected girths

		Girth at breast height (cm)				
		46	76	91	152	290
Sterculia rhinopetala	Age[1]	62	87	97	140	—
	m.a.i.[2]	0.74	0.94	0.94	1.09	—
Triphochiton scleroxylon	Age	—	—	30	69	162
	m.a.i.	—	—	3.27	2.21	1.79
Terminalia superba	Age	—	31	45	84	153
	m.a.i.	—	2.46	2.03	1.8	1.89
Chlorophora excelsa	Age	29	83	109	175	236
	m.a.i.	1.58	0.92	0.84	0.87	1.23

[1]Age in years.
[2]m.a.i.-mean annual increment (in cm) computed as cumulative growth divided by cumulative time for the growth process).
Source: MacGregor (1934), *Oxf. For. Mem.* 18.

Table 18.2 Keay's estimates of ages for trees to attain 2.7 m g.b.h.

Species	Estimated age (years)
Triplochiton scleroxylon	42
Khaya ivorensis	88
Lovoa trichilioides	106
Guarea cedrata	168
Lophira alata	220

Source: Keay (1961), *Nig. For. Inf. Bull.* 11.

Table 18.3 Jones's estimates of growth

	Age to grow to 320 cm girth (years)	Age to grow from 305 to 335 cm (years)
Lovoa trichilioides	140	12.0
Kyaya ivorensis	115	12.0
Entandrophragma cylindricum	110	8.5
Lophira alata	225	27.0
Guarea cedrata	315	45.0

Source: Jones (1956), *J. Ecol.* 44: 83–117.

industrialist or as dictated by machinery available in the developing nations. Using mean annual increments derived from graphs from a selected forest, MacGregor (1934), for instance, derived estimated ages of several species of trees at different girths (Table 18.1). Growth rates are also known to vary within and between tree species. Keay (1956) estimated that *Triplochiton scleroxylon* would take 42 years, while *Lophira alata* would require 220 years to attain the girth of 2.7 m (Table 18.2). Among the mahogany species, Jones (1956) estimated ages between 100 and 320 years for the species in this family to reach a girth of 320 cm in the rainforest of southern Nigeria (Table 18.3). The extremely slow growth rates, albeit not necessarily reliably determined, and the direct effect on growth of the remaining stands of forest should strengthen the concerns of environmentalists and conservationists.

18.1.2 The need for growth studies

The slow growth of the forests is a major concern, but of paramount importance are the extent and magnitude of deforestation and fragmentation. The multinational JARRI project in the Amazon, the extensive infrastructural development in Nigeria (as a result of oil) and the highly profitable charcoal business in Ghana and elsewhere are major assaults on the tropical rainforest.

Bada (1984) has commented that the rate of conversion of the tropical forests in the late 1970s far exceeded 300% of the allowable or acceptable cut. A more glaring picture has been painted by Miller (1991), who indicated that losses of tropical forests amounted to some 17 million hectares per year. Harris (1984) estimated the rate of deforestation as 11.4 million hectares per year in the early 1980s, and suggested as much as 20 million hectares per year at the moment.

Fragmentation involves the breakdown of the forest environment into smaller segments, leading to isolation of biological communities and forcing some organisms to extinction as populations of some species cannot survive in fragmented environments. The opening-up of gaps in the forest system affects the structure of the forest and endangers the survival and growth of shade- tolerant species.

Tree growth should therefore not be seen in isolation of the spatial patterns and inter-tree distances. As foresters approach the new strategy of forest management, plans incorporating the needs of human communities and settlements living in harmony with the forest ecosystem, availability of reliable data on growth, and effective assessment of natural and human interference become more and more important. Some approaches to the study of growth in the tropical rainforests and an illustration using periodically collected data from one of the richest sections of the tropical rainforest in Africa are presented in this chapter.

18.2 QUANTITATIVE ASSESSMENTS

18.2.1 Spatial distribution

Dynamics of trees can be regarded both in the context of radial/vertical movements and of ingrowth/mortality. For effective monitoring of the latter, a spatial distribution map illustrating the exact or relative positions of the trees might ideally be required. Statistical analysis aimed at establishing the nature of the distribution—as to whether it is clustered, random, or uniform/regular—can be carried out on a family, species, or whole-stand basis (Southwood, 1978). Advanced methods of spatial analysis, including nearest-neighbour procedures can also be used, while multivariate statistical procedures may be used to establish various groupings or associations. The use of the Geographic Information System (GIS) technology should facilitate temporal study of the spatial patterns. As a basis for effective monitoring, photographic maps or satellite imageries of the area under study should be initial and necessary acquisitions.

18.2.2 Species diversity

This is an ecological concept and gives an indication of species preferences for certain sites or areas of the forest. Its utility in the study of the dynamics of tree species arises from the fact that changing environments could influence the presence or absence of certain species. GIS can facilitate this aspect, just as for spatial distribution. Statistical measures that may be determined include Simpson's index, Pielou's evenness measure and the Shannon–Weaver index (Southwood, 1978).

18.2.3 Frequency distribution

Species distribution, by sizes or numbers (frequencies), forms an essential component of growth assessment. Distribution is usually positively skewed at the initial stages of stand growth, and then becomes negatively skewed for the overmature forest, (Figure 18.1). Bailey and Dell (1973), Yang *et al.* (1978) and Okojie (1981) have effectively described species size and frequency distribution with the Weibull distribution function:

$$F(x) = 1 - \exp[-(x-a)/b]^c \qquad (18.1)$$

where a is the location parameter, b is a scale parameter and c describes the shape of the curve. Yang *et al.* (1978) have further shown that when c is less

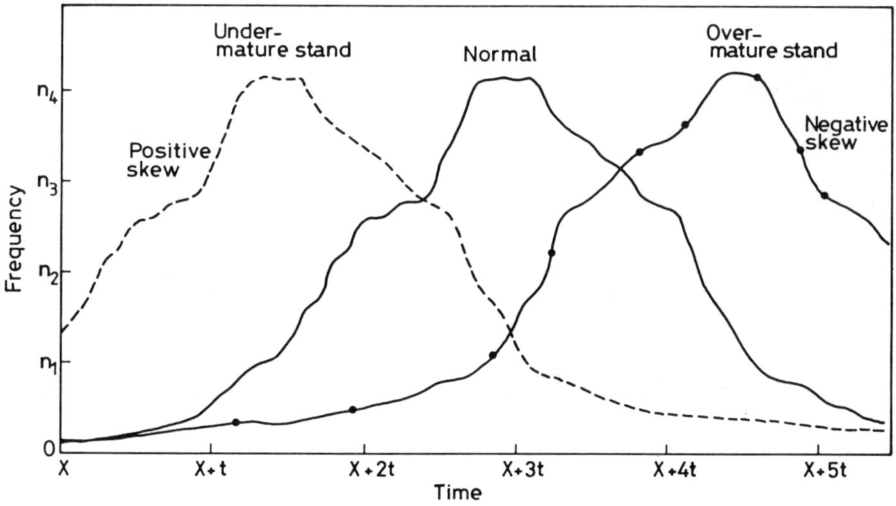

Figure 18.1 Simulated tree-stands dynamics from under- to overmature stage

than 3.6 and b is positive, the distribution shows positive skewness, whereas with $c > 3.6$ a negatively skewed distribution is likely to be obtained. Such a negative skewness is indicative of an overmature and unmanaged forest, or is an indication of the forest approaching the upper limit of its growth potential (Prodan, 1968).

Simple graphical approaches, such as the use of histograms (with frequency or proportions by size classes) and the maintenance of the same interval width over a period of time could give indications of the temporal changing patterns (Figure 18.2), and also enable one to compute the 'time of passage'. The time of passage refers to the time taken by a tree of a given dimension to outgrow the class interval and move on to the next higher class. A study setup to monitor growth or movement patterns will require that trees be marked and their spatial positions be known. Such studies are best carried out in protected areas usually in a Strict or Ecological Natural Reserve. The setting aside of permanent sample

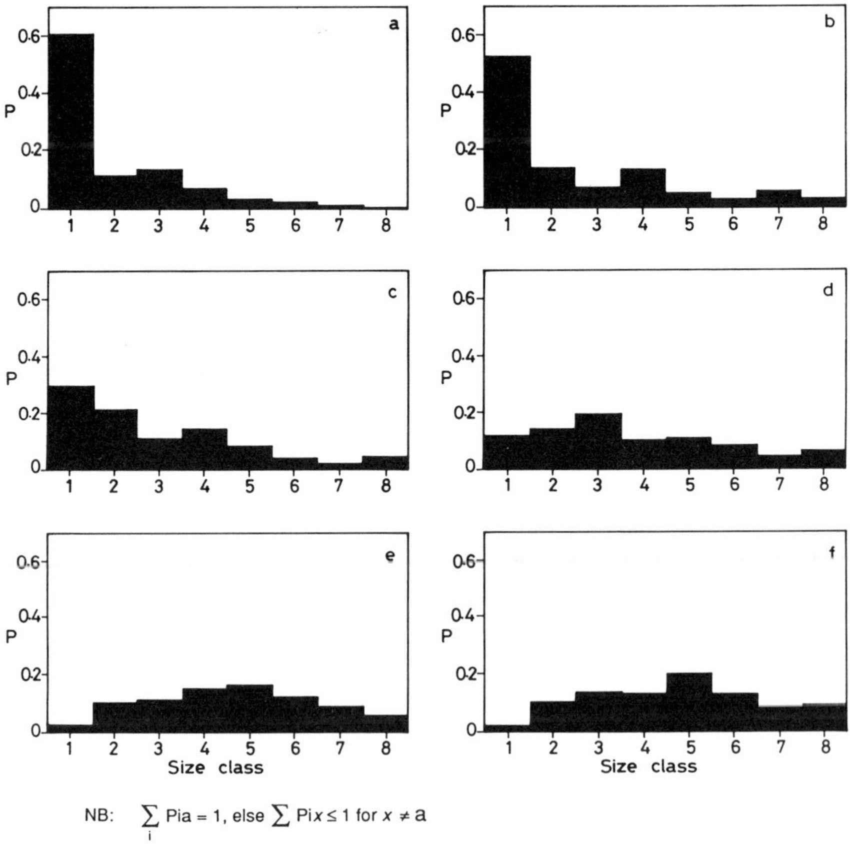

Figure 18.2 Stand distribution by size classes from initial time 'a' to time 'f'

plots or reasonable sizes may also satisfy the requirements for effective growth monitoring.

18.2.4 Growth and yield

Growth is indicative of the change in dimensions of the tree (height or diameter or girth at breast height, which is usually 1.3 m above ground. We denote by d.b.h. the diameter at breast height and by g.b.h. the girth at breast height). Yield is a measure of cumulative growth. Growth and yield assessment can be based on individual trees or be on a whole-stand basis. For tropical forests, whole-stand assessments have rarely been done effectively because of the large number of trees and the relatively few which are considered to be of economic importance. Nevertheless, it has now become increasingly clear that whole-stand assessment should not be ignored as individual growth simulations are very likely to be unrealistic.

Individual tree growth

Logistic or sigmoid curves are adequate to describe the cumulative growth of the individual trees. Simple variations in model form are possible, especially if collection of growth data is not commenced at the very early stages.

Common logistic or growth forms include:

multimolecular

$$Y = a\{1 \pm \exp(-ct)\}^d \tag{18.2}$$

modified multimolecular

$$Y = a\{1 \pm b \cdot \exp(-ct)\}^d \tag{18.3}$$

Verhulst

$$Y = a\{1 \pm \exp(-c(t-g))\}^{-1} \tag{18.4}$$

modified Gompertz

$$Y = a \cdot \exp(-\exp(-c(a-g))) \tag{18.5}$$

modified Weibull

$$Y = a\{1 - \exp(-c\,t^g)\} \tag{18.6}$$

In all cases, the parameter a is indicative of the maximum growth potential and g represents the time (age) at which the maximum current annual increment is maximum (Nokoe, 1978, 1980). The parameters b and c describe the shape of the growth curve. As all models are intrinsically nonlinear, the parameters may best be estimated using iterative nonlinear procedures.

Stand growth

Generation of cumulative stand-growth data involves a long period of observation or monitoring and has rarely been done in the tropics. Thus, proper chronological information may not be available. This problem may be resolved by establishing change in growth as a function of the initial size (mean stand size; example basal area):

$$\frac{\delta B}{\delta t} = f(B_0) \tag{18.7}$$

$$\int_0^t f(B_0)\,\mathrm{d}t \rightarrow \text{cumulative yield from 0 to } t \tag{18.8}$$

where B_0 is, for example, basal area at time 0 (time of first observation), and $\delta B/\delta t$ is the change in growth over time t.

Moser and Hall (1969) used a model of the form;

$$\frac{\delta B}{\delta t} = nB^m - kB, \tag{18.9}$$

which on integration and further manipulation was shown to yield

$$B = [n/k - C{\cdot}\exp(-1(1-m)kt)]^{1/(1-m)} \tag{18.10}$$

This model effectively described an uneven-aged forest stands and can be adapted easily to the tropical forest (Bada, 1984). It is emphasised that description or prediction of stand growth based on proper chronological data should be preferred to the procedure outlined above.

18.2.5 Stand dynamics

Simple formulation

Stand dynamics refers to the movement of trees from one size class to another. At any particular time, the number of trees in a given size class may be given as:

$$\mathrm{SP}_{t+x} = \mathrm{SP}_t + \Sigma_t I_j - \Sigma_t M_j \tag{18.11}$$

where SP_{t+x} is the stand density at time $t + x$, I_j is the ingrowth (movement from a lower size class), and M_j is the mortality (or movement to next higher class) between time $t + x$ and t.

Such straightforward description of stand development can be made specifically for key species. In this case, the preferred variable to observe would be the proportion of the species in the present and the previous class. An issue which should be of mathematical or statistical interest is the time of passage, and the variation in such measure among species and size classes.

The simple formulation ignores elements of change arising from different species mixtures and other uncontrollable factors (such as damage by wind). The stochastic approach, supplemented by empirical evidence, should provide some solutions.

Markov and Leslie-type matrices

If we consider several size classes and the probabilities associated with movement and mortality, Leslie-type matrices may be constructed. Such transition-matrix (Markov) models (Lewis, 1942; Leslie, 1945) have been applied to the study of animal populations over a given time period. Considering the difficulties associated with monitoring animal populations and movements, it thus seems even more feasible to apply the Markov models to tree movements across diameter size classes.

Essentially, the simple deterministic time model could be expressed as:

$$n_{t+1} = M \cdot n_t \qquad (18.12)$$

where n_t is a column vector representing the structure of the population at time t (divided into s stages) and M is a matrix of order S defining the contribution made by each stage (size class) to the other within a unit time period. If n_t is assumed constant, then for period k the structure of the population may be given by the vector

$$n_{t+k} = M^k n_t \qquad (18.13)$$

The resulting matrix has S eigenvalues and eigenvectors satisfying the relationship

$$\lambda n_t = M n_t \qquad (18.14)$$

with λ being an eigenvalue. The natural logarithm of the dominant eigenvalue is the intrinsic rate of natural increase of the population. Expressed as a percentage, the dominant eigenvalue gives an indication of the proportion of the stand that could be harvested without affecting the forest balance (Usher, 1966).

Jeffers (1978) had emphasised that for effective application of the Markov model, the system must be capable of being classified into distinct states, and that transition must take place in discrete intervals. These conditions are unlikely to be accomplished in the forest set-up, but considering that size classes can be measured and tree can be counted accurately, the use of the Markov model should be feasible in the study of forest population dynamics. This has been demonstrated effectively by Usher (1966) and Bada *et al.* (1989). In both studies, transition probabilities were not assumed to be fixed, contrary to a third condition suggested by Jeffers (1978). The utility of the Markov model and associated issues needs to be given serious consideration in the study of tropical forests.

18.3 ILLUSTRATION

18.3.1 The problem

The major issues of concern were the determination of the final population structure given initial (present) frequency (size) distribution of complete enumeration of trees in a known reserve, and the estimation of time taken for a tree to grow from one size class to another. For the forest manager, such information would be a useful guide in deciding which trees could be removed and the frequencies of such removal, while the socio-economic impact on local communities of such removals may also be assessed indirectly by the socio-economist.

18.3.2 The data and initial computations

Six sets of data for 1956, 1959, 1961, 1978 and 1983 were extracted from the growth records of the Usonigbe forest reserve, an extension of the Sapoba forest reserve ($521\,km^2$) in Nigeria. The reserve occupies the south-eastern corner of the Benin division of Nigeria's Edo state. The area is noted for its richness in the slow-growing, but highly valuable, mahogany species.

Four inner plots measuring approximately 120 m × 120 m provided the basic data. Basic data from this reserve were obtained on species counts and their respective d.b.h. sizes. A total of 89 species from 35 families were recorded. To reduce the total number to a reasonable size only trees of 'economic importance' at the time of investigation and with a minimum of 70 representatives in a given size class had been examined. Data were organised by size classes of 10 cm d.b.h. intervals with the lower or minimum d.b.h. class being about 15 cm (approximately 47 cm g.b.h). Information included the total number of trees, the number of trees remaining in each d.b.h. size, and the number that had moved to the next size class. Trees of sizes less than 15 cm were counted and grouped with the seedlings/saplings but not measured for d.b.h.

Close examination of the records on individual tree movements yielded the transition 'probabilities' presented in Table 18.4. Initial stand density, including seedlings and saplings (size class 0), totalled 2230 species.

Table 18.4 Proportion of trees remaining in or moving into next higher class during a given time size class

	0	1	2	3	4	5	6	7	8	9	10
0	0.500										
1	0.100	0.606									
2		0.276	0.559								
3			0.336	0.625							
4				0.275	0.696						
5					0.294	0.859					
6						0.141	0.800				
7							0.200	0.800			
8								0.200	0.825		
9									0.175	0.900	
10										0.100	0.900
11											0.100

18.3.3 Final species structure

To reduce complexities in the determination of the final (or stable) species distribution, we shall assume that within a given time period or iterations in the Markov process, movement is only by one step. Thus, one would not expect any tree in the first size class to move to the third size class during the one-step time period. The assumption is based on the relatively slow growth of the forest in the study area, which is dominated by the mahogany family of trees (see, for example, Tables 18.2 and 18.3).

We apply the Markov process outlined or a simple simulation involving continuous multiplications of probabilities for a given number of iterations. The iterations for the simple formulation is based on that which leads to a converging solution for the Markov process.

The final structure, after 20 iterations or generations, indicates the loss of 1264 species or 43% (Bada *et al.*, 1989) or 1119 species or 51% (present simulation), but considerable improvement in the number of exploitable trees (size class 6 to size class 9). Size classes 10 and 11 represent overmature trees (Table 18.5). The indications are that the forest, even when undisturbed, loses almost half of its species population by the time it attains a stable structure. The final structure also indicates a total of about 400 overmature trees by the time the forest attains a stable structure. Through proper management, these overmature trees could be removed without adversely affecting the growing stock.

Table 18.5 Initial and final population structure

SC	Survival rates	Initial population	Stable structure (a)	Stable structure (b)
0	0.60	1000	0	0
1	0.88	341	1	0
2	0.89	388	2	0
3	0.90	265	12	0
4	0.99	117	42	4
5	1.00	64	237	117
6	1.00	25	187	132
7	1.00	10	152	148
8	1.00	6	105	137
9	1.00	5	65	104
10	1.00	9	23	27
11		0	438	442
Total		2230	1264	1111
Percentage mortality			43%	51%

18.3.4 Determining time-period equivalent in years

Determining time-period equivalents in years is extremely difficult and requires careful monitoring on an individual-tree basis. However, let us consider Tables 18.2 and 18.3 which give estimates of ages at certain dimensions. If we note that Jones's estimates correspond to ages for our size class 10 (as they were for 320 cm g.b.h. or 101 cm d.b.h.), and Keay's estimates could be considerd as size class 8 (80–90 cm d.b.h.), then one might speculate ages for the two classes to fall between 120 and 200 years. We are, however, unable to estimate the ages per time period.

Alternatively, we could use Moser and Hall's equation (18.9) to determine approximate ages associated with the d.b.h. size attained by a specific tree. For the pooled data, the best fit for basal area equivalent to age t was obtained as

$$B_t = \{5.4144 - C \cdot e^{-(0.0047t)}\}^{4.9256} \tag{18.15}$$

where

$$C = 5.4144 - B_0^{0.2030}. \tag{18.16}$$

We notice that B_t, B_0 represent stand basal areas. These are equivalent to the product of the mean tree basal area and the number of trees at times t, o respectively.

To determine the time period or age of a tree of given size, we simply refer to the species frequencies structure, pick up the number of stems in the size classes of interest and divide B_t and B_o by these numbers accordingly. The corresponding B'_t, B'_o values may then be substituted in equations 18.15 and 18.16 for a solution to t.

Applying the above procedure indicated that a tree with initial dimension of 5 cm d.b.h, will take approximately 60 years to reach size 95 cm d.b.h., while that of initial size 10 cm will need 46 years to grow to size 70 cm (Table 18.6).

18.3.5 Assessment of unregulated interference and interventions

Indirect assessments

The rapid loss of species directly affects the quality of human life, especially of those in the rural communities. Losses may be attributed indirectly to disappearance of medicinal plants and natural fodder for livestock, poor quality (or unavailability) of drinking water, erosion of neighbouring agricultural land, etc. Thus, the assessment of interference through tree removal encompasses areas outside traditional forest management. Assessment of indirect effects could be done by the socio-economist and the new breed of forester. Statistical issues include monitoring (and prediction) of soil-fertility status, water-table levels,

Table 18.6 Projections from pooled estimates of basal area

AC1	AC2	AC3	AC4	AC5	AC6	AC7	AC8	AC9	AC10
5	15	25	35	45	55	65	75	85	95
Age	1.12*	12.41	22.26	29.96	36.80	43.14**	49.08	54.73	60.13
BC1	BC2	BC3	BC4	BC5	BC6	BC7	BC8	BC9	BC10
10	30	50	70	90	110	130	150	170	190
Age	15.33	31.52	45.94	57.45	67.86	77.65	86.99	95.99	104.75
CC1	CC2	CC3	CC4	CC5	CC6	CC7	CC8	CC9	CC10
15	20	45	60	75	90	105	120	135	150
Age	15.33	28.13	39.84	49.08	57.45	65.33^{+}	72.82	80.02	86.99

Notes

*Time taken in years for an average tree to move from size class with mean 5 cm d.b.h. to that of 15 cm d.b.h.

**Time taken in years for an average tree to move from size class with mean 5 cm d.b.h. to that of 65 cm d.b.h.

$^{+}$ Time taken in years for an average tree to move from size class with mean 15 cm d.b.h. to 105 cm d.b.h.

fuel-wood supply and availability, and efficient measures for assessing population changes and patterns, and nutritional status of the human and animal communities.

Direct assessment of species structural changes

The selection system of forest management allows for effective utilisation of resources without creating adverse environmental conditions. Thus, one might be interested in knowing the extent to which forest growth and structure can be maintained, while allowing for removal of some trees. Alternatively, the effect of illegal removals can also be simulated.

A simple computer simulation (Appendix 18.1) using initial conditions and transition probabilities given in Table 18.4 is used to illustrate changes in the frequency structure given removals from and entries to the system. From the results of the initial simulation, the forest allowed to grow to its stable state would result in a stand with over 400 trees (30% of initial stand) in the

Table 18.7 Effect on species structure assuming effective removal of overmature trees during simulation period

Interference	0	1	2	3	4	5	6	7	8	9	10	Over-mature
1	0	0	0	0	4	117	132	148	137	104	27	4
2	0	0	0	0	4	117	259	165	214	140	33	4
3	0	0	0	0	4	61	46	37	34	40	13	4
4	0	0	0	0	9	114	113	112	98	80	23	3
5	0	0	0	0	7	131	142	156	141	107	29	4
6	0	0	0	0	7	131	136	138	116	83	21	2
7	0	0	0	0	4	117	132	141	113	63	8	0
8	0	0	0	0	0	3	4	17	50	140	124	130
9	0	0	0	0	4	117	132	148	134	99	26	4
10	0	0	0	0	0	3	4	17	50	152	144	182

Interference codes explained:
1. No interference by man
2. 1 interference during 10th time period, on stage 5 (P succ. 0.05)
3. 1 intervention at time 10 (a) time 0; tree number 1001 (b) stage 5; P succ. 0.05
4. 2 interventions at (a) time 5, stage 0; add 500 trees (b) time 10, stage 5; remove 100 trees time 10, stage 6; remove 59 trees
5. 1 intervention at time 3, stage 0; add 500 trees
6. 2 intervention (a) at time 3, stage 0; add 500 trees (b) at time 10, stage 6; remove 71 trees at time 10, stage 9; remove 10 trees
7. All trees in SC7–9 removed on or before time 10 (20 time simulation output)
8. All trees in SC7–9 removed on or before time 10 (40 time simulaion output)
9. All trees in SC7–9 removed at time 0 (20 time simulation output)
10. All trees in SC7–9 removed at time 0 (40 time simulation output)

overmature class. Proper management, for example, through the selection system could exploit this overmature stock effectively. We shall thus assume that one way or another, overmature stock can be identified and removed during the period loading up to stable structure.

Table 18.7 gives results for 10 different interference or intervention strategies, ignoring the already identified overmature stock. Such simulation output could be useful for management purposes provided the initial data sources (for example, the transition probabilities) are considered reliable.

18.4 CONCLUSIONS

Tropical rainforest is characterised as a complex biological community with an unparalleled diversity and interdependence of its component parts which need to be preserved as changes are irreversible. The forests have suffered greatly due to uncontrolled activities of man, arising mainly out of greed or ignorance of the relatively slow growth rates. The consequences on the environment are obvious: there is a great need for a thorough knowledge of the growth and species/size structure of these forests. More importantly, models for studying the growth and population changes or structure of the tropical rainforests need to be examined, in addition to the use of computers in the simulation of the effects of uncontrolled destruction.

From the illustrated example, which is based on empirical evidence, it could be emphasised that without additional recruitment at the lower size classes, there is a strong likelihood of extinction unless alternative adequate provisions are made to sustain growth and development of the population being managed. The problem of extinction becomes quite crucial when one considers that only about 4% of the tropical forests in Africa, for instance, are protected by legislation, and most are not actually physically protected. Moreover, it needs to be emphasised that damage to the tropical forest is irreversible.

The results from the Markov procedural approach and arbitrarily simulated interventions or interferences indicate substantial mortality in the tropical forest between the initial and the stable population stage. Furthermore, uncontrolled removal of trees could create adverse conditions such as reduction in the potential of shade-tolerant species to grow appreciably and/or the direct effect on soil hardening, increased erodibility and soil temperature which could affect effective survival of seedlings. Hence, efforts should be pursued aimed at supporting the species structure, form, function and interactions in the ecosystems instead of the present commercial (controlled or uncontrolled) and subsistence exploitation (which yield gains only in the short term). One such solution is the setting aside, and the effective enforcement through legislation, of Strict Natural Reserves (SNRs). This will guarantee continuous monitoring and availability of relevant data for future studies. On the other hand, the creation of extractive forests (as set up in Amazonia) would offer long-term use to groups of people who choose to live in harmony with the forest ecosystem, but this may not necessarily guarantee an adequate scientific database for future studies unless data-collection strategies are incorporated.

Although procedures for assessing the effect on human communities as a result of forest removals has not been discussed in this study, they do represent a major area for consideration by environmental statisticians and forest mensurationists.

ACKNOWLEDGMENTS

Most of the illustrated work reported in this chapter was accomplished by a team made up of the author, Dr S. O. Bada of the University of Ibadan and Prof. J. A. Okojie of the University of Agriculture, Abeokuta in Nigeria. The simulation program based on the initial transition probabilities was written with the assistance of Mr Henry H. Meena (programmer and GIS specialist) of the Biomathematics Research Unit, ICIPE, Kenya.

The organisers of the SPRUCE conference in Portugal and the Commission of the European Communities are acknowledged for ensuring the participation of the author and the presentation of this manuscript.

APPENDIX 18.1 BASIC PROGRAM FOR STABLE STRUCTURE SIMULATION

```
10 REM Programmer..........H. Meena
20 REM Date..........Mar 1992
30 REM Program..........Tree Life Stage Simulation
40 REM Institute..........BMRU, ICIPE. Nairobi Kenya.
50 REM SET INITIAL CONDITIONS: SCREEN 0: COLOR 1,2,3
60 STG0PSUCC = .1: STG0PFAIL = .5: STG0INIT = 1000
70 STG1PSUCC = .276: STG1PFAIL = .606: STG1INIT = 341
80 STG2PSUCC = .336: STG2PFAIL = .559: STG2INIT = 388
90 STG3PSUCC = .275: STG3PFAIL = .625: STG3INIT = 265
100 STG4PSUCC = .294: STG4PFAIL = .696: STG4INIT = 117
110 STG5PSUCC = .141: STG5PFAIL = .859: STG5INIT = 64
120 STG6PSUCC = .2: STG6PFAIL = .8: STG6INIT = 25
130 STG7PSUCC = .2: STG7PFAIL = .8: STG7INIT = 10
140 STG8PSUCC = .175: STG8PFAIL = .825: STG8INIT = 6
150 STG9PSUCC = .1: STG9PFAIL = .9: STG9INIT = 5
160 STG10PSUCC = .1: STG10PFAIL = .9: STG10INIT = 9: STG11INIT = 438
170 SIMLEN = 20: REM SIMULATION LENGTH
180 INTERV = 0: INTERVD = 0: DIM INTV(10): CHANGE = 0: FOUT = 2: II = 0
190 OPEN "TREE.PRN" FOR OUTPUT AS #1
200 REM CALL SUBROUTINE TO PRINT CURRENT VALUES
210 REM AND OFFER TO CHANGE THEM
```

```
220 GOSUB 990
230 CLS
240 INPUT "ENTER THE LENGTH OF SIMULATION IN TIME PERIOD (0) TO EXIT",
    SIMLEN
250 IF (SIMLEN < 0) or (SIMLEN > 366) GOTO 240
260 IF (SIMLEN = 0) THEN GOTO 980
270 LOCATE 5,2
280 INPUT "ENTER OUTPUT STEP IN TIME PERIOD (0)) = NO OUTPUT ", FOUT
290 IF (FOUT < 0) OR (FOUT > SIMLEN) THEN GOTO 270
300 IF (FOUT = 0) THEN FOUT = 400: REM ENSURE NOTHING IS PRINTED
310 LOCATE 10,5
320 PRINT "TO INTERVENE, ENTER THE INTERVENTION"
330 LOCATE 11,5
340 PRINT "TIME PERIOD FROM WHICH YOU MAY CHANGE CURRENT"
350 LOCATE 12,5
360 PRINT "VALUES AS WILL BE DISPLAYED"
370 LOCATE 14,1
380 INPUT "ENTER TOTAL NUMBER OF INTERVENTIONS (0) TO SKIP ", INTERV
390 IF (INTERV > 0) AND (NTERV <= SIMLEN)) THEN GOSUB 1560
400 ELSE INTERV = 0
410 REM START THE SIMULATION
420 REM BY SETTING PERIOD = 1
430 II = 1
440 WHILE (II <= SIMLEN)
450 REM CHECK FOR INTERVENTION
460 FOR J = 1 TO INTERV
470 IF (II = INTV(J)) THEN GOSUB 990
480 NEXT J
490 TEMPO = FIX(STG0INIT*STG0PSUCC)
500 STG0INIT = FIX(STG0INIT*STG0PFAIL)
510 REM
520 TEMP1 = FIX(STG1INIT*STG1PSUCC)
530 STG1INIT = FIX(STG1INIT*STG1PFAIL)
540 REM
550 TEMP2 = FIX(STG2INIT*STG2PSUCC)
560 STG2INIT = FIX(STG2INIT*STG2PFAIL)
570 REM
580 TEMP3 = FIX(STG3INIT*STG3PSUCC)
590 STG3INIT = FIX(STG3INIT*STG3PFAIL)
600 REM
610 TEMP4 = FIX(STG4INIT*STG4PSUCC)
620 STG4INIT = FIX(STG4INIT*STG4PFAIL)
630 REM
640 TEMP5 = FIX(STG5INIT*STG5PSUCC)
650 STG5INIT = FIX(STG5INIT*STG5PFAIL)
660 REM
670 TEMP6 = FIX(STG6INIT*STG6PSUCC)
680 STG6INIT = FIX(STG6INIT*STG6PFAIL)
690 REM
```

```
700 TEMP7 = FIX(STG7INIT*STG7PSUCC)
710 STG7INIT = FIX(STG7INIT*STG7PFAIL)
720 REM
730 TEMP8 = FIX(STG8INIT*STG8PSUCC)
740 STG8INIT = FIX(STG8INIT*STG8PFAIL)
750 REM
760 TEMP9 = FIX(STG9INIT*STG9PSUCC)
770 STG9INIT = FIX(STG9INIT*STG9PFAIL)
780 REM
790 TEMP10 = FIX(STG10INIT*STG10PSUCC)
800 STG10INIT = FIX(STG10INIT*STG10PFAIL)
810 STG11INIT = STG11INIT + TEMP10
820 STG1INIT = STG1INIT + TEMP0
830 STG2INIT = STG2INIT + TEMP1
840 STG3INIT = STG3INIT + TEMP2
850 STG4INIT = STG4INIT + TEMP3
860 STG5INIT = STG5INIT + TEMP4
870 STG6INIT = STG6INIT + TEMP5
880 STG7INIT = STG7INIT + TEMP6
890 STG8INIT = STG8INIT + TEMP7
900 STG9INIT = STG9INIT + TEMP8
910 STG10INIT = STG10INIT + TEMP9
920 REM CHECK ON FILE OUTPUT OPTION
930 IF (II MOD FOUT = 0) THEN GOSUB 1640
940 PRINT "END OF PERIOD", II
950 II = II + 1
960 WEND
970 CLOSE #1: PRINT "END OF SIMULATION"
980 END: SYSTEM
990 REM SUBROUTINE TO DISPLAY AND CHANGE CURRENT VALUES
1000 CLS
1010 LOCATE 4,5
1020 PRINT "DISPLAY OF CURRENT STAGE VALUES ON PERIOD" II: PRINT
1030 PRINT "PROBABILITY TO NEXT STAGE PROBABILITY TO REMAIN CURRENT
     TREE NUMBER"
1040 PRINT "STAGE 0" STG0PSUCC SPC(12) "STAGE 0" STG0PFAIL SPC(12) "STAGE
     0" STG0INIT
1050 PRINT "STAGE 1" STG1PSUCC SPC(10) "STAGE 1" STG1PFAIL SPC(10) "STAGE
     1" STG1INIT
1060 PRINT "STAGE 2" STG2PSUCC SPC(10) "STAGE 2" STG2PFAIL SPC(10) "STAGE
     2" STG2INIT
1070 PRINT "STAGE 3" STG3PSUCC SPC(10) "STAGE 3" STG3PFAIL SPC(10) "STAGE
     3" STG3INIT
1080 PRINT "STAGE 4" STG4PSUCC SPC(10) "STAGE 4" STG4PFAIL SPC(10) "STAGE
     4" STG4INIT
1090 PRINT "STAGE 5" STG5PSUCC SPC(10) "STAGE 5" STG5PFAIL SPC(10) "STAGE
     5" STG5INIT
1100 PRINT "STAGE 6" STG6PSUCC SPC(12) "STAGE 6" STG6PFAIL SPC(12) "STAGE
     6" STG6INIT
```

```
1110 PRINT "STAGE 7" STG7PSUCC SPC(12) "STAGE 7" STG7PFAIL SPC(12) "STAGE
     7" STG7INIT
1120 PRINT "STAGE 8" STG8PSUCC SPC(10) "STAGE 8" STG8PFAIL SPC(10) "STAGE
     8" STG8INIT
1130 PRINT "STAGE 9" STG9PSUCC SPC(12) "STAGE 9" STG9PFAIL SPC(12) "STAGE
     9" STG9INIT
1140 PRINT "STAGE 10" STG10PSUCC SPC(12) "STAGE 10" STG10PFAIL SPC(12)
     "STAGE 10" STG10INIT
1150 PRINT"
     STAGE 11" STG11INIT
1160 LOCATE 19,5
1170 INPUT "ENTER STAGE NUMBER TO CHANGE, (-2) TO REDISPLAY, (-1) TO
     EXIT", CHANGE
1180 IF (CHANGE< -2) OR (CHANGE>11) THEN GOTO 1160
1190 IF (CHANGE= -1) THEN GOTO 1550
1200 IF (CHANGE= -2) THEN GOTO 1000
1210 IF (CHANGE=0) THEN INPUT "STAGE 0 SUCCESS TO NEXT STAGE",
     STG0PSUCC
1220 IF (CHANGE=0) THEN INPUT "STAGE 0 PROBABILITY OF REMAINING IN
     STAGE", STG0PFAIL
1230 IF (CHANGE=0) THEN INPUT "STAGE 0 CURRENT NUMBER OF TREES",
     STG0INIT:GOTO 1160
1240 IF (CHANGE=1) THEN INPUT "STAGE 1 SUCCESS TO NEXT STAGE",
     STG1PSUCC
1250 IF (CHANGE=1) THEN INPUT "STAGE 1 PROBABILITY OF REMAINING IN
     STAGE", STG1PFAIL
1260 IF (CHANGE=1) THEN INPUT "STAGE 1 CURRENT NUMBER OF TREES",
     STG1INIT:GOTO 1160
1270 IF (CHANGE=2) THEN INPUT "STAGE 2 SUCCESS TO NEXT STAGE",
     STG2PSUCC
1280 IF (CHANGE=2) THEN INPUT "STAGE 2 PROBABILITY OF REMAINING IN
     STAGE", STG2PFAIL
1290 IF (CHANGE=2) THEN INPUT "STAGE 2 CURRENT NUMBER OF TREES",
     STG2INIT:GOTO 1160
1300 IF (CHANGE=3) THEN INPUT "STAGE 3 SUCCESS TO NEXT STAGE",
     STG3PSUCC
1310 IF (CHANGE=3) THEN INPUT "STAGE 3 PROBABILITY OF REMAINING IN
     STAGE", STG3PFAIL
1320 IF (CHANGE=3) THEN INPUT "STAGE 3 CURRENT NUMBER OF TREES",
     STG3INIT:GOTO 1160
1330 IF (CHANGE=4) THEN INPUT "STAGE 4 SUCCESS TO NEXT STAGE",
     STG4PSUCC
1340 IF (CHANGE=4) THEN INPUT "STAGE 4 PROBABILITY OF REMAINING IN
     STAGE", STG4PFAIL
1350 IF (CHANGE=4) THEN INPUT "STAGE 4 CURRENT NUMBER OF TREES",
     STG4INIT:GOTO 1160
1360 IF (CHANGE=5) THEN INPUT "STAGE 5 SUCCESS TO NEXT STAGE",
     STG5PSUCC
```

```
1370 IF (CHANGE = 5) THEN INPUT "STAGE 5 PROBABILITY OF REMAINING IN
     STAGE", STG5PFAIL
1380 IF (CHANGE = 5) THEN INPUT "STAGE 5 CURRENT NUMBER OF TREES",
     STG5INIT:GOTO 1160
1390 IF (CHANGE = 6) THEN INPUT "STAGE 6 SUCCESS TO NEXT STAGE",
     STG6PSUCC
1400 IF (CHANGE = 6) THEN INPUT "STAGE 6 PROBABILITY OF REMAINING IN
     STAGE", STG6PFAIL
1410 IF (CHANGE = 6) THEN INPUT "STAGE 6 CURRENT NUMBER OF TREES",
     STG6INIT:GOTO 1160
1420 IF (CHANGE = 7) THEN INPUT "STAGE 7 SUCCESS TO NEXT STAGE",
     STG7PSUCC
1430 IF (CHANGE = 7) THEN INPUT "STAGE 7 PROBABILITY OF REMAINING IN
     STAGE", STG7PFAIL
1440 IF (CHANGE = 7) THEN INPUT "STAGE 7 CURRENT NUMBER OF TREES",
     STG7INIT:GOTO 1160
1450 IF (CHANGE = 8) THEN INPUT "STAGE 8 SUCCESS TO NEXT STAGE",
     STG8PSUCC
1460 IF (CHANGE = 8) THEN INPUT "STAGE 8 PROBABILITY OF REMAINING IN
     STAGE", STG8PFAIL
1470 IF (CHANGE = 8) THEN INPUT "STAGE 8 CURRENT NUMBER OF TREES",
     STG8INIT:GOTO 1160
1480 IF (CHANGE = 9) THEN INPUT "STAGE 9 SUCCESS TO NEXT STAGE",
     STG9PSUCC
1490 IF (CHANGE = 9) THEN INPUT "STAGE 9 PROBABILITY OF REMAINING IN
     STAGE", STG9PFAIL
1500 IF (CHANGE = 9) THEN INPUT "STAGE 9 CURRENT NUMBER OF TREES",
     STG9INIT:GOTO 1160
1510 IF (CHANGE = 10) THEN INPUT "STAGE 10 SUCCESS TO NEXT STAGE",
     STG10PSUCC
1520 IF (CHANGE = 10) THEN INPUT "STAGE 10 PROBABILITY OF REMAINING IN
     STAGE", STG10PFAIL
1530 IF (CHANGE = 10) THEN INPUT "STAGE 10 CURRENT NUMBER OF TREES",
     STG10INIT:GOTO 1160
1540 IF (CHANGE = 11) THEN INPUT "STAGE 11 CURRENT NUMBER OF TREES",
     STG11INIT:GOTO 1160
1550 CLS: RETURN
1560 REM SUBROUTINE TO ENTER INTERVENTION TIME PERIOD
1570 CLS: PRINT "TOTAL NUMBER OF INTERVENTION IS", INTERV
1580 PRINT "ENTER TIME PERIOD TO EFFECT INTERVENTIONS"
1590 FOR KI = 1 TO INTERV
1600 PRINT "ENTER TIME PERIOD", KI, "OF INTERVENTION"
1610 INPUT INTV (KI)
1620 NEXT KI
1630 RETURN
1640 REM SUBROUTINE TO PRINT FILE OUTPUT
1650 PRINT #1, STG0INIT; STG1INIT; STG2INIT; STG3INIT; STG4INIT; STG5INIT;
     STG6INIT; STG7INIT; STG8INIT; STG9INIT; STG10INIT; STG11INIT;
1660 RETURN
```

References

The number in parentheses at the end of each reference indicates the first page in the text on which that reference is mentioned.

Alpuim, M. T. (1989) An extremal markovian sequence. *J. Appl. Prob.*, **26**, 219–232. (227)

Anderson, T. W. (1971) *The Statistical Analysis of Time Series*, Wiley, New York. (195)

Anscombe, F. J. (1953) Contribution to the discussion of H. Hotelling's paper. *J. R. Statist. Soc. B*, **15**, 229–230. (319)

ApSimon, H. M. and Davison, A. C. (1986) A statistical model for deriving probability distributions of contamination for accidental releases. *Atmospheric Environment*, **20**, 1249–1259. (124)

ApSimon, H. M., Goddart, A. J. H. and Wrigley, J. (1985) Long-range transport of radio-isotopes. I. The MESOS model. *Atmospheric Environment*, **19**, 99–111. (125)

Arrow, K. J. and Fisher A. C. (1974) Environmental preservation, uncertainty and irreversibility. *Quarterly J. Econ.*, **88**, 312–319. (352)

Ashmore, M. R., Bell, J. N. B. and Brown, I. J. (1990) Air pollution and forest ecosystems in the European Community. An assessment of the current scientific evidence. *Commission of the European Community*, Brussels. (345)

Attenborough, D. (1987) *The First Eden: The Mediterranean World and Man*, Collins, London. (348)

Bada, S. O. (1984) *Growth patterns in an untreated tropical rainforest ecosystem*. Ph.D thesis, University of Ibadan, Nigeria. (382)

Bada, S. O., Nokoe, S. and Okojie, J. A. (1989) Population dynamics selected tree species in a strict Natural Reserve in the moist tropical forest. *Discovery and Innovation*, **1**(3), 71–76. (380)

Baddeley, A. J. and Silverman, B. W. (1984) A cautionary example on the use of second-order methods for analyzing point patterns. *Biometrics*, **40**, 1089–93. (109)

Bailey, R. L. and Dell, T. R. (1973). Quantifying diameter distributions with the Weibull function. *For. Sci.*, **19**, 97–104. (383)

Barnard, G. A. (1963a) The logic of least squares. *J. R. Statist. Soc., Series B*, **25**, 124–127. (101)

Barnard G. A. (1963b) Contribution to the Discussion of Prof. Bartlett's paper. *J. R. Statist. Soc. B*, **25**, 294. (101)

Barnes, F. B. (1954) Storage required for a city water supply. *J. Inst. Eng. Austr.*, **26**(9), 198–203. (208)

Barnett, T. P. (1984) The estimation of 'global' sea level change: a problem of uniqueness. *J. Geophys. Res.*, **89**, 7980–7988. (258)

Bartlett, M. S. (1964) Spectral analysis of two-dimensional point processes. *Biometrika* **51**, 299–311. (97)

Barton, D. E. and David, F. N. (1966) The random intersection of two graphs. In David, F. N. (ed.), *Research Papers in Statistics*, Wiley, New York. (102)

Bayarri, M. J. and DeGroot, M. H. (1986) *Information in selection models*, Technical Report Number 365, Department of Statistics, Carnegie-Mellon University. (55)

Beck, M. and Young, P. C. (1975) A dynamic model for DO-BOD relationships in a non-tidal stream. *Water Research*, **9**, 769–776. (7)

Beck, M. B. and Young, P. C. (1976) Systematic identification of DO-BOD model structure. *J. Env. Eng. Div., American Soc. Civil Eng.*, **102** (EE5), 902–927. (7)

Becker, R. A., Chambers, J. M. and Wilks, A. R. (1988). *The New S Language*, Wadsworth, Pacific Grove. (107)

Beer, T. and Young, P. C. (1983) Longitudinal dispersion in natural streams. *J. Env. Eng. Div., American Soc. Civil Eng.*, **109**, 1049–1067. (7)

Bencala, K. E. and Walters, R. A. (1983) Simulation of solute transport in a mountain pool-and-riffle stream: a transient storage model. *Water Resources Research*, **19**, 718–724. (7)

Beran, J. (1989) A test of location for data with slowly decaying serial correlations. *Biometrika*, **76**, 261–269. (145)

Berman, M. and Diggle, P. J. (1989) Estimating weighted integrals of the second-order intensity of a spatial point process. *J. R. Statist. Soc. B*, **51**, 81–92. (99)

Besag, J. and Newell, J. (1991) The detection of clusters in rare diseases. *J. R. Statist. Soc. A*, **154**, 143–55. (109)

Beven, K. J. and Young, P. C. (1988) An aggregated mixing zone model of solute transport through porous media. *J. Contaminant Hydrology*, **3**, 129–143. (24)

Bhopal, R. S., Diggle, P. J. and Rowlingson, B. S. (1992) Pinpointing clusters of apparently sporadic legionnaires disease. *Br. Med. J.*, **304**, 1022–1027. (104)

Bithell, J. F. (1990) An application of density estimation to geographical epidemiology. *Statist. Med.*, **9**, 691–701. (110)

Bithell, J. F. and Stone, R. A. (1989) On statistical methods for analysing the geographical distribution of cancer cases near nuclear installations. *J. Epidem. Commty Hlth*, **43**, 79–85. (90)

Black, D. (1984) Investigation of the possible increased incidence of cancer in West Cumbria: report of an independent advisory group. HMSO, London. (89)

Blackwell, D. (1951) Comparison of experiments. In *Proceedings of the Second Berkeley Symposium on Mathematical Statistics and Probability*, pp. 93–102. (52)

Blackwell, D. (1953) Equivalent comparisons of experiments, *Ann. Math. Statist.*, **24**, 265–272. (52)

Bloomfield, P. (1992) Trends in global temperature. *Climatic Change*, **21**, 1–16. (159)

Bloomfield, P. and Nychka, D. (1992) Climate spectra and detecting climate change. *Climatic Change*, to appear. (160)

Bonney, G. E. (1987) Logistic regression for dependent binary observations. *Biometrics*, **43**, 951–973. (339)

Boswell, M. T., Burnham, K. P. and Patil, G. P. (1988) Role and use of composite sampling and capture-recapture sampling in ecological studies. In *Handbook in Statistics*, Vol. 6 (Eds. Krishnaiah, P. R. and Rao, C. R.), pp. 469–488, North-Holland, Amsterdam. (313)

Bot, G. P. A. (1983) *Greenhouse climate: from physical processes to a dynamical model*. Ph.D. Dissertation, Agricultural University, Wageningen.

Box, G. E. P. (1975) Statistics and the environment. *Int. Stat. Rev.*, **4**, 279–287. (336)

Box, G. E. P. and Jenkins, G. M. (1970) *Time Series Analysis, Forecasting and Control*. Holden-Day, San Francisco. (4)

Breslow, N. E. (1984). Extra-Poisson variation in log-linear models. *App. Statist.*, **33**, 38–44. (317)

Breslow, N. E. (1990) Tests of hypotheses in overdispersed Poisson regression and other quasi-likelihood models. *J. Amer Statist. Assoc.*, **85**, 565–571. (316)
Brillinger, D. R., Guckenheimer, J., Guttorp, P. and Oster, G. (1980) Empirical modelling of population time series data: The case of age and density dependent vital rates. *Lectures on Mathematics in the Life Sciences*, Vol. 13. American Mathematical Society, Providence R. I. (275)
Brock, W. A., Rothschild, M. and Stiglitz, J. (1988) Stochastic capital theory. In Feiwel, G. (ed.) *Essays in Honour of Joan Robinson*. Cambridge University Press, Cambridge. (352)
Brownie, C. and Robson, D. S. (1976) Models allowing for age-dependent survival rates for band-return data. *Biometrics*, **36**, 419–435. (314)
Brownie, C., Anderson, D. R., Burnham, K. P. and Robson, D. R. (1978, 1985) *Statistical inference from band-recovery data—a handbook.* U.S. Fish and Wildlife Service Resource Publication 1st ed. No. 131, 2nd ed. No. 156. (313)
Burnham, K. P., Anderson, D. R. and Laake, J. L. (1980) Estimation of density for line transect sampling of biological populations. Wildlife Monograph 72. Supplement to *J. Wildlife Management*, **44**. (49)
Burnham K. P., Anderson, D. R., White, G. C., Brownie, C. and Pollock, K. H. (1987) *Design and analysis methods for fish survival experiments based on release-recapture.* American Fisheries Society, Bethesda, Maryland. (313)
California Department of Fish and Game (1987) Status of San Joaquin Drainage Chinook Salmon: Habitat conditions and natural production factors. Region 4, Fresno. Exhibit 15, San Francisco Bay/Sacramento San Joaquin Delta Estuary Water quality/Water Rights Hearings, Phase 1. (272)
Canto e Castro, L. (1992) *Sobre a Teoria Assintótica de Extremos.* Ph.D. Thesis. D.E.I.O., Faculdade de Ciências de Lisboa. (237)
Caughley, G. (1972) Improving the estimates from inaccurate censuses. *J. Wildlife Management*, **36**, 135–140. (47)
Caughley, G. (1974) Bias in aerial survey. *J. Wildlife Management*, **38**, 921–933. (47)
Caughley, G. (1976) Experiments in aerial survey. *J. Wildlife Management*, **40**, 290–300. (47)
Chalabi, Z. S. and Bailey, B. J. (1989) *Simulation of the energy balance in a greenhouse.* Div. Note, DN1516, AFRC Institute of Engineering Research, Silsoe, Bedford. (31)
Chipman, J. S. (1979) Efficiency of least-squares estimation of linear trend when residuals are autocorrelated. *Econometrica*, **47**, 115–128. (195)
Chotai, A., Young, P. C. and Tych, W. (1990) A non-minimum state space approach to true digital control based on the backward shift and delta operator models, appears in M. H. Hamza (ed.), *Proc. IASTED Conference 1990*, Acta Press, Calgary, 1–4. (42)
Clark, C. W. (1972) The dynamics of commercially exploited natural animal populations. *Math. Biosci.*, **13**, 149–164. (307)
Clark, J. E. and Bernard, D. R. (1987) A compound multivariate binomial-hypergeometric distribution describing coded microware tag recovery from commercial salmon catches in Southern Alaska. Alaska Dept. of Fish and Game, *Informational Leaflet* No. 261. (325)
Clarke, H. R. and Reed, W. J. (1989) The tree-cutting problem in a stochastic environment. *J. Econ. Dyn. and Control*, **13**, 569–595. (354)
Clark, H. R. and Reed, W. J. (1990). Land development and wilderness conservation policies under uncertainty: A synthesis. *Nat. Res. Model*, **4**, 11–37. (367)
Clobert, J., Lebreton, J. D. and Allaine, D. (1987) A general approach to survival rate estimation by recaptures or resightings of marked birds. *Ardea*, **75**, 133–142. (313)
Cohen, A. C. (1949) On estimating the mean and standard deviation of truncated normal distributions. *J. Amer. Statist. Assoc.*, **44**, 518–525. (54)
Cohen, A. C. (1950) Estimating the mean and variance of normal populations from singly truncated and doubly truncated distributions. *Ann. Math. Statist.*, **21**, 557–569. (54)

Coles, S. G. and Tawn, J. A. (1990) Statistics of coastal flood prevention. *Phil. Trans. R. Soc. Lond. A*, **332**, 457–476. (193)

Coles, S. G. and Tawn, J. A. (1991) Modelling extreme multivariate events. *J. R. Statist. Soc. B*, **53**, 377–392. (168)

Coles, S. G. and Tawn, J. A. (1992) Statistical methods for multivariate extremes: an application to structural design (with discussion). To appear in *J. R. Statist. Soc. B*. (166)

Conrad, J. M. (1980) Quasi-option value and the expected value of information. *Quarterly J. Econ.*, **94**, 813–820. (352)

Cook-Mozaffari, P. J., Darby, S. C., Doll, R., Forman, D., Hermon, C. and Pike, M. C. (1989) Geographical variation in mortality from leukaemia and other cancers in England and Wales in relation to proximity to nuclear installations, 1969–78. *Br. J. Cancer*, **59**, 476–85 (106)

Cormack, R. M. (1981) Loglinear models for capture-recapture experiments on open populations. In *Mathematical theory of the dynamics of biological populations, II.* (Eds. Hiorns, R. W. and Cooke, D.) pp. 197–215, Academic Press, London. (316)

Cormack, R. M. (1989) Log-linear models for capture-recapture. *Biometrics*, **45**, 395–413. (316)

Cormack, R. M. and Skalski, J. R. (1992) The analysis of coded wire tag returns from commercial catches. *Canadian J Fish. Aquat. Sci.*, **49**, 1816–1825. (316)

Cox, D. R. (1955) Some statistical methods related with series of events (with discussion). *J. R. Statist. Soc. B*, **17**, 129–64. (97)

Cox, D. R. (1958) *Planning of Experiments.* Wiley, London. (318)

Cox, D. R. (1969) Some sampling problems in technology. In Johnson, N. L. and Smith J. R. (eds.) *New Developments in Survey Sampling*, Wiley, New York. (51)

Cox, D. R. (1972a) The statistical analysis of dependencies in point processes. In Lewis, P. A. W. (ed.) *Stochastic Point Processes*, Wiley, New York. (96)

Cox, D. R. (1972b) Regression models and life-tables (with discussion). *J. R. Statist. Soc. B*, **34**, 187–220. (106)

Cox, D. R. (1983) Some remarks on overdispersion. *Biometrika*, **70**, 269–274. (317)

Cox, D. R. (1984) Long-range dependence: A review. In David, H. A. and David, H. T. (eds.) *Statistics: An Appraisal*, Iowa State University Press, Ames, Iowa. (144)

Cox, D. R. (1991) Comments on a paper by L. A. Goodman, *J. Amer. Statist. Assoc.*, **86**, 1121. (68)

Cox, D. R. and Hinkley, D. V. (1974) *Theoretical Statistics*. Chapman and Hall, London. (248)

Cox, D. R. and Isham, V. (1980) *Point Processes*. Chapman and Hall, London. (97)

Cox, D. R. and Lewis, P. A. W. (1966) *The Statistical Analysis of Series of Events.* Chapman and Hall, London. (129)

Cressie, N. (1991) *Statistics for Spatial Data*. Wiley, New York. (91)

Crow, E. L. (1988) Applications in atmospheric sciences. In Crow, E. L. and Shimizu, K. (eds.) *Lognormal Distributions, Theory and Applications*, Marcel Dekker, New York. (112)

Cuzick, J. and Edwards, R. (1990) Spatial clustering for inhomogeneous populations (with discussion), *J. R. Statist. Soc. B*, **52**, 73–104. (100)

Dasgupta, P. (1982) *The Control of Resources.* Basil Blackwell, Oxford. (349)

Davidian, M. and Carroll, R. J. (1988) A note on extended quasi-likelihood. *J. R. Statist. Soc. B*, **50**, 74–82. (315)

Davies, A. M. and Flather, R. A. (1978) Application of numerical models of the north-west European continental shelf and the North Sea to the computation of the storm surges of November to December 1973. *Dt. Hydrgr. Z. Erganzungs., A*, **14** (72 pages). (254)

Davison, A. C. (1984) *A Statistical Model for Contamination due to Long-Range Atmospheric Transport of Radionuclides* Ph.D. Thesis: Imperial College, London. (124)

Davison, A. C. and Smith, R. L. (1990) Models for exceedances over high thresholds (with discussion). *J. R. Statist. Soc. B*, **52**, 393–442. (124)

Day, T. J. (1974) Longitudinal dispersion in natural channels. *Water Resources Research*, **11**, 909–918.

Day, T. J. and Wood, I. R. (1976) Similarity of the mean motion of fluid particles dispersing in a natural channel. *Water Resources Research*, **12**, 655–666.

de Haan, L. (1985) Extremes in higher dimensions: the model and some statistics. In *Proc. 45th Sess. International Statistical Institute, Amsterdam*, paper 26.3. Amsterdam: Int. Statist. Inst. (168)

de Haan, L. (1990) Fighting the arch-enemy with mathematics. *Statistica Neerlandica*, **44**, 45–68. (244)

Dekkers, A. L. M. and de Haan, L. (1989) On the estimation of the extreme-value index and large quantile estimation. *Ann. Statist.*, **17**, 1795–1832. (249)

Dekkers, A. L. M., Einmahl, J. H. J. and de Haan, L. (1989) A moment estimator for the index of an extreme value distribution. *Ann. Statist.*, **17**, 1833–1855. (249)

Delft Hydraulics (1991) Definition of joint extreme condition statistics of hydraulic parameters off the coast of the Netherlands. Rijkwaterstaat, Den Haag. (263)

Delta Committee (1960) *Report of the Delta Committee.* Staatsuitgeverij, Den Haag. (243)

Deriso, R. B. (1980) Harvesting strategies and parameter estimation for an age structured model. *Canadian J. Fish. Aquat. Sci.*, **37**, 268–282. (280)

Diaconis, P. and Efron, B. (1985) Testing the independence of a two-way table: New interpretations of the Chi-square statistic (with discussion). *Ann. Statist.*, **13**, 845–913. (59)

Diggle, P. J. (1983) *Statistical Analysis of Spatial Point Patterns*. Academic Press, London. (91)

Diggle, P. J. (1989) Discussion on cancer near nuclear installations. *J. R. Statist. Soc. A*, **152**, 369–377. (104)

Diggle, P. J. (1990) A point process modelling approach to raised incidence of a rare phenomenon in the vicinity of a pre-specified point. *J. R. Statist. Soc. A*, **153**, 349–62. (104)

Diggle, P. J. and Chetwynd, A. G. (1991) Second-order analysis of spatial clustering for inhomogeneous populations. *Biometrics*, **47**, 1155–63. (100)

Diggle, P. J. and Milne, R. K. (1983) Negative binomial quadrat counts and point processes. *Scand. J. Statist.*, **10**, 257–67. (90)

Diggle, P. J. and Rowlingson, B. S. (1992) A conditional approach to point process modelling of raised incidence near multiple point sources. *Technical Report No. MA93/79*, Department of Mathematics, Lancaster University. (106)

Diggle, P. J., Gates, D. J. and Stibbard, A. (1987) A nonparametric estimator for pairwise-interaction point processes. *Biometrika*, **74**, 763–70. (99)

Diggle, P. J., Gatrell, A. C. and Lovett, A. A. (1990) Modelling the prevelance of cancer of the larynx in part of Lancashire: a new methodology for spatial epidemiology. In Thomas R. M. (ed.) *Spatial Epidemiology*. London Papers in Regional Science No. 21. London: Pion. (105)

Diggle, P. J., Chetwynd, A. G. and Haggkvist, R. (1991) Second-order analysis of space-time clustering. *Tech Report No. MA91/62*. Department of Mathematics, Lancaster University. (103)

Dixon, M. J. and Tawn, J. A. (1992) Trends in U.K. extreme sea levels: a spatial approach. To appear in *Geophys. J. Int.* (254)

Draper, N. H. and Smith, H. (1981) *Applied Regression Analysis*, 2nd edition, Wiley, New York. (192)

Drummer, T. D. and McDonald, L. L. (1987) Size-bias in line transect sampling. *Biometrics*, **13**, 13–21. (47)

Duncan, G. J. and Kalton, G. (1987) Issues of design and analysis of surveys across time. *Int. Statist. Rev.*, **55**, 97–117. (72)

E. A. Engineering, Science and Technology (1991) Submission to Federal Energy Regulatory Commission, Project No. 2299. Appendix 1. (275)

Efron, B. (1986) Double exponential families and their use in generalized linear regression. *J. Amer. Statist. Assoc.*, **81**, 709–721. (59)

Eilers, P. H. C. (1991) Regression in action: Estimating pollution roses from daily averages. *Environmetrics*, **2**, 25–48. (115)

Elliott, P. (1989) *Methodology of Enquiries into Disease Clustering*, Small Area Health Statistics Unit, London. (90)

Elrick, D. E. and French, L. K. (1966) Miscible displacement patterns on disturbed and undisturbed soil cores. *Soil Sci. Soc. Amer. J.*, **30**, 153–156. (24)

Enright, N. J. (1978) The comparative ecology and population dynamics of *Aracaria* species in Papua New Guinea. Ph.D. thesis, Australian National University, Canberra. (380)

Federer, W. T. and Searle S. (1961) Augmented designs for the one-way elimination of heterogeneity. *Biometrics*, **17**, 447–473. (72)

Fienberg, S. E. (1972) The multiple recapture census for closed populations and incomplete 2^k contingency tables. *Biometrika*, **59**, 591–603. (316)

Fienberg, S. E. (1980) *The Analysis of Cross-classified Categorical Data.* 2nd ed. MIT Press, Cambridge, Mass. (317)

Fischer, H. B., List, E. J., Koh, R. C., Imberger, J., Brooks, N. H. (1979) *Mixing in Inland and Coastal Waters*, Academic Press, New York. (6)

Forman, D., Cook-Mozaffari, P., Darby, S., Davey, G., Stratton, I., Doll, R. and Pike, M. (1987) Cancer near nuclear installations. *Nature*, **329**, 499–505. (106)

Freeman, A. M. (1985) Methods for assessing the benefits of environmental programs. In Kneese, A. V. and Sweeney, J. L. (eds.) *Handbook of Natural Resources and Energy Economics*, North-Holland, Amsterdam. (350)

Gardner, M. J. (1991) Investigating childhood leukaemia rates around the Sellafield nuclear plant. *Bull. Int. Statist. Inst.*, **54**(1). (110)

Gardner, M. J., Snee, M. P., Hall, A. J., Powell, C. A., Downes, S. and Terrell, J. D. (1990) Results of a case-control study of leukaemia and lymphoma among young people near Sellafield nuclear plant in West Cumbria. *Br. Med. J.*, **300**, 423–9. (89)

Geweke, J. and Porter-Hudak, S. (1983) The estimation and application of long-memory time series models. *J. Time Series Anal.*, **4**, 221–238. (150)

Gilman, K. (1977) *Movements of heat in soils*. Report No. 44, Institute of Hydrology.

Gilman, E. A. and Knox, E. G. (1991) Temporal-spatial distribution of childhood leukaemias and non-Hodgkin lymphomas in Great Britain. In Draper, G. (ed.) *The Geographical Epidemiology of Childhood Leukaemia and Non-Hodgkin Lymphomas in Great Britain, 1966–83*, HMSO, London. (102)

Godfrey, K. R. and Distefano, J. J. (1985) Identifiability of model parameters. In Barker, H. A. and Young, P. C. (eds.) *Identification and System Parameter Estimation 1985*, Vols 1 and 2, Pergamon, Oxford. (20)

Gomes, M. I. (1981) An i-dimensional limiting distribution function of largest value and its relevance to the statistical theory of extremes. In Taillie, C. *et al.* (eds.) *Statistical Distributions in Scientific Work*, Vol. 6. Reidel, Dordrecht. (230)

Gomes, M. I. (1985) Statistical theory of extremes—comparison of two approaches. *Statistics and Decisions*, **2**, 33–37. (230)

Gomes, M. I. (1990) Statistical inference in an extrema Markovian model. In *Compstat 90*, 257–262, Physica-Verlag. (227)

Gomes, M. I. and Alpuim, M. T. (1986) Inference in a multivariate generalized extreme value model—asymptotic properties of two test statistics. *Scand. J. Statist.*, **13**, 291–300. (240)

Goodman, L. A. (1981) Association models and the bivariate normal for contingency tables with ordered categories. *Biometrika*, **68**, 347–355. (68)

Goodman, L. A. (1991) Measures, models, and graphical displays in the analysis of cross-classified data (with discussion), *J. Amer. Statist. Assoc.*, **86**, 1085–1138. (68)

Gordon, R. D. (1941) Values of Mill's ratio of area to bounding ordinate and of the normal probability integral for large values of the argument. *Ann. Math. Statist.*, **12**, 364–366. (54)

Gornitz, V. and Lebedeff, S. (1987) Global sea-level changes during the last century. In Nummedal, D., Pilkey, O. H. and Howard, J. D. (eds.) *Sea-Level Change and Coastal Evolution*, No. 41. Society for Economic Paleontologists and Mineralogists, Tulsa, Oklahoma, SEPM Special Publication. (258)

Gould, R. P., Minchin, P. R. and Young, P. C. (1988) The effects of sulphur dioxide on phloem transport in two cereals *J. Exp. Bot.*, **39**, 997–1007. (244)

Graff, J. (1981) An investigation of the frequency distributions of annual sea level maxima at ports around Great Britain. *Estuarine Coastal Shelf Sci.*, **12**, 389–449. (316)

Green, P. E. J. and Macdonald, P. D. M. (1987) Analysis of mark-recapture data from hatchery-raised salmon using log-linear models. *Canadian J. Fish. Aquat. Sci.*, **44**, 316–326. (316)

Guerrero-Salazar P. and V. Yevjevich (1975) *Analysis of Drought Characteristics by the Theory of Runs.* Colorado State University Hydrology Papers Series, No. 80, 46 pages. Fort Collins, Colorado, USA. (213)

Gumbel, E. J. (1958) *Statistics of Extremes.* Columbia University Press, New York. (247)

Haltiner, G. J. and Martin, F. L. (1957) *Dynamical and Physical Meteorology.* McGraw-Hill, New York. (180)

Hansen, J. and Lebedeff, S. (1987) Global trends of measured surface air temperature. *J. Geophys. Res.*, **92**, 13345–13372. (159)

Hansen, J. and Lebedeff, S. (1988) Global surface air temperatures: Update through 1987. *Geophys. Res. Lett.*, **15**, 323–326. (159)

Harris, L. D. (1984) *The Fragmented Forest: island biogeography theory and the preservation of biotic diversity.* Univ. of Chicago Press, Chicago. (282)

Harvey, A. C. (1989) *Forecasting, Structural Time Series Models and the Kalman Filter.* Cambridge University Press, Cambridge. (282)

Haslett, J. and Raftery, A. E. (1989) Space-time modelling with long-memory dependence: assessing Ireland's wind power resource (with discussion). *Appl. Statist.*, **38**, 1–50. (150)

Hastie, T. J. and Tibshirani, R. J. (1990) *Generalized Additive Models.* Chapman and Hall, London. (133)

Hastings, W. K. (1970) Monte Carlo sampling methods using Markov chains and their applications. *Biometrika*, **57**, 97–109. (138)

Hazen, Allen (1914) Storage to be provided in impounding reservoirs for municipal water supply. *Trans. Amer Soc. Civil Eng.*, **77**, 1539–1640. (208)

Henderson-Sellars, B., Young, P. C. and Ribeiro da Costa, J. (1990) Water quality models: rivers and reservoirs, in *Proc. 1988 Int. Symp. on Water Quality Modelling of Agricultural Non-Point Sources*; United States Dept. of Agriculture, Agricultural Research Service; 1990; Publication No. ARS-81. (6)

Henry, C. (1974) Investment decisions under uncertainty: The irreversibility effect, *American Economic Review*, **64**, 1006–1012. (352)

Hilborn, R. (1990) Determination of fish movement patterns from tag recoveries using maximum likelihood estimators. *Canadian J. Fish. Aquat. Sci.*, **47**, 635–643. (329)

Hilborn, R. and C. W. Walters (1992) Quantitative fisheries stock assessment: choice, dynamics, and uncertainty. Routledge, Chapman and Hall, Inc., New York. (293)

Hill, B. M. (1975) A simple general approach to inference about the tail of a distribution. *Ann. Statist.*, **3**, 1163–1174. (231)

Hodge, I. (1984) Uncertainty, irreversibility and the loss of agricultural land, *J. Agric. Econ.*, **35**, 191–202.

Hornberger, G. M. and Spear, R. C. (1983) An approach to the analysis of behaviour and sensitivity in the environmental systems. In Beck, M. B. and van Straten, G. (eds.) *Uncertainty and Forecasting of Water Quality*. Springer-Verlag, Berlin. (5)

Hosking, J. R. M. (1981) Fractional differencing. *Biometrika*, **68**, 165–176. (159)

Hsing, T. (1987) On the characterisation of certain point processes. *Stoch. Proc. Appl.*, **26**, 297–316. (247)

Hsing, T. (1988) On the extreme order statistics for a stationary sequence. *Stoch. Proc. and their Appl.*, **29**, 155–169.

Hsing, T. (1989) *Estimating parameters of rare events*. Preprint, Texas A and M University.

Hsing, T., Husler, J. and Leadbetter, M. R. (1988). On the exceedance point process for a stationary sequence. *Probab. Theory and Related Fields*, **78**, 97–112. (247)

Hurst, H. E. (1951) Long-term storage capacity of reservoirs. *Trans. Amer. Soc. Civil Eng.*, **116**, 770–808. (209)

Hurst, H. E. (1956) Methods of using long-term storage in reservoirs. *Proc. Inst. Civil Engrs.*, **1**, 519–543. (209)

Hurst, H. E. (1957) A suggested statistical model of some time series which occur in nature. *Nature*, **180**, 494. (209)

Hurst, H. E., Black, R. P. and Simaika, Y. M. (1965) *Long-Term Storage*. Constable. London. (210)

Husler, J. (1986) Extreme values on non-stationary random sequences. *J. Appl. Probab.*, **23**, 937–950. (247)

Hydraulics Research Station (1980) Design of seawalls allowing for wave overtopping. *HRS Report*, No. 924. (262)

Institution of Civil Engineers (1989) *Floods and reservoir safety: an engineering guide*. Inst. Civ. Eng., London. (163)

Intergovernmental Panel on Climate Change (1990) *Climate Change: The IPCC Scientific Assessment*. Cambridge University Press, Cambridge. (142)

Jakeman, A. J., and Young, P. C. (1980) Towards optimal modelling of translocation data from tracer studies. *Proc. 4th Biennial Conf. Simulaton Soc. of Australia.*, 248–253. (23)

Jakeman, A. J., Littlewood, I. G. and Whitehead, P. G. (1990) Computation of the instantaneous unit hydrograph and identifiable component flows with application to two small upland catchments, *J. Hydrology*, **117**, 275–300.

Jeffers, J. N. R. (1978) *An introduction to Systems Analysis with ecological applications*. Edward Arnold, London. (387)

Joe, H., Smith, R. L. and Weissman, I. (1992) Bivariate threshold methods for extremes. *J. R. Statist. Soc. B*, **54**, 171–183. (168)

Johnson, N. L. and Kotz, S. (1970) *Continuous Univariate Distributions-2*, Wiley, New York. (54)

Johnson, J. (1963) *Econometric Methods*, McGraw-Hill, New York. (20)

Jones, E. E. (1956) Econological studies on the rainforest of Southern Nigeria. IV. (continued). The plateau forest of the 774 Okomu Forest Reserve. *J. Ecol.*, **44**, 83–117. (380)

Jumikis, A. R. (1966) *Thermal Soil Mechanics*, Rutgers University Press, New Brunswick, New Jersey.

Jury, W. A. (1982) Simulation of solute transport using a transfer function model. *Water Resour. Res.*, **18**, 363–368. (24)

Jury, W. A. and Sposito, G. (1986) Field calibration and validation of solute transport models for the unsaturated zone. *J. Soil Sci. Soc. Amer.*, **49**, 1331–1341. (24)

Kalman, R. E. (1960) A new approach to linear filtering and prediction problems. *Transactions of ASME, Journal of Basic Engineering*, **82D**, 35–45. (275)

Kamen, M. I. and N. L. Schwartz (1981) *Dynamic Optimization: The calculus of variations and optimal control in economics and management*. North-Holland, New York.

Karl, T. R., Heim, R. R. and Quayle, R. G. (1991a) The greenhouse effect in central North America: If not now, when? *Science*, **251**, 1058–1061. (156)

Karl, T. R., Kukla, G., Razuvayev, V. N., Changery, M. J., Quayle, R. G., Heim, R. R., Easterling, D. R. and Fu, C. B. (1991b) Global warming: evidence for asymmetric diurnal temperature change. *Geophysical Research Letters*, **18**, 2253–2256. (156)

Kaufmann, P. R., Herlihy, A. T., Elwood, J. W., Mitch, M. E., Overton, W. S., Sale, M. J., Messer, J. J., Cougan, K. A., Peck, D. V., Reckhow, K. H., Kinney, A. J., Christie, S. J., Brown, D. D., Hagley, C. A. and Jager, H. I. (1988) *Chemical Characteristics of Streams in the Mid-Atlantic and Southeastern United States. Volume I: Population Descriptions and Physico-Chemical Relationships.* U.S. Environmental Protection Agency Washington, D.C. (EPA/600/3-88/021a). (72)

Keay, R. W. J. (1961) Increment in Okomu Forest Reserve. *Bull. Nig. For. Info.*, **11**. (380)

Kendall, M. and A. Stuart (1979) *The Advanced Theory of Statistics*, Vol. 2: *Inference and relationship.* 4th ed. Macmillan, New York. (299)

Kio, P. R. O. (1978) *Stand development in naturally regenerated forest in South Western Nigeria.* Ph.D. thesis, University of Ibadan. (380)

Kish, L. (1987) *Statistical Design for Research.* Wiley, New York. (72)

Kitagawa, G. (1987) Non-gaussian state-space modeling of nonstationary time series (with discussion). *J. Amer. Statist. Assoc.*, **82**, 1032–1065. (282)

Kitagawa, G. (1991) A nonlinear smoothing method for time series analysis. *Statistica Sinica* **1**, 371–388. (285)

Klemes, V. (1979) Storage mass-curve analysis in a system-analytic perspective. *Water Resour. Res.*, **15**(2), 359–370. (209)

Knabe, W., Pohlmann, H. and Urfer, W. (1989) Statistische Überprüfung der Silizium-Hypothese in der Waldschadensforschung. *Forstarchiv*, **60**, 223–227. (344)

Knabe, W., Urfer, W., and Venne, H. (1990) Die Variabilität der Immissionsresistenz von Fichtenherkünften-Ein Beitrag zum IUFRO-Fichtenprovenienzversuch 1964/68. *Silvae Genetica*, **33**, 8–17. (345)

Kneese, A. V. and Schulze W. D. (1985) Ethics and environmental problems. In Kneese, A. V. and Sweeney, J. L. (eds.) *Handbook of Natural Resource and Energy Economics*, North-Holland, Amsterdam. (351)

Knox, G. (1963) Detection of low intensity epidemicity. *Br. J. Prev. Soc. Med.*, **17**, 121–7. (91)

Knox, G. (1964) Epidemiology of childhood leukaemia in Northumberland and Durham. *Br. J. Prev. Soc. Med.*, **18**, 17–24. (91)

Koch, G. G., Landis, J. R., Freeman, J. L., Freeman, D. H. and Lehnen, R. G. (1977) A general methodology for the analysis of experiments with repeated measurement of categorical data. *Biometrics* **33**, 133–158. (339)

Kraijenhoff, D. A. and Moll, J. R. (1986) *River Flow Modelling and Forecasting* (Water Science and Technology Library). Reidel, Dordrecht.

Kuo, C., Lindberg, C. and Thomson, D. J. (1990) Coherence established between atmospheric carbon dioxide and global temperature. *Nature*, **343**, 709–714. (160)

Lamb, H. H. (1991) *Historic Storms of the North Sea, British Isles and Northwest Europe.* Cambridge University Press, Cambridge. (243)

Landers, D. H., Eilers, J. M., Brakke, D. F., Overton, W. S., Kellar, P. E., Silverstein, M. E., Schonbrod, R. D., Crowe, R. E., Linthurst, R. A., Omernik, J. M., Teague, S. A. and Meier, E. P. (1987) *Characteristics of Lakes in the Westen United States. Volume I. Population Descriptions and Physico-Chemical Relationships.* U.S. Environmental Protection Agency, Washington, D.C. (EPA/600/3-86/054a). (72)

Lawson, A. (1989) Discussion on cancer near nuclear installations. *J. R. Statist. Soc. A*, **152**, 374–5. (104)

Lawson, A (1991) Score tests for detection of spatial trend in morbidity data. (unpublished). (104)

Leadbetter, M. R. and S. Nandagopalan (1989) On exceedance point processes for stationary sequences under mild oscillation restrictions. In Hüsler, J. and Reiss, R.-D. (eds.) *Extreme Value Theory.* Springer-Verlag, New York. (226)

Leadbetter, M. R., Lindgren, G. and Rootzen, H. (1983) *Extremes and Related Properties of Random Sequences and Series,* Springer-Verlag, New York. (246)

Leadbetter, M. R., Weissman, I., De Haan, L. and H. Rootzén (1989) *On clustering of high values in statistically stationary sequences.* Technical Report 253, Center for Stochastic Processes, Univ. North-Carolina. (237)

Lennon, G. W. (1963) A frequency investigation of abnormally high tidal levels at certain west coast ports. *Proc. Inst. Civ. Engs.,* **25**, 451–484. (263)

Leslie, P. H. (1945) On the use of matrices in certain population mathematics. *Biometrika,* **33**, 183–212. (387)

Lewis, E. G. (1942) On the generation and growth of a population. *Sankhya,* **6**, 93–96. (387)

Linacre, E. and Hobbs, J. (1977) *The Australian Climate Environment.* Wiley, Brisbane. (259)

Linthurst, R. A., Landers, D. H., Eilers, J. M., Brakke, D. F., Overton, W. S., Meier, E. P., and Crowe, R. E. (1986) *Characteristics of Lakes in the Eastern United States. Volume I. Population Descriptions and Physico-Chemical Relationships.* U.S. Environmental Protection Agency, Washington, D.C. (EPA/600/4-86/007a). (72)

Ljung, L. and Soderstrom, T. (1983) *Theory and Practice of Recursive Estimation.* MIT Press, Cambridge, Mass. (15)

Long, C. W., Ossiander, F. J., Ruehle, T. E. and Mathews, W. (1975) *Survival of coho salmon fingerlings passing through operating urbines with and without perforated bulkheads.* Northwest Fisheries Center Report, NMFS, Seattle. (319)

Longman, K. A. and Jenik, J. (1974) *Tropical Forest and its Environment.* Longman London. (379)

McCullagh, P. and Nelder, J. A. (1989) *Generalized Linear Models.* 2nd ed. Chapman and Hall, London. (315)

MacGregor, W. D. (1934) Silviculture of the mixed decidous forests of Nigeria with special reference to the South-Western provinces. *Oxf. For. Mem.,* **18**. (380)

Machin, D. (1975) Note 385: On a design problem in growth studies. *Biometrics,* **31**, 749–753. (75)

Mäler, K.-G. (1985) Welfare economics and the environment. Kneese, A. V. and Sweeney, J. L. (eds.) *Handbook of Natural Resource and Energy Economics,* North-Holland, Amsterdam. (350)

Manley, G. (1974) Central England temperatures: monthly means 1659 to 1973. *Quart. J. R. Met. Soc.,* **100**, 389–405. (142)

Marshall, R. J. (1991) A review of methods for the statistical analysis of spatial patterns of disease. *J. R. Statist. Soc. A,* **154**, 421–41. (90)

Mendelssohn, R. (1988) Some problems in estimating population sizes from catch-at-age data. *Fishery Bulletin* **86**, 617–630. (275)

Mervart, J. (1972) Growth and mortality rates in the natural high forest of Western Nigeria. *Nig. For. Info. Bull.,* **22**. (380)

Middleton, R. H. and Goodwin, G. C. (1990) *Digital Control and Estimation: A Unified Approach,* Prentice Hall, Englewood Cliffs, N.J. (41)

Miller, K. D. (1991) *Biodiversity and the Forestry Profession.* An H. R. MacMillan lecture in forestry, Univ. of British Columbia, Vancouver, Canada 7 Nov. 1992. 31 pp. (384)

Morrison, D. F. (1970) The optimal spacing of repeated measurements. *Biometrics,* **26**, 281–290. (75)

Morton, R. (1988) Analysis of generalized linear models with nested strata of variation. *Austral. J. Statist.,* **30A**, 215–224. (317)

Moser, J. W. and Hall, O. F. (1969) Deriving growth and yield functions for uneven-aged forest stands. *For. Sci.*, **15**(2), 183–188. (386)

Muirhead, C. and Darby, S. (eds.) (1989) Royal Statistical Society Meeting on Cancer Near Nuclear Installations. *J. R. Statist. Soc. A*, **152**, 305–84. (90)

Myers, N. (1979) *The Sinking Ark: A new look at the problem of disappearing species.* Pergamon Press, Oxford. (349)

Myers, N. (1983) *A Wealth of Wild Species: Storehouse for human welfare.* Westview Press, Boulder, CO. (349)

Nandagopalan, S. (1990) *Multivariate Extremes and Estimation of the Extremal Index.* Ph.D. Thesis. Technical Report 315, Center for Stochastic Processes, Univ. North-Carolina. (237)

Natural Environment Research Council (1975) *Flood Studies Report* (*5 volumes*). NERC, London. (165)

Nelder, J. A. and Lee, Y. (1992) Likelihood, quasi-likelihood and pseudo-likelihood: some comparisons. *J. R. Statist. Soc. B*, **54**, 273–284. (316)

Nelder, J. A. and Pregibon, D. (1987) An extended quasi-likelihood function. *Biometrika*, **74**, 221–232. (317)

Nokoe, S. (1978) Demonstrating the flexibility of the Gompertz function as a yield model using mature species data. *Commonwealth Forestry Review*, **57**(1), 35–42. (385)

Nokoe, S. (1980) Non-linear models fitted to stand volume-age data compare favourably with British Columbia Forest Service hand-drawn volume-age curves. *Canadian J. For. Res.* **10**(3), 304–307. (385)

Norton, J. P. (1986) *An Introduction to Identification*, Academic Press, New York. (15)

Ohser, J. and Stoyan, D. (1981) On the second-order and orientation analysis of planar stationary point processes. *Biom. J.*, **23**, 523–33. (96)

Okojie, J. A. (1981) Selecting an appropriate method for estimating the parameters for describing stem-diameter distribution in even-aged plantations. *Niger. J. For.*, **11**(2), 24–31. (383)

Olshen, R. A. (1967) Asymptotic properties of the periodograph of a discrete stationary process. *J. Appl. Prob.*, **4**, 508–528. (144)

Openshaw, S., Craft, A. W., Charlton, M. and Birch, J. M. (1988) Investigation of leukaemia clusters by use of a geographical analysis machine. *The Lancet* **1**, 272–3. (110)

Orlob, G. T. (ed.) (1983) *Mathematical Modelling of Water Quality: Streams, Lakes and Reservoirs.* Wiley, Chichester. (6)

Otto, M. C. and Pollock, K. H. (1990) Size bias in line transect sampling: A field test, *Biometrics*, **46**, 239–245. (47)

Overton, W. S., White, D. and Stevens, D. L., Jr. (1991). *Design Report for EMAP, the Environmental Monitoring and Assessment Program.* U.S. Environmental Protection Agency, Washington, D.C. (EPA/600/3-91/053).

Page, E. S. (1954) Continuous inspection schemes. *Biometrika*, **41**, 100–14. (110)

Pariwono, J. I., Bye, J. A. T. and Lennon, G. W. (1986) Long-period variations of sea-level in Australasia. *Geophys. J. R. Astr. Soc.*, **87**, 43–54. (259)

Patil, G. P. (1991) Encountered data, statistical ecology, environmental statistics, and weighted distribution methods. *Environmetrics*, **2**, 377–423. (46)

Patil, G. P. and Rao, C. R. (1977) Weighted distributions and a survey of their applications. In Krishnaiah, P. R. *Application of Statistics* (ed.) North Holland, Amsterdam. (46)

Patil, G. P. and Rao, C. R. (1978) Weighted distributions and size-biased sampling with applications to wildlife populations and human families. *Biometrics*, **34**, 179–184. (46)

Patil, G. P. and Taillie, C. (1979) Species abundance models ecological diversity, and the canonical hypothesis. *Bull. Int. Statist. Inst.*, **44**, 1–23. (53)

Patil, G. P. and Taillie, C. (1988) Weighted distributions and the effects of weight functions on Fisher information, Unpublished manuscript, Pennsylvania State University, Center for Statistical Ecology and Environmental Statistics, Department of Statistics, University Park, PA.

Patil, G. P. and Taillie, C. (1989) Probing encountered data, meta analysis and weighted distribution methods. In Dodge, Y. (ed.) *Statistical Data Analysis and Inference*, Elsevier Science Publishers, North Holland. (46)

Patil, G. P., Taillie, C. and Wigley, R. L. (1979) Transect sampling methods and their application to the deep-sea red crab. In Cairns, J., Patil, G. P. and Waters, W. E. (eds.) *Environmental Biomonitoring, Assessment, Prediction, and Management—Certain Case Studies and Related Quantitative Issues*, International Co-operative Publishing House, Fairland, Maryland. (47)

Patil, G. P., Taillie, C. and Wigley, R. L. (1980) Transect sampling methods and their applications to the deep-sea red crab. In Diemer, F. P., Vernberg, F. J. and Mirkes, D. Z. (eds.) *Advanced Concepts in Ocean Measurements for Marine Biology*. University of South Carolina Press. (47)

Patil, G. P., Babu, G. J., Hennemuth, R. C., Myers, W. L., Rajarshi, M. B. and Taillie, C. (1988) Data-based sampling and model-based estimation for environmental resources. In Krishnaiah, P. R. and Rao, C. R. (eds.) *Handbook of Statistics*, Vol. 6, Elsevier Science Publishers B. V., North Holland. (46)

Paulsen, S. G., Larson, D. P., Kaufmann, P. R., Wittier, T. R., Baker, J. R., Peck, D. V., MuGue, J., Hughes, R. M., McMullen, D., Stevens, D., Stoddard, J. L., Lazorchak, J., Kinney, W., Selle, A. R. and Hjort, R., (1991) *The Environmental Monitoring and Assessment Program (EMAP)—Surface Waters Monitoring and Research Plan*—Fiscal Year 1991. U.S. Environmental Protection Agency, Washington, D.C. (EPA/600/3-91/022). (84)

Payne, C. D. (1987) *The GLIM System, Release 3.77*. Numerical Algorithms Group, Oxford. (316)

Pegram, G. G. S., Salas, D. C., Boes, J. D. and Yevjevich V. (1980). *Stochastic Properties of Water Storage*. Colorado State University Hydrology Papers Series, No. 100. Fort Collins, Colorado, USA. (213)

Pickands, J. (1971) The two-dimensional Poisson process and extremal processes. *J. Appl. Prob.*, **8**, 745–756. (246)

Pickands, J. (1975) Statistical inference using extreme order statistics. *Ann. Statist.*, **3**, 119–131. (240)

Pierce, D. A. and Schafer, D. W. (1986) Residuals in generalized linear models. *J. Amer. Statist. Assoc.*, **81**, 977–998. (319)

Prandle, D. and Wolf, J. (1978) The interaction of surge and tide in the North Sea and River Thames. *Geophys. J. R. Astr. Soc.*, **55**, 203–216. (252)

Prescott, P. and Walden A. T. (1980) Maximum likelihood estimation of the parameters of the generalized extreme value distribution. *Biometrika*, **67**, 723–724. (239)

Preston, F. W. (1948) The commonness, and rarity, of species. *Ecology*, **29**, 254–283. (53)

Preston, F. W. (1962) The canonical distribution of commonness and rarity. *Ecology*, **43**, 186–215, 410–432. (53)

Prodan, M. (1968) *Forest Biometrics*, Pergamon Press, Oxford. (384)

Provis, D. G. and Radok, R. (1979) Sea-level oscillations along the Australian coast. *Aust. J. Mar. Freshwater Res.*, **30**, 295–301. (254)

Pugh, D. T. (1987) *Tides, Surges and Mean Sea Level*. Wiley, Chichester. (244)

Pugh, D. T. and Vassie, J. M. (1980) Applications of the joint probability method for extreme sea level computations. *Proc. Inst. Civ. Eng.*, **69**, 959–975. (244)

Quang, P. X. (1991) A nonparametric approach to size-biased line transect sampling. *Biometrics*, **47**, 269–279. (47)

Quinn, T. J. (1979) The effects of school structure on line transect estimators of abundance. In Patil, G. P. and Rosenzweig, M. L. (eds.) *Contemporary Quantitative Ecology and Related Ecometrics*, International Co-operative Publishing House, Fairland, Maryland. (47)

Quinn, T. J. (1985) Line transect estimators for schooling populations. *Fisheries Research*, **3**, 183–199. (47)

Raktoe, B. L., Hedayat, A. and Federer, W. A. (1981). *Factorial Designs*. Wiley, New York. (78)

Ramesh, N. I. (1992) *Statistical Inference for Some Classes of Cox Processes*. D.Phil. thesis, University of Oxford. (129)

Ramsey, F. L. (1979) Parametric models for line transect surveys. *Biometrika*, **66**, 505–512.

Ramsey, F. L., Wildman, V. and Engbring, J. (1987) Covariate adjustments to effective area in variable area wildlife surveys. *Biometrics*, **43**, 1–11. (47)

Rao, J. N. K. and Graham, J. E. (1964) Rotation designs for sampling on repeated occasions, *J. Amer. Statist. Assoc.*, **59**, 492–509. (72)

Rao, P. S. C., Ralston, D. E., Jessup, R. E. and Davidson, J. M. (1980) Solute transport in aggregated porous media: theoretical and experimental evaluation. *J. Soil Sci. Soc. Amer.*, **40**, 1139–1146.

Rao, P. V., Portier, K. M. and Ondrasik, J. A. (1981) Density estimation using line transect sampling. In *Estimating Numbers of Terrestrial Birds*. Ralph, C. J. and Scott, J. M. (eds.) Cooper Ornithological Society, Lawrence, Kansas. (47)

Read, T. R. C. and Cressie, N. A. C. (1988). *Goodness-of-fit Statistics for Discrete Multivariate Data*. Springer-Verlag, New York. (318)

Reed, D. W. and Anderson, C. W. (1992) A statistical perspective on reservoir flood standards. In Parr, N. J., Charles, J. A. and Walter, S. (eds.) *Water Resources and Reservoir Engineering*. Thomas Telford, London. (164)

Reed, W. J. (1984) The effects of the risk of fire on the optimal rotation of a forest, *J. Environ. Econ. & Manag.*, **11**, 180–190. (370)

Reed, W. J. (1992) The decision to conserve or harvest old-growth forest, *Ecol. Econ.* (to appear). (354)

Reed, W. J. and D. Errico (1985). Optimal harvest scheduling at the forest level in the presence of the risk of fire, *Can. J. For. Res.*, **16**, 266–278.

Resnick, S. I. (1987) *Extreme Values, Regular Variation and Point Processes*. Springer, New York. (168)

Ribeiro, D. A. and Costa, J. (1992) *Characterisation of Transport and Mixing Phenomena in the Ave River*, Ph.D. Thesis, Centre for Research on Environmental Systems, University of Lancaster. (23)

Richards, P. W. (1966) *The Tropical Rainforests*. Cambridge Univ. Press, London.

Ricker, W. E. (1954) Stock and recruitment. *J. Fish. Res. Board Canada*, **11**, 559–623. (307)

Ricker, W. E. (1975) Computation and interpretation of biological statistics of fish populations. *Bull. Fish. Res. Board Canada*, **191**. (280)

Ripley, B. D (1976). The second-order analysis of stationary point processes. *J. Appl. Prob.*, **13**, 255–66. (93)

Ripley, B. D. (1981) *Spatial Statistics*. Wiley, New York. (91)

Ripley, B. D. (1982) Edge effects in spatial stochastic processes. In Ranneby, B. (ed.)*Statistics in Theory and Practice. Essays in Honour of Bertil Matérn*. Umea: Swedish University of Agricultural Sciences. (91)

Robson, D. S. and Youngs, W. D. (1971) *Statistical analysis of reported tag-recaptures in the harvest from an exploited population*. Cornell University Biometrics Unit, BU-369-M, Ithaca, New York. (314)

Ross, S. M. (1983). *Introduction to Stochastic Dynamic Programming.* Academic Press, New York. (352)

Rowlingson, B. S. and Diggle, P. J. (1991) *Splancs: spatial point pattern analysis code in S-plus.* Tech. Report No. MA91/63, Mathematics Department, Lancaster University. (108)

Salas, J. D. and V. Yevjevich (1972). *Stochastic Structure of Water Use Time Series.* Colorado State University Hydrology Papers Series, No. 52. Fort Collins, Colorado, USA. (211)

Salas, J. D., Delleur, J. D., Yevjevich, V. and W. L. Lane (1985) *Applied Modeling of Hydrologic Time Series.* Water Resources Publications, Littleton, Colorado, USA., (211)

Samarov, A. and Taqqu, M. (1988) On the efficiency of the sample mean in long-memory noise. *J. Time Series Anal.*, **9**, 191–200. (145)

Samarov, A., Smith, R. L. and Taqqu, M. S. (1992), Estimating the intensity of long-range dependence through regression. To appear. (143)

Samuelson, P. A. and Scott, A. (1975) *Economics, Fourth Canadian Edition.* McGraw-Hill Ryerson, Toronto. (350)

Schnute, J. (1985) A general theory for analysis of catch and effort data. *Canadian J. Fish. Aquat. Sci.*, **42**, 414–429. (280)

Schnute, J. (1987) Data uncertainty, model ambiguity, and model identification. ("Mirror, mirror on the wall, what model's fairest of them all?"). *Natural Resource Modeling*, **2**, 159–212. (294)

Schnute, J. (1989) The influence of statistical error on stock assessment: illustrations from Schaefer's model. In Beamish, R. J. and McFarlane, G. A. (eds.) *Effects of ocean variability on recruitment and an evaluation of parameters in stock assessment models.* Can. Spec. Publ. Fish. Aquat. Sci. 108. (294)

Schnute, J. T. (1991) The importance of noise in fish population models. *Fish. Res.*, **11**, 197–223. (235)

Schnute, J. T., Richards, L. J. and Cass, A. J. (1989) Fish survival and recruitment: investigations based on a size-structured model. *Canadian J. Fish. Aquat. Sci.*, **46**, 743–769. (294)

Schnute, J. T., Mulligan, T. J. and Kuhn B. R. (1990) An errors-in-variables bias model with an application to salmon hatchery data. *Canadian J. Fish. Aquat. Sci.*, **47**, 1453–1467. (294)

Schoklitsch, A. (1923) *Graphische Hydraulik.* Teubner, Leipzig and Berlin.

Scott, J. M., Ramsey, F. L. and Kepler, C. B. (1986) *Forest bird communities of the Hawaiian islands. Their dynamics, ecology and conservation*, Allen Press, Lawrence, Kansas, Studies in Avian Biology, Vol. 9. (47)

Seber, G. A. F. (1962) The multi-simple single recapture census. *Biometrika*, **49**, 339–349. (313)

Seber, G. A. F. (1970) Estimating time-specific survival and reporting rates for adult birds from band returns. *Biometrika*, **57**, 313–318. (313)

Seber, G. A. F. (1976) *The Estimation of Animal Abundance*, Hafner, New York. (49)

Self, S. G. and Liang, K.-Y. (1987) Asymptotic properties of maximum likelihood estimators and likelihood ratio tests under nonstandard conditions. *J. Amer. Statist. Assoc.*, **82**, 605–610. (134)

Shen, H. W., Yevjevich, V., Obeysekera, J. T. B. and Decoursey, D. G. editors, (1985) *Multivariate Analysis of Hydrologic Processes.* Proceedings of Fourth International Hydrology Symposium, July 15–17, 1985, Colorado State University, Fort Collins, Colorado, USA. (213)

Shennan, I. (1989) Holocene crustal movements and sea-level changes in Great Britain, *J. Quart. Sci.*, **4**, 77–89. (257)

Shennan, I. and Woodworth, P. L. (1992). A comparison of late Holocene and twentieth century sea-level trends from the UK and North sea region. To appear in *Geophys. J. Int.* (258)

Simonoff, J. S. and Tsai, C. L. (1986) Jackknife-based estimators and confidence regions in nonlinear regression, *Technometrics*, **28**, 103–112. (189)

Skalski, J. R. (1990) A design for long-term status and trends. *Journal of Environmental Management*, **30**, 139–144. (72)

Skopp, J. and Warrick, A. W. (1974) A two phase model for miscible displacement of reactive solutes in soils. *Soil Sci. Soc. Amer. Proc.*, **38**, 545–550.

Smettem, K. R. J. (1984) Soil water residence time and solute uptake: 3. Mass transfer under simulated winter rainfall conditions in undisturbed soil cores. *J. Hydrology*, **67**, 235–248. (24)

Smith, A. F. M. and West, M. (1983) Monitoring renal transplants: an application of the multi-process Kalman filter. *Biometrics*, **39**, 867–78. (110)

Smith, J. A. and Karr, A. F. (1983) A point process model of summer season rainfall occurrences. *Water Resources Research*, **19**, 95–103. (127)

Smith, J. Q. (1988) *Decision Analysis. A Bayesian approach.* Chapman and Hall, London.

Smith, R. (1987) A two equation model for contaminant dispersion in natural streams. *J. of Fluid Dynamics*, **178**, 257–277. (23)

Smith, R. L. (1984a) Contribution to the Discussion of the paper by Stern, R. D. and Coe, R. (1984) A model-fitting analysis of daily rainfall data (with discussion) *J. R. Statist. A*, **147**, 1–34. (127)

Smith, R. L. (1984b) Threshold methods for sample extremes. In Tiago de Oliveira, J. (ed.) *Statistical Extremes and Applications*. Reidel, Dordrecht. (127)

Smith, R. L. (1986) Extreme value theory based on the *r* largest annual events. *J. Hydrology*, **86**, 27–43. (230)

Smith, R. L. (1988) Extreme values for dependent sequences via the Stein–Chen method of Poisson approximation. *Stoch. Proc. Appl.*, **30**, 317–327.

Smith, R. L. (1989) Extreme Value Analysis of environmental time series: an application to trend detection in ground-level ozone. *Statistical Science*, **4**, 367–393. (225)

Smith, R. L. (1992) Regional estimation from spatially dependent data. *Appl. Statist.*, In press. (186)

Smith, R. L., Tawn, J. A. and Coles, S. E. (1993) Markov chain models for threshold exceedances. In preparation. (171)

Smith, S. G. and Skalski, J. R. (1990) *Analysis of survival using the program SURPH.* Tech. Report, CQS, Seattle. (313)

Soderstrom, T. and Stoica, P. (1989) *System Identification* (Systems and Control Engineering Series). Prentice Hall, Hemel Hempstead. (15)

Southwood, T. R. E. (1978) *Ecological Methods.* Chapman and Hall, London and New York. (382)

Speed, T. P. and Ligon, F. K. (1993) An environment-dependent stock recruitment model for chinook salmon in the San Joaquin river system. Manuscript in preparation. (272)

Stedinger, J. R. and Tasker, G. D. (1985) Regional hydrologic analysis 1: ordinary, weighted, and generalized least squares compared, *Water Resour. Res.*, **21**, 1421–1432. (193)

Stein, M. L. (1991). A new class of estimators for the reduced second moment measure of point process. *Biometrika*, **78**, 281–6. (96)

Stoyan, D., Kendall, W. S., and Mecke, J. (1987) *Stochastic Geometry and its Applications.* Akademie-Verlag, Berlin. (91)

Stram, D. O., Wei, L. J. and Ware, J. H. (1988) Analysis of repeated ordered categorical outcomes with possibly missing observations and time-dependent covariates. *J. Amer. Statist. Assoc.*, **83**, 631–637. (344)

Sudler, Charles, E. (1927) Storage required for the regulation of stream flow. *Trans. Amer. Soc. Civil Eng.*, **91**, 622–660. (208)
Sullivan, P. (1992) A Kalman filter approach to catch-at-length analysis. *Biometrics*, **48**, 231–257. (275)
Sutcliffe, J. V. (1978) *Methods of flood estimation: a guide to the Flood Studies Report.* Institute of Hydrology, Report No. 49. (165)
Suthons, C. T. (1963) Frequency of occurrence of abnormally high sea levels on the east and south coasts of England. *Proc. Instn. Civ. Eng.*, **25**, 433–450. (243)
Tase, N. (1976) *Area-Deficit-Intensity Characteristics of Droughts.* Colorado State University Hydrology Paper Series, No. 87. Fort Collins, Colorado, USA. (215)
Tawn, J. A. (1988) An extreme value theory model for dependent observations. *J. Hydrol.*, **101**, 227–250. (230)
Tawn, J. A. (1992) Estimating probabilities of extreme sea levels. *Appl. Statist.*, **41**, 77–93. (250)
Tawn, J. A. and Mitchell, W. M. (1992) A spatial analysis of Australian extreme sea levels. Submitted for publication. (258)
Tawn, J. A. and Vassie, J. M. (1989) Extreme sea levels: the joint probabilities method revisited and revised. *Proc. Inst. Civ. Engrs. Part 2*, **87**, 429–442. (250)
Tawn, J. A. and Vassie, J. M. (1991) Recent improvements in the joint probability method for estimating extreme sea levels. In Parker, B. B. (ed.) *Tidal Hydrodynamics*, Wiley, New York. (250)
Taylor, G. I. (1954) The dispersion of soluble matter in solvent flowing slowly through a tube. *Proc. R. Soc. Lond., Series A.* **223**, 446–468. (6)
Thompson, W. A. (1988) *Point Process Models with Applications to Safety and Reliability.* Chapman and Hall, London. (356)
Tiago de Oliveira, J. and Gomes, M. I. (1984) Two test statistics for choice of univariate extreme models. In Tiago de Oliveira (ed.) *Statistical Extremes and Applications.* Reidel, Dordrecht. (240)
Turnbull, B. W., Iwano, E. J., Burnett, W. S., Howe, H. L. and Clark, L. C. (1990) Monitoring for clusters of disease: application to leukaemia incidence in upstate New York. *Am. J. Epidem.*, **132**, supplement, 136–43. (110)
Turnock, B. J., and Quinn, T. J. (1991) The effect of responsive movement on abundance estimation using line transect sampling. *Biometrics*, **47**, 701–715. (47)
Udink Ten Cate, A. J. (1983) *Modeling and (adaptive) control of greenhouse climates.* Ph.D. Dissertation, Agricultural University, Wageningen.
Urfer, W. (1988) Multivariate Verfahren zur Untersuchung von Wald-Ökosystemen. *Allg. Stat. Arch.*, **72**, 11–23. (344)
Urfer, W. (1989) Statistical evaluation of forest ecosystem measurements. In Bresser, A. H. M. and Mathy, P. (eds.) *Monitoring Air Pollution and Forest Ecosystem Research.* Commission of the European Communities, Brussels.
Urfer, W. and Pohlmann, H. (1992) Statistical analysis of spatially and serially correlated data. In Franke, J. and Roeder, A. (eds.) *Mathematical Modelling of Forest Ecosystems.* Verlag Sauerländer, Frankfurt/Main. (345)
Urfer, W., Knabe, W., Schwarzenbach, F. H., Woltering, F. and Schulte, M. (1990) Auswertung von Bonituren der Immissionsökologischen Dauerbeobachtungsflächen in Nordrhein-Westfalen. *Allg. Forst- und Jagd Zeitung*, **161**, 71–78. (342)
Usher, M. B. (1966) A matrix approach to the management of renewable resources, with special reference to selection forests. *J. Appl. Ecol.*, **3**, 183–212. (387)
Valentine, E. M. and Wood, I. R. (1979) Experiments in longitudinal dispersion with dead zones. *J. Hydraul. Div., Amer. Soc. Civ. Eng.*, **105**, 999–1016. (7)
Van der Made, J. W., Helmers, R., Dekkers, A. L. M., de Haan, L., Daan, H., van Malde, J.,

de Ronde, J. G. and Dillingh, D. (1989). *Frequentielijnen kust—tussentijds verslag 2.* Nota GWIO 89.001, Rijkwaterstaat, Den Haag. (244)

Vannerberg, C. and Holmstedt, G. (1989) *Spridning av NO_2 från en naturgaseldad villapanna.* Report of Department of Fire Safety Engineering, Lund University. (112)

Varlet, H. (1923) Etude graphique des condition d'exploration dur reservoir de regularisation. *Annales des ponts et chaussees*, **IV**, 61–79. (209)

Wallis, S. G., Blakeley, C. and Young, P. C. (1987) A microcomputer-based fluorometric data logging and analysis system, *Jnl. Inst. Water Eng. and Scientists*, **41**, 122–134. (7)

Walls, S. G., Young, P. C. and Beven, K. J. (1989) Experimental investigation of the aggregated dead zone model for longitudinal solute transport in stream channels. *Proc. Ins. Civ. Eng. (U.K.)*, **87**, (Part 2), 1–22. (7)

Ware, J. H., Lipsitz, S. and Speizer, F. E. (1988) Issues in the analysis of repeated categorical outcomes. *Statist. Med.*, **7**, 95–107. (337)

Wedderburn, R. W. M. (1974) Quasi-likelihood functions, generalized linear models, and the Gauss-Newton method. *Biometrika*, **61**, 439–447. (315)

Weissman, I. (1978) Estimation of parameters and large quantiles based on the k largest observations. *J. Amer. Statist. Assoc.*, **73**, 812–815. (230)

Weyman, D. R. (1975) *Runoff Processes and Streamflow Modelling*, Oxford University Press, Oxford.

White, G. C. (1983) Numerical estimation of survival rates from band-recovery and biotelemetry data. *J. Wildlife Mgmt.*, **47**, 716–728. (313)

Whitehead, P. G., Young, P. C. and Hornberger, G. H. (1979) A systems model of stream flow and water quality in the Bedford-Ouse river, I: Streamflow modelling, *Water Research*, **13**, 1155–1169.

Whittle, P. (1982) *Optimization over Time. Dynamic programming and stochastic control*, Vol. 1. Wiley, Chichester.

Wigley, R. L., Theroux, R. B. and Murray, H. E. (1975) Deep-sea red crab. *Marine Fisheries Review*, **37**, 1–21. (48)

Williams, D. A. (1975) The analysis of binary responses from toxicological experiments involving reproduction and teratogenicity. *Biometrics*, **31**, 949–952. (59)

Williams, G. W. (1984) Time-space clustering of disease. In Cornell, R. G. (ed.) *Statistical Methods for Cancer Studies.* Marcel Dekker, New York. (90)

Woltering, F. (1992) Modellbildung bei Langzeitstudien—Regressive logistische Modelle zur Analyse von Erhebungen auf Dauerbeobachtungsflächen. In *Schriften aus der Forstlichen Versuchsanstalt.* Vol. 106. Verlag Sauerländer, Frankfurt/Main. (342)

Woodworth, P. L. (1987) Trends in UK mean sea level, *Marine Geodesy*, **11**, 57–87. (244)

Woodworth, P. L. (1990) A search for accelerations in records of European mean sea level, *Int. J. Clim.*, **10**, 129–143. (244)

World Meteorological Organization (1988) *Hydrological aspects of combined effects of storm surges and heavy rainfall on river flow.* Operational Hydrology Report No. 30. WMO, Geneva. (166)

Yajima, Y. (1988) On estimation of a regression model with long-memory stationary errors. *Ann. Statist.*, **16**, 791–807. (146)

Yajima, Y. (1991) Asymptotic properties of the LSE in a regression model with long-memory stationary errors. *Ann. Statist.*, **19**, 158–177. (146)

Yang, R. C., Kozak, A. and Smith, J. H. G. (1978) The potential of Weibull-type functions as flexible growth curves. *Canadian J. For. Res.*, 9(4), 424–431. (383)

Young, P. C. (1965) Process parameter estimation and self adaptive control. In Hammond, P. H. (ed.) *Theory of Self Adaptive Control Systems.* Plenum Press, New York. (18)

Young, P. C. (1970) An instrumental variable method for real-time identification of a noisy process. *Automatica*, **6**, 271–287. (18)

Young, P. C. (1975) Recursive approaches to time-series analysis, *Bull. Inst. Math. and Applic.*, **10**, 209–224.

Young, P. C. (1978) A general theory of modelling for badly defined dynamic systems. In Vansteenkiste, G. C. (ed.) *Modelling, Identification and Control in Environ-Systems.* North Holland, Amsterdam. (5)

Young, P. C. (1981) Parameter estimation for continuous-time models: a survey, *Automatica*, **17**, 23–39. (18)

Young, P. C. (1983) The validity and credibility of models for badly defined systems. In Beck, M. B. and van Straten, G. (eds.) *Uncertainty and Forecasting of Water Quality.* Springer-Verlag, Berlin. (5)

Young, P. C. (1984) *Recursive Estimation and Time Series Analysis* (Communication and Control Engineering Series). Springer-Verlag, Berlin. (15)

Young, P. C. (1985) The instrumental variable method: a practical approach to identification and system parameter estimation. In Barker, H. A. and Young, P. C. (eds.) *Identification and System Parameter Estimation 1985*, Vols 1 and 2. Pergamon, Oxford. (17)

Young, P. C. (1986) Time-series methods and recursive estimation in hydrological systems analysis. In Kraijenhoff, D. A. and Moll, J. R. (eds.) *River Flow Modelling and Forecasting.* Reidel, Dordrecht.

Young, P. C. (1989) Recursive estimation, forecasting and adaptive control. In Leondes, C. T. (ed.) *Control and Dynamic Systems.* Academic Press, San Diego. (18)

Young, P. C. (1990) Systems methods in the evaluation of environmental problems. In Harrison, R. M. (ed.) *Pollution—Causes*, Effects and Control, Royal Society of Chemistry, London. (23)

Young, P. C. and Benner, S. (1991) *micro*CAPTAIN Handbook, Version 2.0, *Centre for Research on Environmental Systems, I.E.B.S., Univ. of Lancaster.* (21)

Young, P. C. and Beven, K. J. (1991) Comments on computation of the instantaneous unit hydrograph and identifiable component flows with application to two small upland catchments. *Journal of Hydrology*, **129**, 389–396.

Young, P. C. and Jakeman, A. J. (1980) Refined instrumental variable methods of recursive time-series analysis, part 3: extensions, *Int. J. Control*, **31**, 741–764. (18)

Young, P. C. and Minchin, P. E. H. (1991) Environmetric time-series analysis: modelling natural systems from experimental time-series data, *Int. J. Biol. Macromolecules*, **13**, 190–201.

Young, P. C. and Wallis, S. G. (1985) Recursive estimation: A unified approach to identification, estimation and forecasting of hydrological systems. *Applied Mathematics and Computation*, **17**, 299–334.

Young, P. C., Wallis, S. G. (1986) *The Aggregated Dead Zone (ADZ) model for dispersion in rivers.* BHRA Int. Conf. on Water Quality Modelling in the Inland Natural Environment. (7)

Young, P. C. and Wallis, S. G. (1992) Solute transport and dispersion in channels. To appear. In Beven, K. J. and Kirby, M. J. (eds.) *Channel Networks*. Wiley, Chichester. (6)

Young, P. C. Jakeman, A. J. and McMurtrie, R. (1980) An instrumental variable method for model order identification. *Automatica*. **16**, 281–294. (17)

Young, P. C., Chotai, A. and Tych, W. (1991) Identification, estimation and control of continuous-time systems described by delta operator models. In Sinha, N. K. and Rao, G. P. (eds.) *Identification of Continuous-Time Systems*, Kluwer Academic Publishers, Dordrecht. (17)

Zeger, S. L., Liang, K.-Y. and Albert, P. S. (1988) Models for longitudinal data: a generalized estimating equation approach. *Biometrics*, **44**, 1049–1060. (317)

Zwiers, F. W. and Ross, W. H. (1991) An alternative approach to the extreme value analysis of rainfall data. *Atmosphere-Ocean*, **29**, 437–461. (190)

Zygmund, A. (1959) *Trigonometric Series, Volume I.* Second edition, Cambridge University Press, Cambridge. (145)

Index

Active mixing volume. *See* AMV model
Active mixing zone. *See* AMZ model
Adaptive management 15
Advection dispersion equation (ADE) model 6, 7, 24
Advective time-delay 9
ADZ model 6–14, 20–2, 24, 37, 38, 42, 43
  derivation 6–11
  generalisation of 29–38
  potential value of 23
  practical examples 23–9
  practical results 21–3
  transfer function representation 11–14
  wider implications of 23
Aggregated dead zone model. *See* ADZ model
Air pollution
  exposure times from point source 123–38
  quantiles estimation in 111–22
  see also Nitrogen oxides (NO_x) emission
Akaike's AIC criterion 341, 342
AMV model 5, 11, 20, 29–39
  heat flow in soils 30–8
AMZ model 11, 24–5, 38
Annual cycles 202
AR(1) model 303, 304, 306
Argand diagram 131
ARIMA models 159, 160, 211
ARMA models 143, 211, 304
Astronomic cycles 202
Asymptotic variance matrix 105
Australia, sea-level data analysis 258–61
Autoregressive (AR) model 156–7, 211
Autoregressive-moving-average models. *See* ARMA models

Backward-shift operator 41, 42
Bernoulli trials 94, 106
Best linear unbiased estimate (BLUE) 146
Beta-binomial model 60 (BTB) 58, 61
Beta distribution 58
Bilogistic model 182
Binomial mixture model 58
Binomial parameter 58
Binomial response probability 61–2
Bivariate distributions 65
Bivariate overdispersion modelling 62–9
Bivariate sighting-functions 48–51
Bivariate weight function 67–8
Bootstrap methodology 238

Cancers, childhood 101
Cancers of the larynx 106
Carbon dioxide 142
Central England temperature data 142, 147–9
Certainty-equivalence procedure 352
Childhood cancers 101
Childhood leukaemia 89, 102, 108, 110
Chinook salmon (*Oncorhynchus tshawytscha*) 267–71, 297–301, 323–9
  see also Fish populations; Salmon population
Chlorofluorocarbons (CFCs) 142
Chorley-Ribble area 106
Cluster detection problem 110
Cluster point process 93

Cluster sizes 236
Clusters and clustering 91, 109
Clusters of exceedances 236
Coho salmon 319
Coloured noise 16
Conditional intensity 129
Conditional intensity functions 131, 138
Convective-dispersion equation (CDE) 24
Counts 90
Covariance density 92
Cox process 97
CUSUMS 110

Daily cycles 202
Dams
  safety of 163
  see also Reservoirs
Data-based models 4, 5, 20, 35, 37
Data interpretation 46
Data quality 90, 110
De Bilt, annual precipitation maxima 184–6, 191, 193
Declustering 249, 252
Deconvolution problem 99
Deep-sea red crab 47–8
Dependence models 182
Design flood 165, 166
Design initial reservoir level 165
Design wind speed 165
Developmental toxicity data 59
Diaconis-Efron DEF, bivariate form of 63–6
Dirichlet model 182
Discrete differential operator 41
Discrete-time simulation model 272, 275
Disease distribution 89
Disease incidence 106, 110
Dispersion modelling 7, 23
Dispersive characteristics 11
Dispersive fraction 9–10
Double binomial (DBN) model 60, 61
Double binomial response probability 61–2
Double bivariate normal distribution 66–7
Double exponential family (DEF) 59

Dye-tracer experiment 8, 21

Economic cycles 202–3
Economic modelling techniques 3
Edge-corrections 96
Edge-effects 95, 103
Effective half-width 49
Effective residence time 11
Efron's approximation formulae 61–2
Ekman spiral 180
Eliassen, S. 208
EMAP 72–4
Encounter bias in transect sampling 46–52
Encounter sampling 45–69
Environment monitoring, design for 72
Environmental epidemiology 89–91
  software for 107–9
Environmental modelling systems 3–43
Events 91–2
Exceedances of high levels and of upper order statistics 226–9
Expected present value (EPV) 357, 359, 361, 362, 369, 370–2, 375
Exposure times to air pollution from point source 123–38
Extra-distance factors 47
Extremal index 225, 226, 229–31, 247
  estimation of 236–9
  mean values and standard errors of estimators of 238
Extreme sea levels 243–63
  application of spatial model 254–61
  Australian data analysis 258–61
  direct methods 247–50
  estimation of marginal behaviour 247
  incorporating waves in analysis 261–3
  indirect methods 250–2
  separate site methods 247–52
  spatial modelling 252–4
  UK data analysis 254–8
Extreme-value analysis 225
Extreme-value modelling 245–7
Extreme-value-theory models 245

Fickian diffusion equation 6
First-order intensity function 92, 94

Fish populations
  ambiguous inferences from data 293–309
  AR(1) model 303, 304, 306
  ARMA(1,1) model 304
  harvest strategy 307
  Kalman model 304–6
  Lingcod CPUE data 303
  recapture data on tagged fish 311–31
    analysis of deviance 327, 328
    approaches to 312–13
    chinook salmon in South-Eastern Alaska waters 323–9
    effect of experimental dam manipulation on local survival 319–23
    extensions 329–30
    generalised linear models 315–19
    gillnet fishery 328–9
    interval estimates 318
    measures of goodness-of-fit 317–18
    model selection 318
    multinomial models 313–15
    population of interest 317
    replicates 317
    residual analysis 318
    small-number problems 318
    software packages 313
  serial correlation 301–7
  statistical models 294
  stock and recruitment 307–9
  see also Salmon population
Fisher information and estimation efficiency 52
Fisher information matrices 52
Flexible linear model 74
Flood damage 244
Flood defences 243
Flood safety 164
  see also Reservoirs
Flood Studies Report 165
Flooding, wave effects 261
Forestry
  amenity benefits 351
  Ancient World's attitude 348
  categorical data analysis 337–41
  certainty-equivalence procedure 364–7, 373–6
  climatological and ecological factors 335–45
  conservation-development decision model 367–76
  conversion to agricultural land 350
  cost-benefit analysis in decision problems 350–2
  cost-benefit methodology 376
  cost-benefit prescription 356–7, 369–70
  critical ratio 364
  crown transparency 336, 341
  data analysis 336–7
  data structure 345
  determination of sample sizes 345
  developed-land valuation uncertainty 371
  economic analysis with respect to conservation 350
  ecosystem measurements 344–5
  genetic aspects 345
  history of destruction 347–51
  logging industry 349, 351–2, 377
  logging rights 350
  logistic regression analysis 341–4
  logistic regression model 339–41
  longitudinal studies 336–41, 345
  marginal models 336–9
  old-growth 367
  optimal development rule 371
  optimal harvest rule 357–9
  reasons for continued deforestation 349
  service flows 368
  single-stand model for harvest-conservation decision problem 354–67
  size-independent land-service flows 375–6
  social valuations 368
  stochastic optimal control 376
  suboptimality of cost-benefit procedure 359–64, 373
  transitional models 337
  uncertainty and conservation or destruction 347–78
  weighted least squares analysis 341–4

Forestry (*cont.*)
wilderness valuation uncertainty 371
zero development costs and no benefits from scientific
breakthroughs 372
see also Tropical rainforests
Fraga hydrometric station 239–41
Fréchet model 231

General circulation models (GCMs) 141–3, 156
Generalised extreme-value distribution. *See* GEV model
Generalised least squares (GLS) method 193–5
Generalised Pareto distribution 249
Geographical Information Systems (GIS) 107
Geometric Brownian motion 354–5
Geryon quinquedens Smith 48
GEV model 225, 227, 229, 230, 239, 246–8, 254, 260
Geweke-Porter-Hudak procedure 150
Glasshouse microclimate model 31–3
GLIM program 316, 320, 321, 330
Global warming
and detection of trends in temperature series 142–3
long-range dependence 141–61
Great Plains, USA 214–15, 217, 218
Greenhouse gases 142
and temperature effects 142
Gumbel classical approach 229, 230, 239
Gumbel distribution 185, 191
Gumbel double-exponential scale 208
Gumbel estimator 231–5, 241
Gumbel parameters 187, 188, 192–5

Hall, L.S. 208
Hazen, A. 208, 209
Health hazards 89
Heat flow in soils 30–8
Hessian matrix 187
Heuristic estimation algorithms 18
High levels 225
High quantiles 225
Hill estimator 231–5, 240
Historical Climatological Network (HCN) 142, 155–7
Homogeneous Poisson process 130
Hurst, H.E. 210, 213
Hydrological spatio-temporal processes 203–4
Hydrology
statistical methods 204
see also Water resources

Identifiability problems 20, 31, 36–7
Incidence ratio 109
Independent *k* upper order statistics 233
Index of regular variation 225
Indicator function 95
Information collection 46
Inhomogeneous Poisson process 96, 104
Instantaneous discount rate 355
Instrumental cross-product matrix (ICPM) 17, 19
Intensity function 105
Intergovernmental Panel on Climate Change (IPCC) 142, 143, 156, 157, 159
Iodine-131, log exposures to wet deposition 125–7
Isotropy 92

Jack-knife method 188–9
JARRI project 382
Joint Probability Studies for Reservoir Flood Safety 164, 166, 250
statistical methodology 167–73
strategy of 166–7

K-function 92–4, 100–4
bivariate 94
effect of random perturbation on 97
k-homogeneous model 259
Kalman filter 282, 284, 285
Kalman model 304–6
Kernel estimator 106
Klemes, V. 208, 209
Knox's statistic 103

Laplace operator 40
Least squares estimate (LSE) 146
Legionnaires' disease 104

Leslie-type matrices 387–8
Leukaemia 89, 102, 108, 110
Likelihood inference 106
Likelihood ratio statistic 134
Likelihood ratio test 240
Line-transect sampling 46–7, 49, 51
Linear-in-the-parameters 15, 16
Linear jack-knife method 189–90
Linear regression model 272, 274
Log-likelihood 105
Log-normal distribution 112, 115
  of species abundance 53–8
LOGIST procedure 339
Logistic model 182
Long-range dependence
  and estimation of linear trend 143–7
  English data 149–55
  global data 157–60
  global warming 141–61
  US data 155–7
Long-range periodicities 202
Lophira alata 381
Lunar cycle 202

Markov chain 127, 130
Markov models 387–8
Markov-modulated Poisson processes 127–31
  second-order properties 129–30
Markov process 390
Markovian max-autoregressive sequences 226
MATLAB numerical computational package 32
Max-autoregressive model 241
Maximum likelihood equations 233
Maximum likelihood estimation 54, 55, 230, 231
Maximum likelihood estimator 105
Maximum likelihood method 185–6
Mean sea levels 243, 244, 259
Mean squared error 105
Mechanistic models 4, 5, 7, 20, 21, 35, 38
MESOS model 124–7
Methane 142
MicroCAPTAIN program 15, 21, 35
Milankovitch cyclicities 202
Minimal sufficient statistics (MSS) 314
Model order identification 18
Model specification 46
Monte Carlo test 101, 104, 106
Multivariate dependent structure 230
Multivariate extremes 167
Multivariate sighting function 50

Natural-gas furnaces 111–22
Negative bilogistic model 182
Negative definiteness 52
Negative logistic model 182
NERC 21
NEVN (normalised error variance norm) 18–19
Newton-Raphson algorithm 233
Newton-Raphson procedure 189
Nitrogen oxides (NO_x) emission 111–22
  correction for drift 113–14
  critical level 118
  data analysis 116–18
  data collection 114
  distribution with fitted normal-plus-log-normal density 118
  effect of measurement error on empirical extreme quantiles 118–21
  empirical distribution 117
  influence of wind direction 115
  measuring system 112–13
  normal and extreme 112–15
  regression-corrected measurements 114
  statistical model 115–18
Nitrous oxide 142
Nonstationary models 14
Norway spruce (*Picea abies*) 335, 345
Nuisance function 106

Object-factors 47
Object-specific factor 50
Oncorhynchus kisutch 319
Oncorhynchus tshawytscha. *See* Chinook salmon

Palm distributions 171, 180, 181
Pareto excesses 231
Peaks over threshold. *See* POT
Periodic-autoregressive (PAR) model 211

Periodic-autoregressive-moving-average (PARMA) model 211
Periodic-stochastic processes 202
Persistently exciting input 20
Perturbation distribution 98
Physically motivated considerations 19
Picea abies. *See* Norway spruce
Point process 249
  bivariate 94
  characterisation 247
  data analysis 131
  data-perturbation 97–9
  first- and second-moment properties 91–4
    of bivariate stationary, isotropic point process 95
  in pseudo-polar coordinates 177
  independent stationary, isotropic 94
  isotropic 92
  models 89–110, 168–9, 181
    parameter estimates 176
  pseudo-polar coordinates 178
  raised incidence near one or more point sources 104–7
  second-order stationary 92
  stationary space-time 103
Point sources 104–7
  exposure times to air pollution from 123
  multiple 106
Poisson distribution 93
Poisson log-likelihood 329
Poisson point process modelling 106
Poisson process 93, 226, 246, 249
Poisson regression modelling 106
Polygonal regions 108
Positive definiteness 52
POT estimator 229, 231–5, 241
POT-ALL 241
Precipitation. *See* Rainfall depth-duration-frequency curves
Prediction error 16
Prefilter parameters 17
Prefiltered variables 16
Primary variables 167–8
  dependence between extremes of 168–9
  dependence models for 173
Probable maximum flood 165
Quantile estimation
  in airborne pollution 111–22
  rainfall depth-duration-frequency curves 195–6
Quantile relations in disturbed distributions 118–21

Rainfall depth-duration-frequency curves 183–97
  annual maxima at De Bilt 184–6, 191, 193
  bootstrap method 190
  estimation of correlation between parameter estimates 186–92
  jack-knife method 188–9
  linear jack-knife method 189–90
  mathematical description of change of Gumbel parameters with duration 192–5
  quantile estimation 195–6
  Smith's method 187–8
Random labelling 94
Random perturbations 97
Random phenomena 201
Random sighting-functions 51–2
Random thinning 101
Rare events 225–41
r-coefficient 218
Recapture data on tagged fish. *See* Fish populations
Recursive fixed-interval smoothing 18
Recursive parameter estimation 14–18
Red crab 47–8
Reduced second-moment measure 92
Reference location 90
Reference time 90
Regressive logistic regression models 339
Relative informativeness of weighted observations 69
RELEASE 313
Reservoir safety 163–82
  catchment wetness 172
  current method of assessment 164–6
  dependence between rain and wind 173–8
  dependence structure for standardised margins 167
  reservoir fullness 172

snowmelt 172
structure function 166
wind speed and direction 180
see also Joint Probability Studies for Reservoir Flood Safety
Reservoirs Act 1975 163
Residence time 10, 37, 42
Residual surface 103–4
Response error 16
Return period 225
Revised joint probabilities method (RJPM) 250–2
Rhodamine WT 10
River Conder 8, 21–3
River Douro 239–41
RJPM model 260
Rotating panel 72
Run lengths 236
Runs of consecutive exceedances 236

Salmon population 267–92
Beverton-Holt curves 280, 290
chinook salmon 267–71, 297–301, 323–9
coho salmon 319
decline of population size 268, 270–1
density-dependent mortality of eggs or juveniles 290–1
Deriso curves 280, 290
discrete-time simulation model 272, 275
escapement in San Joaquin river system 269
EV model 299–300
juvenile mortality 268, 271
linear regression model 272, 274
models and methods 271–89
predicted and observed series for catch 287, 289
predicted and observed series for escapement 287, 288
Ricker curves 280, 290
San Joaquin River System 267–71, 275
spring flow and population abundance 272
spring flow in determination of recruit numbers 289–90
state-space model 275–89
sums of squared one-step prediction errors (SSPE) 287–9
see also Chinook salmon; Coho salmon; Fish populations
Sampling designs 71–85
augmentation 72, 80–3, 85
comparison of 73, 74, 79–83, 85
discussion 84
modelling approach 74–8
precision of 79
principal difference between 84
rotating panel 72, 79, 84–5
serially alternating 72, 79, 84–5
San Joaquin River System 267–71, 275
Scaling factor 50
Scotland 104
Sea levels 243–63
annual maximum 247
observations 244
Seascale 89
Sea-wall design 262
Second-order intensity function 92, 94
Secondary variables 167–8, 171
Sellafield 89, 110
Semi-parametric approach 239
Serial correlation in fisheries data 301–7
Serially alternating design 72
Short-range periodicities 202
Sighting-distance bias 48
Sighting frequency 48
Sighting functions 48–52
Simplified refined instrumental variable (SRIV) procedure 15
SIMULAB program 32
Soft data 46
Soil energy-balance equations 32–5
Soil energy-balance model 31, 34
Soil tracer experiment 24–9
Soils
heat flow in 30–8
solute transport in 24–9
Solute transport in soils 24–9
Southern Oscillation effects 259
S P (semi-parametric) estimator 235, 241
Space-time incidence ratio 110
Space-time interaction 91, 102–4
Spatial clustering 91, 100–2

Spatial databases 110
Spatial heterogeneity 91, 96
Spatial modelling 252–4
Spatial phenomena 91
Spatial point process 90
Species abundance, log-normal distribution of 53–8
Splancs 107–9
Split panel designs 72
S-plus 108
SRIV algorithm 17–18
SRIV methods 15, 21, 26
SRIV model 27, 29
Stationarity 92, 99–100
Stationary Cox process 100
Stationary Poisson process 93
Statistical analysis, point processes 91–100
Statistical extremes 229
Statistical functions 107
Statistical Sciences Inc. 108
Steady-state gains (SSGs) 10, 37, 38
Stochastic dynamic programming (SDP) 352, 371
Stochastic models, exposure times to air pollution from point source 123–38
Stochastic optimal control 376
Storm events 164–5
Storm-Tide Warning Service 243
Structural models 20
Sudler, Charles E. 208
SURGE 313
SURPH 313
Survey factors 47, 50
SURVIV 313
Survivor function 355–6

Temperature effects and greenhouse gases 142
Teratological binary data, overdispersion and heterogeneity of 58–62
Test statistics 100–1, 104
TF models 5, 7, 11–14, 39–43
  backward-shift 41
  discrete differential operator 41
  example 42–3
  identification and estimation 14–21
  order identification 18–20
  time-derivative 40
Tidal predictions 258
Tide gauges 244
Tide-surge interaction 252
Time-derivative operator 40
Time-series analysis 14
Time-variable parameters (TVP) 14
Total reach volume 11
Tracer experiments 9
Transect sampling, encounter bias in 46–52
Transfer function models. *See* TF models
Trinary data 63
Trinomial distribution 66
Triplochiton scleroxylon 381
Tropical rainforests 379–99
  amenity benefits 351
  assessment of species structural changes 393–4
  assessment of unregulated interference and interventions 391–4
  BASIC program for stable structure simulation 395–9
  data sets 388
  diversity of species 349
  frequency distribution 383–5
  growth and yield assessment 385
  individual tree growth 385
  major issues of concern 388
  need for growth studies 381–2
  quantitative assessments 382–8
  regions 379–80
  simulated tree-stands dynamics from under-to overmature stage 383
  spatial distribution 382
  species diversity 383
  species population structure 390
  stand distribution by size classes 384
  stand dynamics
    Leslie-type matrices 387–8
    Markov models 387–8
    simple formulation 386–8
  stand-growth data 386
  studies on growth or dynamics of 380
  time-period equivalents in years 391
  tree species' ages at selected girths 380
  see also Forestry

United Kingdom, sea-level data
analysis 254–8
Upper quantiles
comparison of estimators of 231–6
estimation of 229–31

Varlet, Henri 209
Visibility function 50

Water cycle 201
Water resources 201–44
ageing and renovation of systems and structures 221
and environmental protection 220
application of statistics 208–10
areal models of distribution of parameters of point time series 214–15
as random phenomena 201
changes in priority of use of 222
changing risks in function of new information and demands 221
cleaning polluted environments 219–20
climatic changes and environment 220
complexity of problems 210–11
conflicts of interests in development 222
correlation analysis 206–7
drinking water industrially produced 223
frequency distribution analysis 205–6
future applications of statistics 219–24
improvement of existing systems 221
input, state and output processes 212
local spatial dependence 217–18
long-range sampling 204–5
marketing 222
modelling of spatial hydrologic processes 213–14
multivariate analysis 213
past applications of statistics 204–10
periodicity and stochastic modelling 211–12
pollution 219
present application of statistics 210–18
probability distribution analysis 205–6
ranges and runs as important, derived hydrologic variables 213
reallocation of water rights 219
regional time dependence 215–17
regression analysis 206–7
residual mass curve 209
river-flow mass curve 209
selection of statistical methods 204
solutions of future problems 223–4
state of water problems 204
statistics and future problems 219
stochastic modelling of hydrologic time processes 212
temporal stochastic models 211
testing statistical hypotheses 207–8
year as critical subsample 204–5
Water time series 202–4
Wave effects in extreme sea levels 261–3
Wave-tank modelling 262
Weekly cycles 202–3
Weibull distribution function 383
Weighted bivariate normal correlation model 67–8
Weighted distribution methods 45–69
informativeness of 52–8
Welfare economics 350
White noise 16
Wilderness, conservation-development decision model 367–76

YIC statistic 19–22

Index compiled by G. Jones